Asheville-Buncombe
Technical Community College
Learning Resources Center
340 Victoria Rd.
Asheville, NC 28801

DISCARDED

DEC 1 2 2024

Molecular Biology of Parasitic Protozoa

Frontiers in Molecular Biology

SERIES EDITOR

B. D. Hames

Department of Biochemistry and Molecular Biology
University of Leeds, Leeds LS2 9JT, UK

AND

D. M. Glover

Cancer Research Campaign Laboratories
Department of Anatomy and Physiology
University of Dundee, Dundee DD1 4HN, UK

TITLES IN THE SERIES

1. Human Retroviruses
Bryan R. Cullen
2. Steroid Hormone Action
Malcolm R. Parker
3. Mechanisms of Protein Folding
Roger H. Pain
4. Molecular Glycobiology
Minoru Fukuda and Ole Hindsgaul
5. Protein Kinases
Jim Woodgett
6. RNA–Protein Interactions
Kyoshi Nagai and Iain Mattaj
7. DNA–Protein: Structural Interactions
David Lilley
8. Mobile Genetic Elements
David Sherratt
9. Chromatin Structure and Gene Expression
Sarah C. R. Elgin
10. Cell Cycle Control
Chris Hutchinson and D. M. Glover
11. Molecular Immunology (Second Edition)
B. David Hames and David M. Glover
12. Eukaryotic Gene Transcription
Stephen Goodbourn
13. Molecular Biology of Parasitic Protozoa
Deborah F. Smith and Marilyn Parsons

Molecular Biology of Parasitic Protozoa

EDITED BY

Deborah F. Smith

Department of Biochemistry
Imperial College of Science, Technology and Medicine, London SW7 2AZ, UK

and

Marilyn Parsons

Seattle Biomedical Research Institute,
4 Nickerson Street, Seattle, WA 98109-1651 USA
and
Department of Pathobiology, School of Public Health and
Community Medicine, University of Washington, Seattle WA 98195 USA

OXFORD UNIVERSITY PRESS
Oxford New York Tokyo

Oxford University Press, Walton Street, Oxford OX2 6DP
Oxford   New York
Athens   Auckland   Bangkok   Bombay
Calcutta   Cape Town   Dar es Salaam   Delhi
Florence   Hong Kong   Istanbul   Karachi
Kuala Lumpur   Madras   Madrid   Melbourne
Mexico City   Nairobi   Paris   Singapore
Taipei   Tokyo   Toronto
and associate companies in
Berlin   Ibadan

Oxford is a trade mark of Oxford University Press

Published in the United States
by Oxford University Press Inc., New York

© *Oxford University Press, 1996*

All rights reserved. No part of this publication may be
reproduced, stored in a retrieval system, or transmitted, in any
form or by any means, without the prior permission in writing of Oxford
University Press. Within the UK, exceptions are allowed in respect of any
fair dealing for the purpose of research or private study, or criticism or
review, as permitted under the Copyright, Designs and Patents Act, 1988, or
in the case of reprographic reproduction in accordance with the terms of
licences issued by the Copyright Licensing Agency. Enquiries concerning
reproduction outside those terms and in other countries should be sent to
the Rights Department, Oxford University Press, at the address above.

This book is sold subject to the condition that it shall not,
by way of trade or otherwise, be lent, re-sold, hired out, or otherwise
circulated without the publisher's prior consent in any form of binding
or cover other than that in which it is published and without a similar
condition including this condition being imposed
on the subsequent purchaser.

A catalogue record for this book is available from the British Library

Library of Congress Cataloging-in-Publication Data
Molecular biology of parasitic protozoa/edited by Deborah F. Smith
and Marilyn Parsons.—1st ed.
(Frontiers in molecular biology; 13)
Includes bibliographical references.
1. Protozoa, pathogenic.   2. Protozoan diseases—molecular
aspects.   3. Medical protozoology.   I. Smith, Deborah F.
II. Parsons, Marilyn.   III. Series.
[DNLM: 1. Protozoa.   2. Molecular Biology—methods. QX 50 M7177
1996]
QR251.M675   1996   616'.016–dc20   95-39520
ISBN 0 19 963602 8 (Hbk)
ISBN 0 19 963601 X (Pbk)

Typeset by Footnote Graphics, Warminster, Wilts
Printed and bound in Great Britain by
Butler & Tanner Ltd, Frome and London

Contents

List of contributors   xii

Abbreviations   xv

1 Introduction   1
DEBORAH F. SMITH and MARILYN PARSONS

2 Trypanosomatid genetics   6
JOHN SWINDLE and ANDREW TAIT

    1 Introduction   6
    2 Ploidy and genome size   7
    3 Genetic analysis   8
       3.1 Genetic exchange   8
       3.2 Genetic exchange in natural populations   10
    4 Molecular karyotype   11
       4.1 *Trypanosoma cruzi* karyotype   12
       4.2 *Leishmania* spp. karyotype   12
       4.3 *Trypanosoma brucei* karyotype   14
    5 Gene organization   14
    6 Requirements for gene expression   16
    7 DNA-mediated transformation systems   18
    8 Transient transfection systems   19
       8.1 Transient expression vectors   19
       8.2 Reporter genes   19
       8.3 Promoters and mRNA processing   20
       8.4 Methods of transformation   21
       8.5 Applications of transient transfection   21
    9 Stable transformation   22
       9.1 Selectable markers   22
       9.2 Integrative transformation   22
       9.3 Episomal transformation   24
       9.4 Inducible gene expression   24

10 Conclusion		25
References		26

3 The three genomes of *Plasmodium*   35
JEAN E. FEAGIN and MICHAEL LANZER

1 Introduction	35
2 Nuclear genomic organization	36
2.1 Polymorphisms at chromosome ends	37
2.2 Repetitive elements in subtelomeric domains	38
2.3 Chromatin structure and breakage sites	39
3 Biological ramifications of chromosomal polymorphism	41
3.1 Chromosomal rearrangements and antigenic variation	41
3.2 Sexual recombination and genetic diversity	42
4 Subtelomeric organization and chromosome pairing	42
5 The extrachromosomal DNAs	43
5.1 The mitochondrial genome	43
5.2 Mitochondrial expression and function	44
5.3 The putative plastid genome	45
5.4 Expression and function of the 35kb DNA	47
5.5 Extrachromosomal DNA inheritance	48
6 Conclusions	48
References	48

4 *Toxoplasma* as a model genetic system   55
L. DAVID SIBLEY, DAN K. HOWE, KIEW-LIAN WAN, SHAHID KHAN, MARTIN A. ASLETT and JAMES W. AJIOKA

1 Introduction	55
1.1 Advantages of *Toxoplasma* as a model genetic system	55
1.2 Life cycle of *Toxoplasma*	55
1.3 Human infection and pathogenesis	56
2 Genomic organization	57
2.1 The genomes	57
2.2 The genes	57
2.3 Repetitive DNAs	58
2.4 Gene expression	58
2.5 Stage-specific gene expression	59
3 Population genetic structure	59

    3.1 Strain-specific antigenic markers   59
    3.2 Polymorphic isoenzyme and DNA markers   59
    3.3 The clonal population structure of *Toxoplasma*   60
4 Classical genetic analyses   60
    4.1 Tools for *in vitro* analysis   60
    4.2 Use of drug resistance markers in genetic crosses   61
    4.3 RFLP linkage studies   61
5 Genome mapping   62
    5.1 Physical mapping, construction of contigs and mapping of cDNAs   63
    5.2 Physical mapping of parasite genomes   64
    5.3 Database development   64
6 Molecular genetics   65
    6.1 DNA transfection   66
    6.2 Stable transformation   66
    6.3 Targeted gene disruption   67
7 Future developments   68
References   68

5 Kinetoplast DNA: structure and replication   75

PAUL T. ENGLUND, D. LYS GUILBRIDE, KUO-YUAN HWA, CATHARINE E. JOHNSON, CONGJUN LI, LAURA J. ROCCO and AL F. TORRI

1 Introduction   75
2 Structure of a kDNA network   76
    2.1 The isolated network   76
    2.2 Organization of the network *in vivo*   77
    2.3 Conditions required for network formation   78
3 Replication of the kDNA network   78
    3.1 Free minicircles   78
    3.2 Structure of a replicating network   79
    3.3 Distribution of newly synthesized minicircles around the network periphery   81
    3.4 The spinning kinetoplast model   81
    3.5 Changes in minicircle valence during replication   83
    3.6 Final stages of network replication   83
4 A closer look at the replication of minicircles and maxicircles   85
    4.1 Minicircle replication   85
    4.2 Maxicircle replication   85
References   86

6 Developmental regulation of gene expression in African trypanosomes — 88
ETIENNE PAYS and LUC VANHAMME

1 Introduction — 88
  1.1 A model: *Trypanosoma brucei* — 88
  1.2 Transcription units of the genes for the major stage-specific antigens of *T. brucei* — 90
  1.3 Possible levels of control for gene expression — 92
2 Promoters and the control of transcription initiation — 93
  2.1 The few promoters known — 94
  2.2 Regulation of promoter activity — 95
  2.3 Influence of the chromosomal context — 97
3 Transcription elongation and processing of the primary transcripts — 98
  3.1 RNA elongation — 98
  3.2 *Trans*-splicing and polyadenylation — 98
  3.3 The untranslated regions and RNA amounts — 100
4 RNA translation and protein stability — 100
  4.1 Translational controls — 100
  4.2 Protein stability — 100
5 Antigenic variation and novel mechanisms of gene regulation — 101
  5.1 *In situ* (in)activation of VSG expression sites — 103
  5.2 VSG gene rearrangements — 103
  5.3 Programming of VSG expression — 104
  5.4 Point mutations — 105
  5.5 Evolution of VSGs — 106
6 Conclusions — 107
Acknowledgements — 107
References — 107

7 *Trans*-splicing in trypanosomatid protozoa — 115
ELISABETTA ULLU, CHRISTIAN TSCHUDI and ARTHUR GÜNZL

1 Introduction — 115
2 Mechanism of *trans*-splicing — 115
  2.1 *Trans*-spliceosomal U-snRNPs — 116
  2.2 U-snRNA interactions in *trans*-splicing — 118
3 The substrates of *trans*-splicing — 119

	3.1 The spliced leader RNA: structure and function	119
	3.2 Structure of the pre-mRNA	121
4	Functional role and consequences of *trans*-splicing	124
5	The physiology and regulation of *trans*-splicing	125
6	Evolutionary considerations	128
	References	129

8 RNA editing: post-transcriptional restructuring of genetic information — 134

STEPHEN L. HAJDUK and ROBERT S. SABATINI

1	Introduction	134
	1.1 RNA editing defined	135
	1.2 Mechanisms of RNA processing	135
	1.3 Kinetoplastid RNA editing	136
2	General aspects of kinetoplastid RNA editing	140
	2.1 The mitochondrial genome of trypanosomes	140
	2.2 Guiding of RNA editing	143
3	Biochemical mechanisms of RNA editing	144
	3.1 Cleavage–ligation mechanism	144
	3.2 Transesterification mechanism	145
	3.3 TUTase addition mechanism	147
4	Involvement of mitochondrial RNPs in RNA editing	148
5	Function and origin of RNA editing	150
	5.1 Role of RNA editing in regulation of gene expression	150
	5.2 RNA editing is developmentally regulated	152
	5.3 Evolutionary origin of kinetoplastid RNA editing	152
	References	153

9 Biogenesis of specialized organelles: glycosomes and hydrogenosomes — 159

JÜRG M. SOMMER, PETER J. BRADLEY, C. C. WANG and PATRICIA J. JOHNSON

1	Introduction	159
2	The glycosome	160
	2.1 Morphology	160

2.2 Function	160
2.3 Biogenesis	161
2.4 Glycosomal targeting signals	161
2.5 Glycosomal versus peroxisomal import	167
2.6 Mutational analysis of glycosome assembly	167
2.7 Glycosomal protein import as a potential target for chemotherapy	168
3 The hydrogenosome	**168**
3.1 Trichomonad hydrogenosomes	168
3.2 Morphology	169
3.3 Function	170
3.4 Biogenesis	171
3.5 Hydrogenosome-like organelles in other organisms	174
4 Conclusions	**174**
Acknowledgements	**175**
References	**175**

10 Mechanisms of drug resistance in protozoan parasites 181

DYANN F. WIRTH and ALAN COWMAN

1 Emergence of drug resistance and its impact	181
2 Role of intrinsic genetic and biological components	182
3 Impact of drug resistance on public health and disease	182
4 Common themes of resistance in protozoan parasites	183
5 Amplification and drug resistance	183
6 Mutations in target enzymes	185
7 Folate antagonists	185
8 Drug transport	187
9 Quinine-like antimalarials	191
10 Genetic characterization of chloroquine resistance	192
11 Drug resistance in *Toxoplasma gondii*	194
12 Metronidazole resistance	195
13 Conclusions and perspectives	196
References	196

11 Glycosyl-phosphatidylinositols and the surface architecture of parasitic protozoa 205
MALCOLM J. McCONVILLE

1 Introduction	205
2 Structure of protozoan GPI protein anchors	205
3 Function of GPI protein anchors in the protozoa	207
3.1 Subcellular trafficking	208
3.2 Protein turnover	209
3.3 Lipase-mediated release	210
3.4 Maintenance of surface coats	210
3.5 Modulation of host signal transduction pathways	211
4 Protein-free GPI glycolipids in the trypanosomatids	212
4.1 Structure of the GIPLs and LPG in *Leishmania* parasites	212
4.2 Function of *Leishmania* LPG and GIPLs	213
4.3 Distribution of free GPIs in other protozoan parasites	217
5 Metabolism and biosynthesis of protozoan GPIs	218
5.1 Protein anchor biosynthesis	218
5.2 Biosynthesis of protein-free GPIs in *Leishmania*	220
6 Conclusion	222
References	222

Index 229

Contributors

JAMES W. AJIOKA
University of Cambridge, Department of Pathology, Tennis Court Road, Cambridge CB2 1QP, UK.

MARTIN A. ASLETT
University of Cambridge, Department of Pathology, Tennis Court Road, Cambridge CB2 1QP, UK.

PETER J. BRADLEY
Department of Microbiology and Immunology, University of California, Los Angeles, CA 90024-1747, USA.

ALAN COWMAN
Immunoparasitology Unit, The Walter and Eliza Hall Institute of Medical Research, Post Office, The Royal Melbourne Hospital, Victoria, Australia.

PAUL T. ENGLUND
Department of Biological Chemistry, Johns Hopkins University Medical School, 725 N. Wolfe Street, Baltimore, MD 21205, USA.

JEAN E. FEAGIN
Seattle Biomedical Research Institute, 4 Nickerson Street, Seattle, WA 98109-1651, USA and Department of Pathobiology, School of Public Health and Community Medicine, University of Washington, Seattle, WA 98195, USA.

D. LYS GUILBRIDE
Department of Biological Chemistry, Johns Hopkins Medical School, Baltimore, MD 21205, USA.

ARTHUR GÜNZL
Department of Internal Medicine, Section of Infectious Diseases, 805 LCI, 333 Cedar Street, Yale University School of Medicine, New Haven, CT 06520-8022, USA.

STEPHEN L. HAJDUK
Department of Biochemistry and Molecular Genetics, University of Alabama at Birmingham, Birmingham, AL 35294, USA.

DAN K. HOWE
Department of Molecular Microbiology, Washington University School of Medicine, St Louis, MO 63110, USA.

KUO-YUAN HWA
Department of Biological Chemistry, Johns Hopkins Medical School, Baltimore, MD 21205, USA.

CATHARINE E. JOHNSON
Department of Biological Chemistry, Johns Hopkins Medical School, Baltimore, MD 21205, USA.

PATRICIA J. JOHNSON
Department of Microbiology and Immunology, University of California, Los Angeles, CA 90024-1747, USA.

SHAHID KHAN
University of Cambridge, Department of Pathology, Tennis Court Road, Cambridge CB2 1QP, UK.

MICHAEL LANZER
Zentrum für Infektionsforschung, University of Würzburg, Röntgenring 11, D-97070 Würzburg, Germany.

CONGJUN LI
Department of Biological Chemistry, Johns Hopkins Medical School, Baltimore, MD 21205, USA.

MALCOLM J. McCONVILLE
Department of Biochemistry, University of Melbourne, Parkville 3052, Australia.

MARILYN PARSONS
Seattle Biomedical Research Institute, 4 Nickerson St., Seattle WA 98109-1651, USA and Department of Pathobiology, School of Public Health and Community Medicine, University of Washington, Seattle, WA 98195, USA.

ETIENNE PAYS
Département de Biologie Moléculaire, Laboratoire de Parasitologie Moléculaire, Université Libre de Bruxelles, Rue des Chevaux 67, B-1640 Rhode-Saint-Genèse, Belgium.

LAURA J. ROCCO
Department of Biological Chemistry, Johns Hopkins Medical School, Baltimore, MD 21205, USA.

ROBERT S. SABATINI
Department of Biochemistry and Molecular Genetics, University of Alabama at Birmingham, Birmingham, AL 35294, USA.

L. DAVID SIBLEY
Department of Molecular Microbiology, Washington University School of Medicine, St Louis, MO 63110, USA.

JÜRG M. SOMMER
Department of Pharmaceutical Chemistry, University of California, San Francisco, CA 94143-0446, USA.

DEBORAH F. SMITH
Department of Biochemistry, Imperial College of Science, Technology and Medicine, London SW7 2AZ, UK.

JOHN SWINDLE
Department of Microbiology and Immunobiology, University of Tennessee, Memphis, 858 Madison Ave, Memphis, TN 38163, USA.

ANDREW TAIT
Wellcome Unit of Molecular Parasitology, University of Glasgow, Bearsden Road, Glasgow G61 1QH, UK.

AL F. TORRI
Department of Biological Chemistry, Johns Hopkins Medical School, Baltimore, MD 21205, USA.

CHRISTIAN TSCHUDI
Department of Internal Medicine, Section of Infectious Diseases, 805 LCI, 333 Cedar Street, Yale University School of Medicine, New Haven, CT 06520-8022, USA.

ELISABETTA ULLU
Department of Internal Medicine, Section of Infectious Diseases, 805 LCI, 333 Cedar Street, Yale University School of Medicine, New Haven, CT 06520-8022, USA.

LUC VANHAMME
Département de Biologie Moléculaire, Laboratoire de Parasitologie Moléculaire, Université Libre de Bruxelles, Rue des Chevaux 67, B-1640 Rhode-Saint-Genèse, Belgium.

KIEW-LIAN WAN
University of Cambridge, Department of Pathology, Tennis Court Road, Cambridge CB2 1QP, UK.

C. C. WANG
Department of Pharmaceutical Chemistry, University of California, San Francisco, CA 94143-0446, USA.

DYANN F. WIRTH
Tropical Public Health, Harvard School of Public Health, 665 Huntington Ave, Boston, MA 02115, USA.

Abbreviations

AIDS	Acquired immune deficiency syndrome
BAC	Bacterial artificial chromosome
CAT	Chloramphenicol acetyl transferase
CHO	Chinese hamster ovary
cM	Centimorgan
COI	Cytochrome *c* oxidase subunit I
COIII	Cytochrome *c* oxidase subunit III
CYb	Apocytochrome *b*
DHRF/TS	Dihydrofolate reductase/thymidylate synthase
DHPS	Dihydropteroate synthase
ESAG	Expression site-associated genes
EST	Expressed sequence tags
FACS	Fluorescence-activated cell sorting
FP	Ferriprotoporphyrin IX
GIPL	Glycoinositol phospholipid
GPI	Glycosylphosphatidylinositol
gPGK	Glycosomal phosphoglycerate kinase
gRNA	Guide RNA
GUS	β-glucuronidase
HGPRTase	Hypoxanthine–guanine phosphoribosyltransferase
IL-1	Interleukin 1
kDNA	Kinetoplast DNA
LPG	Lipophosphoglycan
LSU	Large subunit (of rRNA)
MDH	Malate dehydrogenase
MDR	Multidrug resistance
MSA-1	Merozoite surface antigen 1
ORF	Open reading frame
PAG	Procyclin-associated gene
PARP	Procyclic acidic repetitive protein (procyclin)
PCR	Polymerase chain reaction
PEPCK	Phosphoenolpyruvate carboxykinase
PfEMP1	*Plasmodium falciparum* erythrocyte membrane protein 1
PFGE	Pulsed-field gel electrophoresis
Pgh1	P-glycoprotein homologue
PI	Phosphatidylinositol
PMSF	Phenylmethylsulphonyl fluoride
pol II	RNA polymerase II

pPy	Polypyrimidine
RFLP	Restriction fragment length polymorphism
RNP	Ribonucleoprotein particle
SL	Spliced leader
SSU	Small subunit (of rRNA)
STS	Sequence-tagged sites
Tg DB	*Toxoplasma* database
TNF	Tumour necrosis factor
TR	Trypanothione reductase
TUTase	Terminal uridylyl transferase
UTR	Untranslated region
VSG	Variable surface glycoprotein
YAC	Yeast artificial chromosome

1 | Introduction

DEBORAH F. SMITH and MARILYN PARSONS

The eukaryotic parasites are a diverse group of organisms that range from some of the most ancient single-celled species (e.g. *Giardia lamblia*) to the cytologically complex metazoan helminths. Despite this extreme diversity in body form, all endoparasites share the common features of living within and exploiting a host organism. This exploitation has resulted in the evolution of some remarkable adaptations at the biochemical level: many parasites have dispensed with whole metabolic pathways, while developing novel strategies for the evasion of host defences. In addition, complex mechanisms have evolved to facilitate transmission of parasites to new hosts. Typically, parasites take advantage of the relationships between two species to facilitate transmission: for example, the mosquito takes a bloodmeal from a malarious individual and thereby becomes infected with the parasite. Transmission is not passive, but rather involves parasite development. Therefore most parasites undergo complex, multiphasic life cycles that can involve extracellular and intracellular stages and, for unicellular parasites, maintenance as actively dividing populations or as dormant, non-dividing cysts.

Evidence suggests that a preponderance of the evolutionary history of eukaryotes occurred when organisms existed solely as single cells or protists (see Fig. 1). It should come as no surprise, therefore, that there is a huge amount of diversity within the protozoa, despite the fact that in the past they were grouped together as a single phylum. It is clear now that these organisms, including the parasitic protozoa, are polyphyletic. This is reflected at both the molecular and cytological levels. For example, certain protozoa lack mitochondria (*Giardia lamblia*, *Trichomonas vaginalis*, *Entamoeba histolytica*, microsporidia). Many others possess but a single mitochondrion, while higher eukaryotic cells often possess hundreds. Another example is at the level of RNA processing, where kinetoplastids show the unusual properties of polycistronic transcription and *trans*-splicing as well as RNA editing. The more ancient species, *G. lamblia*, even shows a relic of its close relationship with prokaryotes by the presence of the Shine Dalgarno sequence on its ribosomal RNA (3).

These unusual approaches to survival, coupled with a range of novel biochemical features explain, in part, the fascination of parasitic organisms to the biologist. However, the economic and medical importance of parasitic infections adds an extra dimension to their study. The consequences of parasitic disease for human and animal health and economic growth around the world have captured the attentions of a diverse group of scientists. Many parasites cause devastating diseases in otherwise healthy, immunocompetent humans (see Table 1). In addition, parasitic

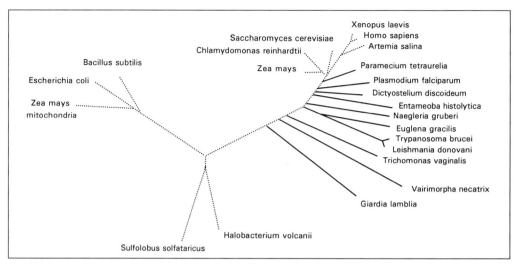

Fig. 1 Multi-kingdom tree. This unrooted tree (adapted from Ref. 1) is based on the analysis of 16S rRNA-like sequences. The length of each line indicates the evolutionary distance between organisms with the protozoan branches shown by solid lines. Among the earliest branches are the diplomonads (e.g. *G. lamblia*), microsporidians (e.g. *Vairimorpha necatrix*), and trichomonads (*T. vaginalis*, not shown); the branching order of these has not been firmly established (2).

infections can be of importance in the progression and resolution of other infectious diseases, most notably AIDS (acquired immune deficiency syndrome). There has been a steady increase in the number of *Leishmania*-HIV co-infections over the past 10 years, for example. In one study in Spain, 80% of leishmaniasis cases were in immunodepressed patients, 60% of which were HIV positive (4). Better understanding of these organisms and their interactions with their hosts will improve the human condition at all corners of the globe.

In this book, we have chosen to focus on the parasitic protozoa, causative agents of a broad spectrum of diseases. This specialization omits the large number of other protozoan species that are not parasitic but gives a 'window' on the diversity of the kingdom as a whole. Also excluded from this volume is any discussion of the metazoan parasites, the worms and the ectoparasites. The molecular and cellular biology of metazoan parasites is more closely related to the organisms they infect than to the protozoan parasites, and thus can be more appropriately discussed in conjunction with the molecular biology of higher eukaryotes.

The protozoan organisms described in the chapters that follow include trypanosomatid species of the order Kinetoplastida (*Trypanosoma brucei, Trypanosoma cruzi, Leishmania*—causative agents of West African sleeping sickness, Chagas' disease and leishmaniasis respectively); species of the phylum Apicomplexa (the hematozoan *Plasmodium* spp. and the coccidian *Toxoplasma gondii*—causative agents of malaria and toxoplasmosis, respectively); and *T. vaginalis*, causative agent of sexually transmitted disease infecting 180 million people worldwide. Collectively, these parasites cause disease in most areas of the world. An initial impression of the detrimental effects of these infections can be obtained from the data gathered by the

Table 1 Distribution of parasitic diseases caused by the organisms discussed in this book

Parasite	Disease	Distribution	Population at risk	Cases (deaths) p.a.
Plasmodium spp.	Malaria	90 countries (sub-tropical/tropical regions)	2400 million	[a,b] 3–500 million (1.5–2.7 million)
Toxoplasma	Toxoplasmosis	Worldwide	5430 million (total world)	[c] ?
Trichomonas vaginalis	Trichomoniasis	Worldwide	5430 million (total world)	[d] 128 million
Leishmania spp.	Leishmaniasis	88 countries (sub-tropical/tropical regions)	367 million	[b,e] 0.4 million
Trypanosoma brucei	African sleeping sickness	36 countries (sub-Saharan Africa)	55 million	[b,f] >0.05 million
Trypanosoma cruzi	Chagas' disease	21 countries (South and Central America)	90 million	[b] 3 million

p.a. per annum
[a] World Malaria Situation in 1992. (1994) Reprinted from World Health Organization (WHO) Wkly. Epidem. Rec. 42-44.
[b] Control of Tropical Diseases, Progress Report 1 (1994) WHO, Geneva.
[c] Zuber, P. and Jacquier, P. (1995) Schweiz Medizinsche Wochenschrift 1995, 125 (suppl. 65): 195-225.
[d] Gerbase, A. C. (1995) Global Programme on AIDS, WHO, Geneva.
[e] Ashford, R. W., Desjeux, P. and deRaadt, P. (1992) Parasitology Today 8, 104-105.
[f] Kuzoe, F. A. (1993) Acta Tropica 54, 153-162.

World Health Organization (WHO) and others on mortality and morbidity in infected populations (Table 1).

The topics chosen for review in this book focus on some, but not all, of the most rapidly moving areas in current research in molecular parasitology. These focal areas have largely been determined by the extent of research on a given group of organisms which in turn reflects, in part, the availability of suitable material for analysis. Some of the protozoa are relatively easy to grow in culture, while others are more difficult if not impossible to maintain *in vitro*. Even given adequate numbers of organisms, there may be added complications: intracellular stages may be difficult to release from host cell material; or sexual cycles, if they occur, may yield only limiting numbers of cells that are not amenable to investigation. Of necessity, therefore, certain organisms have become the models. In recent years, however, the advent of molecular biological techniques has promoted the study of a much wider group of organisms. Use of the polymerase chain reaction, for example, has allowed useful molecular data to be generated from very small amounts of material (5,6). Technical advances in the development of genetic transfection systems have also led to rapid progress in the study of some but not all parasitic organisms. These developments are reflected in the pages of this book.

Molecular investigations of the trypanosomatids, apicomplexans and trichomonads have revealed novel aspects of gene regulation and expression that are of intrinsic interest to the student of eukaryotic lifestyle. Additionally, these studies have provided seminal research applicable to a broader understanding of molecular systems of higher eukaryotes. One example is in the analysis of glycophosphatidyl inositol anchored proteins, first investigated in detail in the trypanosome system, but found to be of relevance to many mammalian cell surface proteins. Similarly, the description of RNA editing as a genetic process in trypanosomatids led the way for the description of RNA editing in human disease. These topics are discussed in Chapters 11 and 8, respectively.

Treatments for most parasitic diseases are difficult, for reasons of drug toxicity, efficacy or cost. Novel aspects of parasite molecular and cellular biology are consequently not only of intrinisic interest, but may be of importance in developing new methods of chemotherapy. Therefore, this book contains discussions of unique, highly specialized parasite organelles and novel mechanisms of gene regulation. Topics directly at the interface between the laboratory and medicine are also included; Chapter 10 describes the current status of research on drug resistance in malaria parasites and *Leishmania*. An understanding of the mechanisms that promote genome plasticity and gene amplification, as well as knowledge of gene regulatory mechanisms, will be critical in the design of strategies to avoid the emergence of extensive drug resistance in these and other parasites. Parasite genome structures, including the phenomena of high genome plasticity and unusual organellar genomes, are discussed in Chapters 2 to 5.

Space limitations have led us to omit several research areas that are important to the field of parasitology, but less germane to the molecular biology of protozoan parasites. Among these are the study of host responses and pathogenesis. Also not covered are two areas of more applied research that have benefited from investigations into parasite molecular biology. The first of these is the identification, expression, and production of candidate antigens for vaccination studies. This has been a focal area of malaria research, chiefly because of the overwhelming need for an effective vaccine against a disease that causes more than 400 million clinical cases and between 1 and 3 million deaths each year (7). At the time of writing, however, none of the antigens tested have been shown to give long-term protection against the human malarias. In fact there are, as yet, no effective vaccines for any of the parasitic diseases discussed in these pages. The reasons for this lack of success are complex, but a major factor must be the ingenuity of the parasites in evading host interventions in their developmental and proliferative cycles. In particular, the elegant mechanisms of immune evasion by antigenic variation (characteristic of the life cycle of *T. brucei* (8), but probably of importance too in the survival of the malaria parasite (9–12)) have so far confounded the vaccinologists. The second area is the generation of recombinant DNA and antibody probes for use in epidemiological surveys. These molecules have been increasingly important in determining the incidence and spread of the major parasitic infections and in the development of public health strategies worldwide. Readers are referred to several recent reviews for consideration of these issues (13-15).

Despite the rapid explosion in our knowledge of parasite molecular mechanisms, discussed in detail in the following pages, it is necessary to end this introduction on a note of caution. Although the numbers of cloned parasite genes and isolated parasite proteins are ever increasing, the number of tools available for molecular analysis remains fairly limited compared with many other eukaryotic systems. The development of parasite molecular genetics, although rapid, still lags behind the sophisticated advances made in yeast and mammalian genetics. As this balance is redressed, we expect that more novel attributes of the molecular biology of parasites will be revealed. In the long term, however, the successful exploitation of fundamental research in molecular biology for the treatment and prevention of parasitic infections will require not only 'cutting edge' research but the provision of adequate resources for drug and vaccine development, testing and distribution. Effective treatment for most parasitic diseases may still be a long way off.

References

1. Sogin, M. L. (1989) Evolution of eukaryotic microorganisms and their small subunit ribosomal RNAs. *Am. Zool.*, **2**, 487.
2. Leipe, D. D., Gunderson, J. H., Nerad, T. A., and Sogin, M. L. (1993) Small subunit ribosomal RNA of *Hexamita inflata* and the quest for the first branch in the eukaryotic tree. *Mol. Biochem. Parasitol.*, **59**, 41.
3. Sogin, M. L., Gunderson, J. G., Elwood, H. J., Alonso, R. A., and Peattie, D. A. (1989) Phylogenetic meaning of the kingdom concept: an unusual ribosomal RNA from *Giardia lamblia*. *Science*, **243**, 75.
4. Alvar, J. (1994) Leishmaniasis and AIDS co-infection: the Spanish example. *Parasitol. Today*, **10**, 160.
5. Schriefer, M. E., Sacci, J. B., Wirtz, R. A., and Azad, A. F. (1991) Detection of polymerase chain reaction-amplified malarial DNA in infected blood and individual mosquitoes. *Exp. Parasitol.*, **73**, 311.
6. de Bruijn, M. H. and Barker, D. C. (1992) Diagnosis of New World leishmaniasis: specific detection of species of the *Leishmania braziliensis* complex by amplification of kinetoplast DNA. Acta Trop., **52**, 45.
7. Sturchler, D. (1989) Malaria, how much is there worldwide? *Parasitol., Today*, **5**, 39.
8. Van der Ploeg, L. H. T., Gottesdiener, K., and Lee, M. G.-S. (1992) Antigenic variation in African trypanosomes. *Trends Genet.*, **8**, 452.
9. Roberts, D. J., Craig, A. G., Berendt, A. R., Pinches, R., Nash, G., Marsh, K., and Newbold, C. I. (1992) Rapid switching to multiple antigenic and adhesive phenotypes in malaria. *Nature*, **357**, 689.
10. Baruch, D. I., Pasloske, B. L., Singh, H. B., Bi, X., Ma, X. C., Feldman, M., Taraschi, T. F., and Howard, R. J. (1995) Cloning the *P. falciparum* gene encoding PfEMP1, a malarial variant antigen and adherence receptor on the surface of parasitized human erythrocytes. *Cell*, **82**, 77
11. Su, X.-Z., Heatwole, V. M., Wertheimer, S. P., Guinet, F., Herrfeldt, J. A., Peterson, D. S., Ravetch, J. A., and Wellems, T. E. (1995) The large diverse gene family *var* encodes proteins involved in cytoadherence and antigenic variation of *Plasmodium falciparum*-infected erythrocytes. *Cell*, **82**, 89.
12. Smith, J. D., Chitnis, C. E., Craig, A. G., Roberts, D. J., Hudson-Taylor, D. E., Peterson, D. S., Pinches, R., Newbold, C. I., and Miller, L. H. (1995) Switches in expression of *Plasmodium falciparum var* genes correlate with changes in antigenic and cytoadherent phenotypes of infected erythrocytes. *Cell*, **82**, 101.
13. Hide, G. and Tait, A. (1991) The molecular epidemiology of parasites. *Experientia*, **47**, 128.
14. Day, K. P., Koella, J. C., Nee, S., Gupta, S., and Read, A. F. (1992) Population genetics and dynamics of *Plasmodium falciparum*: an ecological view. *Parasitology*, **104**, S35.
15. Blackwell, J. M. (1992) Leishmaniasis epidemiology: all down to the DNA. *Parasitology*, **104**, S19.

2 | Trypanosomatid genetics

JOHN SWINDLE and ANDREW TAIT

1. Introduction

The genetics of the trypanosomatid group of protozoa have been the subject of considerable research over the last 15 years. The trypanosomatids include a number of highly pathogenic parasitic species which infect a very wide range of hosts including man and domestic livestock. In terms of molecular genetics, the most intensively studied are *Leishmania* and trypanosomes. This chapter will primarily focus on *Trypanosoma brucei, Trypanosoma cruzi,* and a few *Leishmania* species. Two fundamental aspects of the genetics of these parasites will be considered: first, a basic description of their genomes and genome organization together with a review of our current state of knowledge on classical genetics; and second, a review of modern transfection-based molecular genetics. The lack of readily available classical genetic systems, which have been used so effectively in yeast, *Drosophila*, and *Caenorhabditis elegans,* has been a major restraint, although the availability of reverse genetics has circumvented some of these limitations.

Trypanosomatids have two genomes, one within the nucleus and another within the mitochondrion. The latter, called the kinetoplast, constitutes some 10–30% of the total cellular DNA (1). The nuclear genomes are of low complexity (2). While these organisms are largely diploid, the question of ploidy is complex (see discussion below). Classical approaches to defining ploidy by cytological studies of mitotic chromosomes have been precluded by the lack of chromosome condensation at mitosis (3, 4). Development of electrophoretic techniques (pulsed-field gel electrophoresis, PFGE) capable of separating large DNA molecules of up to 10Mb (5) has allowed the characterization of the genomes of the trypanosomatids. These 'molecular karyotypes' permit the localization of gene markers by DNA hybridization and, potentially, allow the determination of ploidy. The data on gene organization in trypanosomatids are based on the analysis of a relatively small number of genes when compared to the number characterised in yeast, insects and mammals. Nevertheless, a general feature appears to be multiple copies of genes either in tandem arrays or in several different chromosomal localizations—the most extreme case being the *T. brucei* variable surface glycoprotein (VSG) genes, which occur as 1000 copies per genome (6).

2. Ploidy and genome size

Analyses of the renaturation kinetics of total DNA have shown that the genomes of the trypanomatids are quite small (4–6 x *E. coli*) and contain highly repetitive, middle repetitive, and single copy sequences. *Leishmania donovani* (7) and *T. brucei* (2) have around 60–65% single copy sequences while *T. cruzi* appears to have a higher level of repetitive sequences and only 30% single copy sequences (8). Chemical analysis of the DNA content of a known number of cells, or microfluorometry of single nuclei stained with Feulgen or other DNA binding dyes, have provided measures of the total nuclear DNA content. These values are approximately twofold higher than the estimates based on kinetic complexity analysis of DNA content per haploid genome and so suggest that the trypanosomatids are diploid (Table 1).

Analysis of iso-enzyme variation and restriction fragment length polymorphisms (RFLPs) of single copy genes in cloned lines of a range of trypanosomatid species shows patterns of variation that are consistent with a diploid genome (9–13). Perhaps the most direct evidence, albeit from a relatively small set of genes, is provided by the results of 'gene knockout' experiments in *T. brucei* and *Leishmania*. It is clear that gene deletion (by homologous recombination) requires two rounds of disruption and Southern blotting of single and double knockouts with appropriate gene probes demonstrates the loss of first one allele followed by the second (14, 15). Taken with the DNA content and complexity data, it is clear that the genomes of the trypanosomatids are largely diploid. One notable exception to diploidy is found in the VSG genes of *T. brucei*. Available evidence from gene cloning and Southern blotting of restriction digests shows that there is only one allele per diploid genome (16). As described below, some of these genes are found on smaller chromosomes that are thought to be aneuploid, while a smaller number are on larger chromosomes. In the latter case, the genes can be considered allelic by virtue of their homologous chromosome location despite extensive divergence in the VSG genes themselves.

The ploidy of different life cycle stages has been examined in *T. brucei*, where measurements of DNA content between the bloodstream and the metacyclic forms

Table 1 DNA content and ploidy

	Single copy genes (% genome)	Kinetic complexity[c] (x 10^7 bp)	Kinetoplast DNA (% total)	DNA content (pg)	
				A	B
T. cruzi (Y strain)[7]	23	12	16	0.28	0.13
T. brucei (427)[1]	68	2.5	6	0.091	0.041
Leishmania donovani[13]	60	4.5–6.5	15	0.1	0.06

A – Measured by cytophotometry
B – Derived from kinetic complexity
C – Haploid value

(in the salivary glands of the vector) have been undertaken. Conflicting results have been obtained, with one report indicating similar DNA contents in both stages (17) and another suggesting that metacyclic forms are haploid (18). However when individual metacyclic forms (derived from trypanosomes heterozygous at various loci) were used to infect mice, the resulting bloodstream stage trypanosomes were still heterozygous (19). Thus, the trypanosomes did not pass through a haploid stage, thereby establishing that the metacyclic stage must be diploid. Analysis of the DNA content of the epimastigote and trypomastigote stages of *T. cruzi* (20), using fluorescence-activated cell sorting (FACS) analysis of cells stained with propidium iodide, has shown that the epimastigote stage (in several stocks) contains some 2–12% less DNA. The molecular basis of these changes remains unknown, but suggests no major difference in ploidy.

Using similar techniques, strain differences in DNA content of the order of 5–10% have been shown in *T. brucei* spp, *T. vivax*, and *T. congolense* (19, 21). There are also differences between species, with the largest variation being found between *T. congolense* and *T. brucei* spp. This may reflect differing chromosome sizes or copy numbers of aneuploid chromosomes (see below). Much larger differences in DNA content (up to 70%) have been reported between strains of *T. cruzi* (22, 23). These occur between different isolates but also arise spontaneously during the culture of a cloned line (23). The mechanisms and molecular basis for these differences and changes in DNA content remain obscure, although changes in chromosome number or amplification of large regions of the genome could be involved.

The base composition and codon usage of those genes sequenced to date predict average GC contents of 49% (*T. brucei*), 56% (*T. cruzi*) and 64% (*L. major*) (24, 25). Recent reports provide evidence for two modified bases in *T. brucei*, namely 5-hydroxymethyl uracil and a novel modified base, β-D-glycosyl hydroxymethyl uracil (26, 27). The latter is of low abundance, is present only in the bloodstream stage of the life cycle, and is associated with silent telomeric VSG genes (27).

3. Genetic analysis

The question of the existence of genetic exchange together with its occurrence and role in natural populations has been and continues to be the subject of debate (9, 11, 13, 28, 29). To review this area, it is convenient to consider two aspects separately: first, the evidence for the existence of genetic exchange, and second, the arguments and data concerning its frequency and role in natural populations.

3.1 Genetic exchange

At present there is incontrovertible evidence that *T. brucei* spp. can undergo genetic exchange (9, 30) based on laboratory crosses, while there is indirect evidence that some *Leishmania* species can undergo genetic exchange (31) but no evidence for the process in *T. cruzi* (13).

In *Leishmania*, the evidence for genetic exchange rests on three sets of observations. First, fusion of promastigote cells in culture has been observed microscopically in *L. infantum* and *L.tropica* (31), although marker analysis of the products of fusion has not yet been undertaken. Second, data on the DNA content of individual amastigotes in macrophages have shown that two different populations of parasite exist, one with the same DNA content as the promastigote stage and the other with twice the DNA content, suggesting that at least fusion between cells can occur (32). Third, analysis of isolates from a small area in Saudi Arabia has identified a cloned line which is an apparent hybrid between the two sympatric species of *L. major* and *L. arabica* (33). Based on an analysis of six RFLPs, five iso-enzyme patterns, and PFGE karyotyping, this clone has all the characterisitics of a heterozygote formed by gene exchange between the two species. These latter observations are difficult to explain on any other basis than that of genetic exchange (although this strain could represent the product of a fusion, followed by DNA loss). On the other hand, attempts to demonstrate genetic exchange directly in *Leishmania*, by laboratory infection of the sandfly vector with two genetically marked stocks of *L. major*, have been unsuccessful (34). In summary, genes can or have been exchanged between species and within strains of *Leishmania* but the mechanism involved and its relationship to a classical genetic system remain obscure.

In *T. brucei* spp., crosses can be made by infecting tsetse flies with two different stocks and then isolating hybrid progeny after the trypanosomes have completed the developmental stages in the fly. The main features of the system have been reviewed recently (9, 30). The analysis of cloned and uncloned trypanosome lines derived from tsetse flies infected with two stocks of *T. brucei* has shown that genetic exchange is non-obligatory (35, 36) and has also provided evidence for both cross- and self-fertilization (36–39). Marker analysis shows that the products of mating are the equivalent of an F1 in a broadly classical Mendelian system. Analysis of the segregation and recombination of alleles at several loci, coupled with data on the segregation of homologous chromosomes (separated by PFGE), is consistent with *T. brucei* being diploid and undergoing meiosis (37–41). However, it is unclear whether meiosis precedes mating (to generate haploid nuclei or cells) or whether it occurs after fusion of two diploid cells (9, 30). The life cycle stage at which mating takes place is the subject of conflicting reports (40, 42). It is clear that the metacyclic stage (in the salivary glands) is a product of mating, as parasite lines produced from single metacyclic stage trypanosomes derived from flies infected with two parasite stocks are recombinant (9, 30, 37); the determination of the life cycle stage involved awaits further investigation.

While much of the data to date are consistent with a conventional Mendelian system, no formal proof in terms of segregation ratios has been obtained. In addition, a proportion of the progeny clones have approximately 1.5 times the normal DNA content (39, 43, 44). Since those clones studied in detail show trisomy for the specific chromosomes analysed (44), it is reasonable to assume that such progeny clones are triploid. The initial interpretation of the raised DNA content led to a fusion model without meiosis (43). However, the occurrence of a considerable number of progeny clones with normal DNA content that show segregation and recombination between alleles and chromosomes (39) argues in favour of meiosis, with the increased DNA content arising as a result of chromosomal non-disjunction.

A number of aspects of this system of genetic exchange require further research in order to elucidate the basic mechanisms and frequency. The ability to use selective markers enhances the amenability of the system to study (40, 41) and, although genetic analysis is unlikely to be used routinely, it could be an invaluable tool, when coupled with genome analysis, in the identification of genes determining traits of clinical and epidemiological significance.

3.2 Genetic exchange in natural populations

Analysis of isoenzyme variation in a *T. brucei* population has been interpreted as providing evidence for panmixia and random mating (45), since the allele frequencies observed fit with those predicted by the Hardy-Weinberg equilibrium. Statistical tests have shown that this fit could be due to chance and that the number of expected recombinant genotypes was higher than the number observed (46). Although this suggests that the population was not panmictic, the variation observed cannot be accounted for on a purely asexual mutational model.

The role of genetic exchange in the generation of the observed diversity between strains in natural populations has recently received renewed attention with the publication of a clonal theory to explain the population structure of the parasitic protozoa (47). This theory extends findings from the analysis of *T. cruzi* population variation (11) to a wide range of parasitic protozoa, including *Leishmania* spp., *Trypanosoma brucei* spp., and *Trypanosoma congolense*. Essentially, using available data on iso-enzyme variation, a set of criteria for recombination were tested statistically for deviations from an idealized panmictic population in which there is no selection and no bias introduced by sampling methodology (the latter may be a significant factor, as analysis of these parasite isolates was dependant on their amplification by *in vitro* culture or infection of laboratory hosts). These criteria included Hardy-Weinberg predictions, representation of multi-locus genotypes, heterozygosity, absence of segregation genotypes, and linkage disequilibrium (48). When *T. brucei brucei* populations isolated from West and East Africa were analysed separately, the data did not show significant deviations from expectation for many of the criteria (47). However, if the data from the two populations were combined, significant deviations from a model predicting regular crossing were demonstrated. The fact that these populations are several thousand miles apart and could represent isolated populations was not considered. Analysis of the data from *T. cruzi* and *Leishmania* spp. showed significant deviations from a model of frequent genetic exchange and panmixia, suggesting that these parasites have a clonal population structure in which genetic exchange is rare or absent (47, 48). In the analysis of similar questions with bacterial populations, it has been argued (28) that linkage disequilibrium and the predominance of a few multilocus genotypes can arise in essentially sexual populations as a result of an epidemic population structure.

Using statistical methods which test whether the rapid expansion of a few genotypes in a population can 'mask' a predominantly sexual population, it is possible to determine whether a population is clonal or has an 'epidemic' population structure in which recombination is frequent but occasionally a particular subset of genotypes predominate. Application of these tests to a large iso-enzyme data set from *T. brucei* (49) and *T. cruzi* (50) led to the conclusion that, while *T. cruzi* was clonal, the *T. brucei* population had an 'epidemic' structure in which recombination was frequent (28). These conclusions are in agreement with the data presented by Tibayrenc *et al.* (47) for *T. b. brucei* although not with the interpretation presented by those authors.

In summary, the available evidence suggests that the population structure of *T. cruzi* is essentially clonal while that of the *T. brucei* group of sub-species has an epidemic structure with the role of recombination being variable, depending on the specific epidemiological situation. The situation as regards *Leishmania* spp. is less clear. The available evidence (47, 48) supports a clonal structure, although the statistical analyses of Maynard-Smith *et al.* (28) have not been undertaken.

4. Molecular karyotype

The development of techniques for the separation of large, chromosome-sized DNA fragments by gel electrophoresis has led to the definition of some of the features of the karyotype of the trypanosomatids. A number of different systems have been used for such separations but these will all be referred to by the generic term PFGE. These studies reveal considerable diversity in the size of homologous chromosomes both within a genome and between genomes of different strains and species. No detailed long range restriction maps of these chromosomes have been constructed to date, resulting in a less than ideal state of knowledge about the structural relationships between the different DNA bands identified on gels. While each trypanosomatid species shows certain unique attributes, the following generalizations can be made:

1. The relative staining intensity of individual bands suggests that many contain the DNA from several chromosomes.
2. The number of bands and their size varies between different cloned strains.
3. The karyotypes are the same in different life cycle stages.
4. Homologous chromosomes can differ significantly in size within a genome as well as between strains.

To clearly establish these generalizations, it will be necessary to construct detailed physical maps for each chromosome. Currently, a number of laboratories are undertaking the genome analysis of representative strains of *T. cruzi*, *Leishmania*, and *T. b. brucei*, and the data generated from this work will lead to clarification of the structural relationships between chromosome bands.

4.1 *Trypanosoma cruzi* karyotype

The molecular karyotype of a number of different strains and clones of *T. cruzi* has been examined by PFGE (51–55). Some 20–24 ethidium bromide staining DNA bands have been separated which range in size from 300kb to 2Mb. Different cloned lines from a single isolate show karyotype variation, suggesting that isolates can contain several different genotypes (52). The data on clonal variation in karyotype are provocative. Analyses of cloned derivatives of the Y strain show that clones with similar or identical DNA contents have identical karyotypes, while those differing in DNA content show variation in the staining intensity of bands and altered chromosome mobilities (23, 53). In other strains (51, 52), clones with high DNA content do not show evidence for a larger number of chromosomes. Furthermore, in some strains, DNA content and chromosome profiles change during continuous culture, while in other strains the karyotype is stable. At the present time, the basis for these changes in karyotype and DNA content are unclear.

In order to define the genome organization and the relationship between the chromosome bands separated by PFGE, a range of probes encoding housekeeping, antigen, spliced leader (SL), and ribosomal genes have been used to probe blots of separated chromosomes and restriction enzyme digested total genomic DNA. It is difficult to define a comprehensive 'map' of the genome from these experiments as the parasite stocks, the separation conditions, and the probes used differ between studies. These data are illustrated schematically in Fig. 1 which shows the three different classes of hybridization pattern observed. At present, a more extensive analysis using many markers is needed to define the number of chromosomes and their relationships to each other at the molecular level.

4.2 *Leishmania* spp. karyotype

The karyotype of a wide range of *Leishmania* species has been examined using a number of different PFGE systems (56) and variation in the size of the chromosome bands demonstrated, both between strains of the same species and between different species. The total number of resolved ethidium bromide staining bands is strain and species dependent, within a range of 19–29. The sizes of the bands vary between 250kb and >2Mb with some bands estimated to contain DNA between two and seven chromosomes. Some authors (57–59) have estimated, on the basis of size and intensity of fluorescence of each band, that 25–50% of the DNA content can be accounted for in these separations. In order to define putative homologous chromosomes within a genome and between genomes of different strains and species, as well as to examine the ploidy of *Leishmania* spp., specific gene probes have been mapped to chromosomes separated by PFGE. For example, the genes for ribosomal RNA, the hsp70 heat shock protein, β-tubulin, phosphofructokinase, pyruvate kinase, and ubiquitin (57–63) have been mapped to chromosomes ≥ 1Mb, while the genes encoding the spliced leader RNA, the major surface glycoprotein (gp63), dihydrofolate reductase, and a number of anonymous probes (56–62) have been mapped to the 'mid' sized chromosomes (400–900kb). Without detailed physical

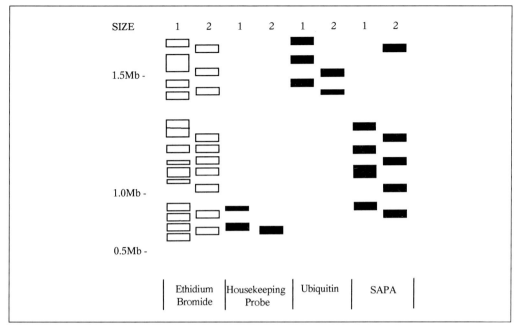

Fig. 1 Diagrammatic representation of the karyotype of two strains (strains 1 & 2) of *T. cruzi* showing the variation in size of homologous chromosomes both within and between strains detected by Southern blots of PFGE separations probed with different classes of gene sequence. This diagram is based on the data presented in Refs 52–54. SAPA, shed acute phase antigen gene.

mapping it is difficult to define, with any certainty, whether all chromosomes are present as pairs of homologues.

A number of chromosome size changes in cloned lines have been reported. These can be conveniently divided into two classes: first, the appearance of small (≤200kb) DNAs of high copy number (57, 63); and second, significant size changes (30–100kb) in specific chromosomes (60, 64). The first class occurs spontaneously during culture and arises during selection or nutrient stress (63). These DNA elements have been termed 'extra-chromosomal' and are derived from chromosomes within the genome. The best studied alterations of the second type are in *L. major*, where it has been shown that amplification of the repeated spliced leader gene located on chromosome 2 causes an increase in size of one homologue from 350kb to 385kb, the latter co-migrating with chromosome 3 (64). Size changes in this chromosome have also been reported in response to drug selection. In *L. infantum*, alterations in chromosome size after sub-cloning have been reported (62). The mechanisms could involve translocations between non-homologous chromosomes, amplification of sub-telomeric sequences, or amplification of chromosome internal sequences (58). Again, detailed mapping studies are needed to define these alterations in detail.

Although these chromosome alterations tend to be highlighted, the main body of data suggest that the *Leishmania* karyotype is, for the most part, stable. Whether the observed chromosome size changes represent the basic mechanisms by which karyotype diversity between strains is generated, or are the result of artificial conditions imposed by *in vitro* culture, is unclear but requires further analysis.

4.3 *Trypanosoma brucei* karyotype

The most studied African trypanosome species in terms of karyotype is *T. brucei* (65, 66) and this discussion will primarily focus on this organism. Chromosomes of *T. brucei* can be divided into three size classes that account for 80% of the genome: the mini-chromosomes (50–150kb) and two larger groups of 150–700kb and 800kb-6Mb (67, 68). There are approximately 100 mini-chromosomes per genome (69) and these are linear, aneuploid molecules containing telomeres, multiple copies of a 177bp repeat and a reservoir of non expressed VSG genes (65). The size and number of these mini-chromosomes vary between species and it has been suggested that this reflects the size of the VSG gene repertoire. The intermediate size chromosomes are also variable in number and size between strains, species, and cloned lines expressing different antigenic variants (70). Like the mini-chromosomes, they have been shown to contain the 177bp satellite sequence, VSG genes, and a VSG expression site, but a detailed analysis of their structure has not been undertaken. Comparisons between cloned lines of trypanosomes expressing different VSG genes have shown that size alterations in these chromosomes can occur as a result of duplicative transposition of VSG genes (71; see Chapter 6).

Ethidium bromide staining of separated chromosomes from stock 427 has identified a total of 13 bands larger than 1Mb (65, 66) in addition to a further band close to the sample slot; this may represent trapped DNA. The clusters of basic copy VSG genes are not distributed on all of the large chromosomes but restricted to a few (three in the case of stock 427), while some of the expression sites for these genes have also been localized to three chromosome bands of which two are of different size to those encoding the clusters of basic copy VSG genes (65). The large chromosomes have been further characterized by the localization of 18 genetic markers (66) and the results are summarized in Fig. 2. The available evidence supports the conclusion that the large (>1Mb) chromosomes are diploid and that homologous chromosomes can differ in size by several 100kb or be of identical size, depending on the chromosome considered and the parasite strain. As a result of such size variations, non-homologous chromosomes can be of the same size in some cases. The nature of the size differences between homologous chromosomes is unclear and, although some can be accounted for by telomere growth and recombination in these regions, this is unlikely to account for the very large size differences seen between some homologues.

5. Gene organization

Single copy genes appear to occur in relatively low frequency in trypanosomatids and tandemly repeated genes represent about half the genes studied to date. This appears to be a general characteristic rather than the sampling of a particular class

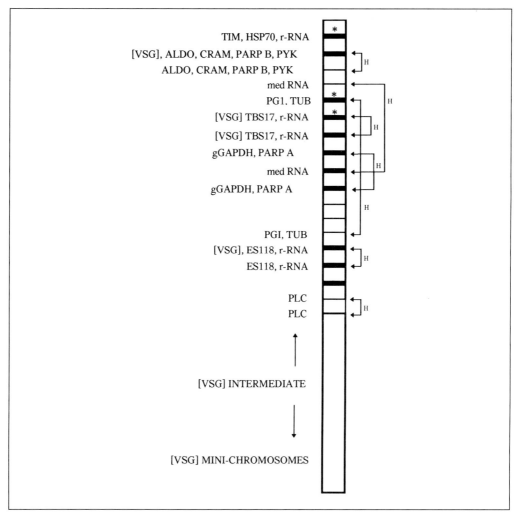

Fig. 2 Diagrammatic representation of the genetic markers localized on PFGE separations of *T. brucei* chromosomes. Arrows marked with an H indicate pairs of homologous chromosomes. Bands represent the ethidium bromide staining pattern; the relative intensity of staining is indicated by the thickness of the band. Gene probes used: TIM, triosephosphate isomerase; HSP70, 70kDa heatshock protein; r-RNA, ribosomal RNA; ALDO, aldolase; [VSG], C-terminal region of the variable surface glycoprotein; CRAM, putative cell surface receptor; PARP B, procyclic acid repetitive protein, locus B; PYK, pyruvate kinase; med RNA, spliced leader repeat sequence; PGI, phosphoglucose isomerase; TUB, α and β tubulin; TBS17, anonymous cDNA clone; gGAPDH, glycosomal glyceraldehyde phosphate dehydrogenase; ES118, expression site for VSG18; PLC, phospholipase C; * 'presumed' band, not actually separated. Data taken from Ref. 70.

of genes (e.g. antigen versus housekeeping). The apparently high proportion of multicopy genes could be explained in two ways: first, as a means of preventing the loss of essential genes, given the propensity of these organisms to translocate and amplify their gene sequences; and second, as a means of determining levels of specific mRNA given the general lack of transcriptional regulation of trypanosomatid gene expression (see Chapter 6).

Five classes of gene organization can be defined (Fig. 3). The main examples of the first class (one to three copies at a single locus) encode the genes for enzymes essential for normal metabolism (10, 12). Typically, for such genes, the organization is similar between the three genera. However, for repeated genes (classes 3 and 4), the organization is not always conserved among genera or even species. For example, the tubulin genes of *T. brucei* are arranged as alternating α-β repeats on a single chromosome (72), while the α and β tubulin genes of *Leishmania* are arranged in separate tandem arrays on different chromosomes (59–61). The converse is found with the rRNA genes which exist as tandem arrays on several chromosomes in *T. brucei* (70) but are arranged in tandem arrays at a single locus in *Leishmania* (57–59) and *T. cruzi* (53). An interesting feature of a number of the multiple copy, single locus genes, is that the copy number may be different on homologous chromosomes. For example, in *T. brucei*, the hexose transporter gene cluster varies in copy number between homologues (73).

In *T. brucei*, two transposon-like elements, *RIME* and *INGI*, have been described and are widely dispersed in the genome (74). A third transposon-like sequence, *SLACS*, has been described which is confined to the spliced leader RNA locus and occurs at a copy number of 2–30 (75). Related sequences, *CZAR* (76) in *T. cruzi* and *CRE1* in *Crithidia fasciculata* (77), have also been located to the same locus. Conser-

Class		Examples		
		T. brucei	*T. cruzi*	*Leishmania*
1. 1-3 copies single locus		PG1, ALD, gGAPDH	ALD, gGAPDH	DHFR, gGAPDH
2. 1-4 copies at two loci		RNA polII, PARP	?	?
3. Multicopy, single locus		UBI, SL, THT, β-TUB	r-RNA, SL^3, UBI	gp63^{1,2}, r-RNA
4. Multicopy, > 1 locus		r-RNA	SL^3	β-TUB^2
5. Gene family, many loci		VSG	gp85	?

Fig. 3 Gene organization in *T. brucei*, *T. cruzi* and *Leishmania*. Examples of the genes corresponding to the different organizational classes are shown. The lines represent chromosomes, with non-homologous chromosomes being indicated by different types of line. Abbreviations: PGI, phosphoglucose isomerase; ALD, aldolase; gGAPDH, glycosomal glyceraldehyde phosphate dehydrogenase; DHFR, dihydrofolate reductase; RNA polII, RNA polymerase II; PARP, procyclic acidic repetitive protein; UBI, ubiquitin; SL, spliced leader RNA gene; r-RNA, ribosomal RNA repeat; THT, trypanosome hexose transporter repeat; β-TUB, β tubulin repeat; gp63, 63kDa surface glycoprotein; VSG, variable surface glycoprotein; gp85, 85kDa surface glycoprotein. [1] gp63 has variable gene organization depending on the species of *Leishmania*; [2] occur as multicopy arrays but also as single copy at other loci, in addition; [3] can be transposed to other chromosomes and loci.

vation of specific domains within these sequences suggest that they represent a family of elements of related function (78). Whether they have a role in translocation of sequences within the genome is an open question, although this is suggested by the high frequency of rearrangement (1% per generation) of the *CRE1* element in *Crithidia fasciculata* (77). Their wider role in the evolution and stability of gene organization in the trypanosomatids remains to be established.

A further mechanism for altering gene organization and, potentially, ploidy is DNA amplification. To date, some form of DNA amplification has been observed in all three main groups of trypanosomatids in response to drug selection (63, 79, 80), nutritional stress, or as a spontaneous event (53, 81). The amplification and transposition events within a chromosome have been outlined in the previous section while those leading to the generation of 'extrachromosomal', self-replicating elements present in multiple copies are dealt with in Chapter 10 of this volume. Although amplification events have been observed in both *T. cruzi* and *T. brucei* spp., these appear to be limited to intrachromosomal amplifications or complete chromosomal duplications. By contrast, DNA amplification seems to occur frequently in *Leishmania* spp. and can generate as much as 5–10% of the total cellular DNA (68). This propensity to amplify circular 'extrachromosomal' elements has been extensively exploited in the generation of *Leishmania* transfection vectors.

6. Requirements for gene expression

For a gene to be expressed in trypanosomatids, whether it be endogenous or introduced by transformation, multiple events must occur. Of these events, transcription initiation is the least understood, principally because few promoters have been identified. Thus far, most of the promoters which have been described are transcribed by RNA polymerase I (82–85) or RNA polymerase III (86–88). The only RNA polymerase II (pol II) promoter so far described is that of the spliced leader gene of *Leishmania tarentolae* (89). Why pol II promoters of protein coding genes have not been identified is unclear but may be related to two facts. The first is that much of the transcription in trypanosomatids appears to be polycistronic (90–93) and the second is that in almost every reported case, the genes on the chromosome are transcribed from the same DNA strand (94–97). Very few examples of either convergent or divergent transcription have been reported (88, 98). With such an organization of transcription units, very few promoters may be needed to transcribe large segments of a chromosome. In an extreme case, one could imagine a single promoter driving transcription of an entire chromosome arm. Although this is an unlikely example, it could explain the scarcity of pol II promoters. An alternative explanation is that promoters of this type do not exist in trypanosomes. Rather, transcription may initiate randomly within the ApT rich regions found between trypanosomatid genes. In either case, it is possible that investigators will have to look beyond the classically defined promoters seen in other eukaryotic organisms to understand transcription initiation in trypanosomatids.

In most cases, transcription of the tandemly repeated genome results in the generation of polycistronic primary transcripts which are processed into monogenic mRNAs via 5′-end *trans*-splicing (99–102) and 3′-end polyadenylation (92, 103–105). *Trans*-splicing essentially involves the joining together of the first 39 nucleotides of the spliced leader primary transcript and the precursor polycistronic transcript. As a result each mRNA is capped at its 5′-end with a common sequence, referred to as the spliced leader or mini-exon RNA. The basic mechanism of *trans*-splicing is highly reminiscent of that seen in *cis*-splicing in higher eukaryotes and the *cis*-acting elements required for *trans*-splicing are beginning to be fairly well characterized. These sequence elements include the ApG 3′ splice-acceptor site, a polypyrimidine tract, and one or more branch points (106–108). *Trans*-splicing is essential for gene expression in trypanosomatids and, because of its central role, it has been one of the most intensively studied molecular mechanisms in these organisms (see Chapter 7 for a full discussion of *trans*-splicing).

7. DNA-mediated transformation systems

DNA-mediated transformation is well developed in each of the major pathogenic species of trypanosomatids, including members of the *T. brucei* complex (84, 109–114), *T. cruzi* (115–119), and *Leishmania* (109, 117, 120–123). In large part, the coming of age of trypanosomatid molecular genetics has been due to the development of these systems. Transformation is rapidly becoming a fundamental tool for the analysis of gene expression and function in these parasites. The mechanisms of transcription (83, 86, 89, 111, 124–128), mRNA processing and mRNA stability (103, 104, 106, 119, 129) have each been the subject of intensive investigation using transformation-based methodologies. More recently, these methods have been used to study protein targeting (130) and function (131), in addition to isolating genes by functional complementation (132).

Essentially, transformation systems can be subdivided into two major categories: transient and stable systems. Transient systems are those in which a plasmid vector is introduced into the parasite and expression of a plasmid-borne reporter gene is measured a short time later. These systems have proven to be very useful for the study of transcription, RNA processing, and protein targeting.

Stable transformation refers to the permanent alteration of the parasite genome. This can occur either as a result of the maintenance of a freely replicating episomal plasmid (113, 117, 133–136) or by insertion of a plasmid or DNA fragment into a chromosome by homologous recombination (109, 112, 114, 116, 118, 119, 121, 137–140). The efficiency of this latter process has made the isolation of targeted gene disruptions and replacements relatively straightforward. Homologous recombination has also made it possible to place non-selected reporter genes and mutations such as mis-sense, nonsense, and frameshift mutations on the chromosome (83, 89, 107, 118, 119, 129). Therefore, essentially all the molecular tools are in place to carry out careful genetic analyses of most cellular processes. The significance of molecular genetic tools is highlighted by the inability, or at best difficulty, in carrying out classical genetic studies in trypanosomatids. Therefore future progress in understanding the mechanisms of gene expression and function will to a large extent be dependent on transformation-based genetics.

8. Transient transfection systems

Transient transformation has proven to be an indispensable component of many molecular analyses in trypanosomatids. In the absence of purified *in vitro* systems, transformation has played an instrumental role in many studies investigating the mechanisms of transcription and RNA processing. Much of our current understanding of the ways in which trypanosomatid mRNAs are matured is the product of experiments in which the activity of wild-type and mutated versions of *cis*-acting sequences have been studied in transiently transfected parasites.

8.1 Transient expression vectors

Although many different transient expression vectors have been used successfully in trypanosomatids, they all commonly consist of a prokaryotic plasmid backbone and reporter gene as well as the trypanosomatid DNA sequences necessary for reporter gene expression. A generic expression vector therefore contains a reporter gene flanked on the 5′ side by sequences required for *trans*-splicing and on the 3′ side by sequences which direct polyadenylation of the reporter gene mRNA. The tandem arrangement of genes in trypanosomatids dictates that the intergenic regions separating protein coding sequences must contain the sequence elements required for polyadenylation and *trans*-splicing of the upstream and downstream genes respectively. It is a relatively straightforward matter, therefore, to include these elements in transformation vectors (Fig. 4).

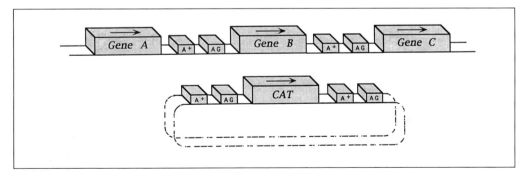

Fig. 4 Map of a typical transient expression vector. Genes on the chromosome are designated as boxes labeled Gene A, Gene B and Gene C. On the chromosome and the plasmid vector, the sequences required for polyadenylation and *trans*-splicing are designated as boxes labeled A+ and AG respectively. The box labeled CAT represents the coding sequence for the chloramphenicol acetyltransferase gene. The plasmid backbone is represented by the dashed lines. The direction of transcription is indicated by the arrows above each gene.

8.2 Reporter genes

As described above, transient transformation systems are based on a reporter gene whose expression can be accurately measured. Reporter genes are usually of prokaryotic origin and encode products for which there is no trypanosomatid analogue. Therefore, there is little if any background activity to obscure the result. Several reporter genes have been used extensively, including the chloramphenicol acetyltransferase (*CAT*) (84, 111, 119, 127, 141–143), β-glucuronidase (*GUS*) (130, 144), firefly luciferase (*LUX*) (130, 145) and β-galactosidase (146–149) genes. The expression of each reporter gene can be measured by relatively simple biochemical techniques. Synthesis of the reporter gene product is an indirect measure of the processes which lead to expression, i.e. transcription, *trans*-splicing, and polyadenylation.

8.3 Promoters and mRNA processing

Of the processes leading to gene expression in trypanosomatids, transcription initiation is clearly the least understood. Although transient transfection has been used to identify RNA polymerase I promoters in African trypanosomes (84, 85, 111, 150) and the RNA polymerase II promoter of the *L. tarentolae* spliced-leader gene (89), little success has been achieved in the search for RNA polymerase II promoters of protein coding genes. Why this has been the case is not completely clear although many of the currently used transient transfection plasmids support promoter independent transcription, making them unsuitable for these experiments. It has been frequently observed that providing the sequence elements necessary for RNA processing, 5'-end *trans*-splicing and 3'-end polyadenylation is sufficient to achieve gene expression. In the absence of *trans*-splicing, however, reporter gene expression is invariably blocked. Although this simple observation has lead to a large body of work in which transient expression systems have been used to define the sequence elements required for efficient *trans*-splicing *in vivo* (104, 106–108, 151, 152), it has complicated efforts at defining promoters. Many mutations that block reporter gene expression, and on initial inspection appear to define promoter elements, in fact define nucleotide sequences involved in *trans*-splicing. Therefore, it has become clear that care must be taken when using transformation systems to identify promoters of transcription in trypanosomatids.

Although much less is known about the mechanism of 3'-end polyadenylation, it is clear that this is required for efficient gene expression. In both *Leishmania* (104) and *T. brucei* (106), transient expression analyses have been used to generate convincing evidence that, in fact, polyadenylation and *trans*-splicing are functionally coupled (see Chapter 7). Using transformation based methodologies, investigators have shown that, within a single intergenic region, the polypyrimidine tract which plays a role in *trans*-splicing of the downstream transcript also dictates the site of 3'-polyadenylation of the upstream transcript. Further systematic analyses have not yet been carried out to identify possible sequences which play a role in polyadenylation. In spite of this, however, many different intergenic regions have been successfully used in expression vectors to provide the sequences required for polyadenylation. It is likely that most, if not all, intergenic regions include these 3'-end processing sequences due to the tandem arrangement of genes in these parasites.

8.4 Methods of transformation

Electroporation is clearly the method used most frequently to introduce DNA into trypanosomatids. A wide range of different conditions have been used, that differ slightly for each parasite and often between different laboratories working with the same parasite (84, 116–118, 122, 126, 137, 141, 143, 153, 154).

8.5 Applications of transient transfection

With further development, transient transfection systems are likely to become useful tools in future investigations of, among other things, promoter function, mRNA processing, gene regulation, and protein targeting. The major drawback of spurious transcription from the plasmid backbone of many constructs (mentioned earlier) could be overcome by including transcription terminators to block the read-through transcription. Although no terminators have been identified in trypanosomatids to date, polyadenylation signals have been successfully used for this purpose in mammalian cell expression systems. However, it remains an open question as to how closely transcription from a plasmid resembles transcription from the native chromosome. Therefore, promoters identified using transient transfection will have to be studied in their native chromosomal context as well.

Although promiscuous transcription may pose a significant problem for the use of transient transfection systems as methods for promoter identification, it is not a serious problem for the analysis of mRNA processing. Transient transfections have been an integral component in many of the studies which have lead to our understanding of mRNA processing, 5'-end *trans*-splicing, 3'-end polyadenylation, and the relationship between the two processes (see chapters 6 and 7). Mutations which disrupt mRNA processing adversely affect reporter gene expression even if transcription initiates from the plasmid backbone. Therefore mutational analyses can be carried out without a complete understanding of transcription initiation.

9. Stable transformation

Stable transformation, one of the most useful techniques available for the study of molecular parasitology, is well developed for each of the pathogenic trypanosomatid species (109, 112, 113, 118, 119, 122, 155, 156). With stable transformation, the parasite genome can essentially be manipulated at will. The only significant constraint is that changes cannot be made which would be lethal to the parasite. Several different methods have been developed to generate stable transformants. As described below, these methods include maintenance of a plasmid as an extrachromosomal element, integration of a circular plasmid into the chromosome by a single recombination event, and integration of a linear DNA fragment by double recombination.

9.1 Selectable markers

Generation of stable transformants necessitates the use of dominant selectable marker genes. Since auxotrophic markers are not available for any of the trypanosomatids, antibiotic resistance genes must be used to select for positive transformants. Several effective dominant markers have been used successfully. The first stable transformants of *T. brucei*, *Leishmania*, and *T. cruzi* were selected on the basis of expression of the aminoglycosidase phosphotransferase gene (*neo*) of the bacterial transposon Tn5, which confers resistance to the antibiotic G418 (109, 114, 116, 140). The *neo* gene is still the most widely used dominant selectable marker although a second marker, the *E. coli* hygromycin B phosphotransferase gene (*hyg*) which confers resistance to hygromycin B, has also been sucessfully used in all three parasite species (121, 137, 156). Less widely used, to date, are the genes confering phleomycin (112, 155) and puromycin (155) resistance.

9.2 Integrative transformation

Sequence-specific integration of plasmids and DNA fragments has been surprisingly straightforward in typanosomatids. Thus far, all integration events described have been mediated by homologous recombination; non-homologous events have not been observed. Therefore, in this regard, trypanosomatids resemble other lower eukaryotes such as *Saccharomyces cerevisiae* rather than higher eukaryotes, in which site-specific targeting is at best inefficient. Integration of both circular (109, 114, 116) and linear DNA molecules (113, 118, 119, 121, 137, 138, 140, 157, 158) has been described. A single recombination event is sufficient for integration of circular molecules while integration of linear molecules requires at least two recombination events (Fig. 5).

Integrative transformation has been used to both disrupt (116, 157) and replace (118–120, 137, 138, 140, 158, 159) targeted genes. These are critically important experimental tools considering the multicopy nature of trypanosomatid genomes. When studying the multicopy gene families, it is often difficult and sometimes impossible to determine if specific members of the gene family are expressed or if expression of all members is coordinated throughout the parasites' life cycle. It is also frequently difficult to determine if mRNAs for each member of the gene family are processed in the same manner. This may turn out to be an important consideration since much of gene regulation in trypanosomatids appears to be post-transcriptional. These difficulties can be overcome by using gene replacement techniques. One member of a multicopy gene family can be replaced with an unrelated reporter gene or a tagged gene copy. By analysing the patterns of expression of the inte-

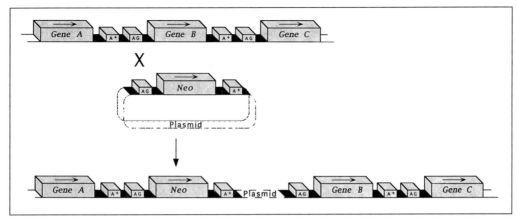

Fig. 5 Stable integrative transformation. The labeling is exactly as described for Figure 4. The target locus is labeled, as are the products of circular plasmid integration (Integrative) and gene replacement mediated by a linear DNA fragment (Replacement) events.

grated sequence, insights can be gained into the pattern of expression of the replaced gene as well as the other members of the gene family. Processing of the reporter transcript can also be studied outside the background imposed by the other gene copies.

Integrative transformation also provides the means to determine if a gene or genes are essential for parasite survival. This is typically shown by demonstrating a positive selection for at least one expressed copy of a gene or gene family. If the gene of interest is essential for parasite survival, then all copies of the gene cannot be replaced unless one copy is expressed from either an episomal plasmid or an alternative chromosomal locus. An example of the latter approach was the demonstration the both calmodulin-ubiquitin associated (*CUB*) genes of *T. cruzi* could be deleted only if a new *CUB* gene was expressed from the calmodulin locus (160).

The minimal requirements for integration are unclear since few systematic studies have been carried out. Although it has been assumed that the free ends of linear DNA molecules are more recombinagenic than closed circular plasmids, this assumption has not been systematically tested. The extent of uninterrupted homology between the transforming DNA and the target sequence necessary to promote integration is also unclear and a systematic study of the requirements for integration has not been undertaken. Different investigators have successfully used anywhere from several thousands of base pairs to less than 100 base pairs of homologous sequence to promote site-specific integration (113, 119). Thus far, investigators have primarily concentrated on generating replacements and disruptions of their gene(s) of interest rather than studying the recombination events which lead to integration. This will certainly change since transfection systems are ideally suited for studying this phenomenon.

Experiments have also been carried out in *T. cruzi* demonstrating that, following electroporation, linear DNA fragments can be converted to circular molecules which can then integrate by a single recombination event (161). In the reported case, the linear fragment was generated by digestion with a single restriction endonuclease, *Mlu*I, giving a linear DNA fragment with compatible ends. When the linear fragment had non-compatible ends, circularization was not observed. Although it is not known if compatible ends are universally required for circularization, it is likely that circularization of linear molecules occurs at some measurable frequency in all trypanosomatid species.

9.3 Episomal transformation

Stable transformation has also been achieved using freely replicating plasmid vectors (117, 120, 135). Although episomal transformation is common in *Leishmania*, it is less so in *T. brucei* and *T. cruzi*. The stability of these transformants is generally contingent upon continued antibiotic selection. In the absence of drug selection, the plasmids are rapidly lost since the *cis*-acting DNA elements necessary for episome stabilization have not been defined in trypanosomatids and therefore would be unlikely to be present in most vectors. Although trypanosomatid telomeres (158, 162, 163) and at least one origin of DNA replication (133) have been characterized, nothing is known about chromosome centromeres, or if in fact these organisms have conventional centromeres. In spite of these shortcomings, episomal transformation has proven to be an invaluable tool in many molecular genetic analyses.

Plasmid vectors used for episomal transformation have the same requirements as those used for transient transfection but also need a selectable marker gene (see above). As with transient transfection, transcription initiation again falls into the realm of uncertainty. Frequently, as long as the sequence elements required for mRNA processing are included in the plasmid, there is no obvious requirement for a defined transcription promoter. Although this at first would seem unlikely, it is possible that spurious transcription initiation from the ApT-rich trypanosome sequences on the plasmid construct, or from the plasmid backbone itself, may be sufficient to establish drug resistance provided that the transcripts can be accurately processed.

Plasmid amplification is one of the common characteristics of episomal transformation for each trypanosomatid species. Normally, after obtaining a transformed parasite or parasite population, the plasmid copy number remains relatively stable in the presence of continued antibiotic selection. Frequently, however, if the concentration of antibiotic is gradually increased there is a corresponding increase in plasmid copy number. It is not difficult to obtain transformants in which the transforming plasmid vector has been amplified a hundred times or more. The plasmid amplification is not stable since once drug selection is decreased or removed the plasmid copy number decreases. The observation that many plasmid vectors can be amplified suggests that most vectors include a hidden trypanosomatid origin of DNA replication.

One of the most useful aspects of episomal transformation is the ability to easily transfer (shuttle) genes back and forth between the parasite and *E. coli*. This is particularly true when one considers the possibility of cloning genes by genetic complementation as has been done in *Leishmania* (130, 132). Genes identified by complementation in the parasite can then be shuttled to *E. coli* where they can be more easily studied. Shuttle vectors typically carry two selectable genes, one for selection in *E. coli* and the other for selection in the parasite, as well as an expression cassette. The expression cassette normally consists of a cloning site(s) flanked by parasite intergenic regions which provide the sequence necessary for mRNA processing. Cosmid vectors, which have been used in *T. cruzi* (164) and *Leishmania* (120), are one of the most versatile types of shuttle vector since large segments of chromosomal DNA (up to approximately 40kb) can be cloned in a single plasmid. Although only in the developmental stages, *Leishmania* artificial chromosomes are also being tested and hold great promise owing to their ability to carry even larger chromosome segments. The full development of cosmid vectors and parasite artificial chromosomes will have tremendous impact on the analysis of gene function in parasites.

9.4 Inducible gene expression

Although inducible expression systems have not been fully developed for *Leishmania* and *T. cruzi*, such a system has recently been described for *T. brucei* (165). This system uses transgenic trypanosomes expressing the tetracycline repressor of *E. coli* to modulate expression from the procyclic acid repetitive protein (PARP) RNA polymerase I promoter of *T. brucei*. In the absence of tetracycline, expression is repressed but can be induced by the addition of low concentrations of the antibiotic to the growth medium. When the tetracycline operator is cloned between the PARP promoter and the reporter gene, expression of the latter can be up-regulated by a factor of 10000. This ability to regulate the expression of transfected genes represents an important advance which will significantly enhance studies of a variety of processes in trypanosomes.

10. Conclusion

There have been significant impediments to genetic analyses of the trypanosomatids. Although models for various molecular processes could be envisioned, it was difficult or impossible to test them *in vivo*. Transformation-based genetics now provides the means to test most models *in vivo*, thus lifting one of the last major obstacles to molecular genetic analysis in the trypanosomatids. The recent development of an inducible gene expression system has removed one of the last limitations to progress. Although this system is currently available only in *T. brucei*, it is likely to be adapted to *Leishmania* and *T. cruzi* within a short time. The remaining major limitation is the inability to carry out random insertional mutagenesis similar to transposon mutagenesis in bacteria. If past experience is a guide, however, this barrier is likely to fall soon as well.

The trypanosomatids are a diverse family of protozoan parasites, many of which cause disease world wide, in both humans and animals. Hundreds of millions of people and countless agriculturally important animals are afflicted and multitudes more are at risk. Knowledge of the mechanisms of gene expression and function will undoubtedly lead to a better understanding of these important pathogens.

References

1. Borst, P., Van der Ploeg, M., Van Hoek, J. F. M., Tas, J., and James, J. (1982) On the DNA content of trypanosomes. *Mol. Biochem. Parasitol.*, **6**, 13.
2. Borst, P., Fase-Fowler, F., Frasch, A. C. C., Hoeijmakers, J. H. J., and Weijers, P. J. (1980) Characterisation of DNA from *Trypanosoma brucei* and related trypanosomes by restriction endonuclease digestion. *Mol. Biochem. Parasitol.*, **1**, 221.
3. Vickerman, K. and Preston, T. M. (1970) Spindle microtubules in the dividing nuclei of trypanosomes. *J. Cell Sci.*, **6**, 365.
4. Solari, A. J. (1980) The 3-dimensional structure of the mitotic spindle in *Trypanosoma cruzi*. *Chromosoma*, **78**, 239.
5. Schwartz, D. and Cantor, C. R. C. (1984) Separation of yeast chromosome sized DNAs by pulse field gradient gel electrophoresis. *Cell*, **37**, 67.
6. Van der Ploeg, L. H. T., Valerio, D., De Lange, T., Bernards, A., Borst, P., and Grosveld, F. G. (1982) An analysis of cosmid clones of nuclear DNA from *Trypanosoma brucei* shows that genes for variant surface glycoproteins are clustered in the genome. *Nucleic Acids Res.*, **10**, 5905.
7. Leon, W., Fonts, D. L., and Manning, J. (1978) Sequence arrangement of the 16S and 26S rRNA genes in the pathogenic haemoflagellate *Leishmania donovani*. *Nucleic Acids Res.*, **5**, 491.
8. Lanar, D. E., Levy, L. S., and Manning, J. E. (1991) Complexity and content of the DNA and RNA in *Trypanosoma cruzi*. *Mol. Biochem. Parasitol.*, **3**, 327.
9. Tait, A. and Turner, C. M. R. Genetic exchange in *Trypanosoma brucei*. *Parasitol. Today*, **6**, 70.
10. Gibson, W. C., Osinga, K. A., Michels, P. A., and Borst, P. (1985) Trypanosomes of subgenus Trypanozoon are diploid for housekeeping genes. *Mol. Biochem Parasitol.*, **16**, 231.
11. Tibayrenc, M., Ward, P., Moya, A., and Ayala, F. J. (1986) Natural populations of *Trypanosoma cruzi*, the agent of Chagas disease, have a complex multiclonal structure. *Proc. Natl Acad. Sci. USA.* **83**, 115.
12. Gibson, W. C. and Miles, M. A. (1986) The karyotype and ploidy of *Trypanosoma cruzi*. *EMBO J.*, **5**, 1299.
13. Tait, A. (1983) Sexual processes in the kinetoplastida. *Parasitology*, **86**, 29–57.
14. Souza, A. E., Bates, P. A., Coombs, G. H., and Mottram, J. C. (1994) Null mutants for the lmcpa cysteine proteinase in *Leishmania mexicana*. *Mol. Biochem. Parasitol.* **63**, 213.
15. Webb, J. R. and McMaster, W. R. (1994) *Leishmania major* HEXBP deletion mutants generated by double targeted gene replacement. *Mol. Biochem. Parasitol.*, **63**, 231.
16. Borst, P. and Cross, G. A. M. (1982) Molecular basis for trypanosome antigenic variation. *Cell*, **29**, 291.
17. Kooy, R. F., Hirumi, H., Moloo, S. K., Nantulya, V. M., Dukes, P., Van der Linden, P. M., Duijudan, W. A. L., Janse, C. J. and Overdulve, J. P. (1989) Evidence for diploidy in metacyclic forms of African trypanosomes. *Proc. Natl Acad. Sci. USA*, **86**, 5469.

18. Zampetti-Bosseler, F., Schweizer, J., Pays, E., Jenni, L., and Steinert, M. (1986) Evidence for haploidy in metacyclic forms of *Trypanosoma brucei*. *Proc. Natl Acad. Sci. USA*, **83,** 6063.
19. Tait, A., Turner, C. M. R., Le Page, R. W. F., and Wells, J. M. (1989) Genetic evidence that metacyclic forms of *Trypanosoma brucei* are diploid. *Mol. Biochem. Parasitol.*, **37,** 247.
20. Nozaki, T. and Dvorak, J. A. (1991) *Trypanosoma cruzi*: flow cytometric analysis of developmental stage differences in DNA. *J. Protozool.*, **18,** 234.
21. Garside, L., Bailey, M., and Gibson, W. C. (1994) DNA content and molecular karyotype of trypanosomes of the subgenus *Nannomonas*. *Acta Tropica*, **57,** 21.
22. Dvorak, J. A., Hall, T. E., Crane, M. J., Engel, J. C., McDaniel, J. P., and Uriegas, R. (1982) *Trypanosoma cruzi*: flow cytometric analysis. I. Analysis of total DNA/organism by means of mithramycin-induced fluorescence. *J. Protozol.*, **29,** 430.
23. McDaniel, J. P. and Dvorak, J. A. (1993) Identification, isolation and characterisation of naturally occurring *Trypanosoma cruzi* variants. *Mol. Biochem. Parasitol.* **57,** 213.
24. Musto, N., Rodrigiez-Maseda, H., and Bernardi, G. (1994) Nuclear genomes of African and American trypanosomes are strikingly different. *Gene*, **141,** 63.
25. Alvarez, F., Robello, C., and Vignali, M. (1994) Evolution of codon usage and base contents in kinetoplastid protozoa. *Mol. Biol. Evol.*, **11,** 790.
26. Gommers-Ampt, J. H., Lutgeriuk, J., and Borst, P. (1991) A novel DNA nucleotide in *Trypanosoma brucei* only present in the mammalian phase of the life cycle. *Nucleic Acids. Res.*, **19,** 1745.
27. Gommers-Ampt, J. H., Van Leewen, F., de Beer, A. L. J., Villegenthart, J. F. G., Dizdaroglu, M., Kowalak, J. A. Crain, P. F., and Borst, P. (1993) β-D-glucosyl-hydroxymethyl uracil: a novel modified base present in the DNA of the parasitic protozoan *T. brucei*. *Cell*, **75,** 1129.
28. Maynard Smith, J., Smith, N. H., O'Rourke, M., and Spratt, B. G. (1993) How clonal are bacteria? *Proc. Natl Acad. Sci. USA.* **90,** 4384.
29. Bastein, P., Blaineau, C., and Pagès, M. (1992) *Leishmania*: sex, lies and karyotype. *Parasitol. Today*, **8,** 174.
30. Sternberg, J. and Tait, A. (1990) Genetic exchange in African trypanosomes. *Trends Genet.* **6,** 317.
31. Lanotte, G. and Rioux, J. A. (1990) Fusion cellulaire chez les *Leishmania* Kinetoplastida, Trypanosomatidae. *C. R. Acad. Sci.*, **310,** Sér III, 285.
32. Kreutzer, R. D., Yemma, J. J., Grocl, M., Tesh, R. B., and Martin, T. I. (1994) Evidence of sexual reproduction in the protozoan parasite *Leishmania*. *Am. J. Trop. Med. Hyg.*, **51,** 301.
33. Kelly, J. M., Law, J. M., Chapman, C. J., van Eys, G. J., and Evans, D.A. (1991) Evidence for genetic recombination in *Leishmania*. *Mol. Biochem. Parasitol.*, **46,** 253.
34. Panton, L. J., Tesh, R. B., Nadean, L. C., and Beverley, S. M. (1991) A test for genetic exchange in mixed infections of *Leishmania major* in the sandfly *Phleobotomus papatasi*. *J. Protozool.*, **38,** 224.
35. Schweizer, J., Tait, A., and Jenni, L. (1988) The timing and frequency of hybrid formation in African trypanosomes during cyclical transmission. *Parasitol. Res.*, **75,** 98.
36. Sternberg, J., Turner, C. M. R., Wells, J. M., Ranford-Cartwright, L. C., Le Page, R. W. F. and Tait, A. (1989) Gene exchange in African trypanosomes: frequency and allelic segregation. *Mol. Biochem. Parasitol.*, **27,** 191.
37. Turner, C. M. R., Sternberg, J., Buchanan, N., Smith, E., Hide, G., and Tait, A. (1990) Evidence that the mechanism of genetic exchange in *Trypanosoma brucei* involves meiosis and syngamy. *Parasitology*, **101,** 377.

38. Gibson, W. C. and Garside, L. H. (1991) Genetic exchange in *Trypanosoma brucei brucei*: variable chromosomal location of housekeeping genes in different trypanosome stocks. *Mol. Biochem. Parasitol.*, **45**, 77.
39. Tait, A., Buchanan, N., Hide, G., and Turner, C. M. R. (1993) Genetic recombination and karyotype inheritance in *Trypanosoma brucei* species. In *Genome analysis of protozoan parasites* (ed. S.P. Morzaria), pp. 93, ILRAD, Kenya.
40. Gibson, W. C. and Whittington, H. (1993) Genetic exchange in *Trypanosoma brucei*: selection of hybrid trypanosomes by introduction of genes conferring drug resistance. *Mol. Biochem. Parasitol.*, **60**, 19.
41. Gibson, W. and Bailey, M. (1994) Genetic exchange in *Trypanosoma brucei*: evidence for meiosis from analysis of a cross between drug resistant transformants. *Mol. Biochem. Parasitol.*, **64**, 241.
42. Schweizer, J. and Jenni, L. (1991) Hybrid formation in the life cycle of *Trypanosoma brucei*: detection of hybrid trypanosomes in a midgut derived isolate. *Acta Tropica*, **48**, 319.
43. Paindavoine, P., Zampetti-Bosler, F., Pays, E., Schweizer, J., Guyaux, M., Jenni, L., and Steinert, M. (1986) Trypanosome hybrids generated in tsetse flies by nuclear fusion. *EMBO J.*, **5**, 3631.
44. Gibson, W. C., Garside, L., and Bailey, M. (1992) Trisomy and chromosome size changes in hybrid trypanosomes from a genetic cross between *Trypanosoma brucei rhodesiense* and *T. b. brucei*. *Mol. Biochem. Parasitol*, **52**, 189.
45. Tait, A. (1980) Evidence for diploidy and mating in trypanosomes. *Nature*, **287**, 536.
46. Cibulskis, R. E. (1988) Origins and organisation of genetic diversity in natural populations of *Trypanosoma brucei*. *Parasitology*, **96**, 303.
47. Tibayrenc, M., Kjellberg, F., and Ayala, F. J. (1990) A conal theory of parasitic protozoa: the population structures of *Entamoeba, Giardia, Leishmania, Naegleria, Plasmodium, Trichomonas* and *Trypanosoma* and their medical and taxonomic consequences. *Proc. Natl. Acad. Sci. USA*, **87**, 2414.
48. Tibayrenc, M. and Ayala, F. J. (1991) Towards a population genetics of micro-organisms: the clonal theory of parasitic protozoa. *Parasitol. Today*, **7**, 228.
49. Mihok, S., Otieno, L. H., and Darji, N. (1989) Population genetics of *Trypanosoma brucei* and the epidemiology of human sleeping sickness in the Kaubie Valley, Kenya. *Parasitology*, **100**, 219.
50. Zhang, Q., Tibayrene, M., and Ayala, F. J. (1988) Linkage disequilibrium in natural populations of *Trypanosoma cruzi* (Flagellate), the agent of Chagas' disease. *J. Protozool.*, **35**, 81.
51. Gibson, W. C. and Miles, M. A. (1986) The karyotype and ploidy of *Trypanosoma cruzi*. *EMBO J.*, **5**, 1299.
52. Engman, D. M., Reddy, L. V., Donelson, J. E., and Kirchoff, L. V. (1987) *Trypanosoma cruzi* exhibits inter and intra strain heterogeneity in molecular karyotype and chromosomal gene location. *Mol. Biochem. Parasitol.*, **22**, 115.
53. Wagner, W. and So, M. (1990) Genomic variation of *Trypanosoma cruzi*: involvement of multicopy genes. *Infect. Immunity*, **58**, 3217.
54. Henriksson, J., Ashmd, L., Macina, R. A., Franke de Cazzulo, B. M., Cazzulo, J. J., Frasch, A. C. C., and Petersson, U. (1990) Chromosomal localisation of seven cloned antigen genes provides evidence of diploidy and further demonstration of karyotype variability in *Trypanosoma cruzi*. *Mol. Biochem. Parisitol.*, **42**, 213.
55. Ayermich, S. and Goldenberg, S. (1989) The karyotype of *Trypanosoma cruzi* Dm 28C; comparison with other strains and trypanosomatids. *Exp. Parasitol.*, **69**, 107.

56. Lighthall, G. K. and Giannini, S. H. (1992) The chromosomes of *Leishmania*. *Parasitol. Today*, **8**, 192.
57. Scholler, J. K., Reed, S. G., and Stuart, K. (1986) Molecular karyotype of species and subspecies of *Leishmania*. *Mol. Biochem. Parasitol.*, **20**, 279.
58. Pagès, M., Bastien, P., Veas, F., Rossi, V., Bellis, M., Winkler, P., Rioux, J. A., and Roizès, G. (1989) Chromosome size and number polymorphisms in *Leishmania infantum* and possible genetic exchange. *Mol. Biochem. Parasitol.*, **36**, 161.
59. Galindo, I. and Ochoa, J. L. R. (1989) Study of *Leishmania mexicana* electrokaryotype by clamped homogeneous electric field electrophoresis. *Mol. Biochem. Parasitol.*, **34**, 245.
60. Giannini, S. H., Curry, S. S., Tesh, R. B., and Van der Ploeg, L. H. T. (1990) Size conserved chromosomes and stability of molecular karyotype in cloned stocks of *Leishmania major*. *Mol. Biochem., Parasitol.*, **39**, 9.
61. Spithill, T. W. and Samaras, N. (1987) Genomic organisation, chromosomal location and transcription of dispersed and repeated tubulin genes in *Leishmania major*. *Mol. Biochem. Parasitol.*, **24**, 23.
62. Blaineau, C., Bastien, P., Rioux, J.-A., Roizès, G., and Pagès, M. (1991) Long range restriction maps of size variable chromosomes in *Leishmania infantum*. *Mol. Biochem. Parasitol.*, **46**, 293.
63. Beverley, S. M. (1991) Gene amplification in *Leishmania*. *Annu. Rev. Microbiol.*, **45**, 417.
64. Iovannisci, D. M. and Beverley, S. M. (1989) Structural alterations of chromosome 2 in *Leishmania major* as evidence for diploidy, including spontaneous amplification of the miniexon array. *Mol. Biochem. Parasitol.*, **34**, 177.
65. Van der Ploeg, L. H. T., Smith, C. L., Polvere, R. I., and Gottesdiener, K. M. (1989) Improved separation of chromosome sized DNA from *Trypanosoma brucei* stock 427–60. *Nucleic Acids Res.*, **17**, 3217.
66. Gottesdiener, K., Garcia-Anoveros, J., Lee, M. G. and Van der Ploeg, L. H. T. (1990) Chromosome organisation of the protozoan *Trypanosoma brucei*. *Mol. Cell. Biol.*, **10**, 6079.
67. Gibson, W. C. and Borst, P. (1986) Size fractionation of the small chromosomes of Trypanozoon and Nennamonas trypanosomes by pulse field gel electrophoresis. *Mol. Biochem. Parasitol.*, **18**, 127.
68. Van der Ploeg, L. H. T., Cornelissen, A. W. C. A., Barry, J. D., and Borst, P. (1984) Chromosomes of kinetoplastida. *EMBO J.*, **3**, 3109.
69. Van der Ploeg, L. H. T., Cornelissen, A. W. C. A., Michels, P. A. M., and Borst, P. (1984) Chromosome re-arrangements in *Trypanosoma brucei*. *Cell* **37**, 77.
70. Weiden, M., Osheim, Y. N., Beyer, A. L., and Van der Ploeg, L. H. T. (1991) Chromosome structure: DNA nucleotide sequence elements of a sub-set of the minichromosomes of the protozoan *Trypanosoma brucei*. *Mol. Cell. Biol.*, **11**, 3823.
71. Van der Ploeg, L. H. T., Cornelissen, A. W. C. A., Michels, P. A. M., and Borst, P. (1984) Chromosome re-arrangements in *Trypanosoma brucei*. *Cell* **39**, 213.
72. Thomashow, L. S., Milhansen, M., Rutter, W. J., and Agabian, N. (1983) Tubulin genes are tandemly linked and clustered in the genome of *Trypanosoma brucei*. *Cell*, **32**, 35.
73. Bringau, F. and Baltz, T. (1994) African trypanosome glucose transporter genes: organisation and evolution of a multigene family. *Mol. Biol. Evolution*, **11**, 220.
74. Kimmel, B. E., Ole-Moiyoi, O. K. and Young, J. R. (1987) Ingi, a 5.2kb dispersed sequence element from *Trypanosoma brucei* that carries half of a small mobile element at either end and has homology with mammalian LINEs. *Mol. Cell. Biol.*, **7**, 1465.
75. Aksoy, S., Williams, S., Chang, S., and Richards, F. F. (1990) SLACS retrotransposon from *Trypanosoma brucei gambiense* is similar to mammalian LINEs. *Nucleic Acids Res.*, **18**, 785.
76. Villanueva, M. S., Williams, S. P., Beard, C. B., Richards, F. F., and Asksoy, S. (1991) A

new member of a family of site specific retrotransposons is present in the spliced leader RNA genes of *Trypanosoma cruzi*. *Mol. Cell. Biol.*, **11,** 6139.
77. Gabriel, A., Yen, T. J., Scwartz, D. C., Smith, C. L., Boeke, J. D., Solluer-Webb, B., and Cleveland, D. W. (1990) A rapidly rearranging retrotransposon within the mini-exon gene locus of *Crithidia fasciculata*. *Mol. Cell. Biol.*, **10,** 615.
78. Aksoy, S. (1991) Site specific retrotransposons of the trypanosomatid protozoa. *Parasitol. Today*, **7,** 281.
79. Stuart, K. D. (1991) Circular and linear multicopy DNAs in *Leishmania*. *Parasitol. Today*, **7,** 158.
80. Navarro, M., Maingon, R., Hamers, R., and Sergovia, M. (1992) Dynamics and size polymorphisms of mini-chromosomes in *Leishmania major* LV-561 cloned lines. *Mol. Biochem. Parasitol.* **55,** 65.
81. Wilson, K., Berens, R. L., Sifri, C. D., and Ullman, B. (1994) Amplification in the inosinate dehydrogenase gene in *Trypanosoma brucei gambiense* due to an increase in chromosome copy number. *J. Biol. Chem.*, **269,** 28978.
82. Zomerdijk, J. C., Kieft, R., and Borst, P. (1992) A ribosomal RNA gene promoter at the telomere of a mini-chromosome in *Trypanosoma brucei*. *Nucleic Acids Res.*, **20,** 2725.
83. Brown, S. D., Huang, J., and Van der Ploeg, L. H. (1992) The promoter for the procyclic acidic repetitive protein (PARP) genes of *Trypanosoma brucei* shares features with RNA polymerase I promoters. *Mol. Cell. Biol.*, **12,** 2644.
84. Jefferies, D., Tebabi, P., and Pays, E. (1991) Transient activity assays of the *Trypanosoma brucei* variant surface glycoprotein gene promoter: control of gene expression at the posttranscriptional level. *Mol. Cell. Biol.*, **11,** 338.
85. Rudenko, G., Le Blancq, S., Smith, J., Lee, M. G., Rattray, A., and Van der Ploeg, L. H. (1990) Procyclic acidic repetitive protein (PARP) genes located in an unusually small alpha-amanitin-resistant transcription unit: PARP promoter activity assayed by transient DNA transfection of *Trypanosoma brucei*. *Mol. Cell. Biol.*, **10,** 3492.
86. Fantoni, A., Dare, A. O., and Tschudi, C. (1994) RNA polymerase III-mediated transcription of the trypanosome U2 small nuclear RNA gene is controlled by both intragenic and extragenic regulatory elements. *Mol. Cell. Biol.*, **14,** 2021.
87. Campbell, D. A., Suyama, Y., and Simpson, L. (1989) Genomic organisation of nuclear tRNAGly and tRNALeu genes in *Trypanosoma brucei*. *Mol. Biochem. Parasitol.*, **37,** 257.
88. Nakaar, V., Dare, A. O., Hong, D., Ullu, E., and Tschudi, C. (1994) Upstream tRNA genes are essential for expression of small nuclear and cytoplasmic RNA genes in trypanosomes. *Mol. Cell. Biol.*, **14,** 6736.
89. Saito, R. M., Elgort, M. G., and Campbell, D. A. (1994) A conserved upstream element is essential for transcription of the *Leishmania tarentolae* mini-exon gene. *EMBO J.*, **13,** 5460.
90. Johnson, P. J., Kooter, J. M., and Borst, P. (1987) Inactivation of transcription by UV irradiation of *T. brucei* provides evidence for a multicistronic transcription unit including a VSG gene. *Cell*, **51,** 273.
91. Muhich, M. L. and Boothroyd, J. C. (1988) Polycistronic transcripts in trypanosomes and their accumulation during heat shock: evidence for a precursor role in mRNA synthesis. *Mol. Cell. Biol.*, **8,** 3837.
92. Huang, J. and van der Ploeg, L. H. (1991) Maturation of polycistronic pre-mRNA in *Trypanosoma brucei*: analysis of trans splicing and poly(A) addition at nascent RNA transcripts from the hsp70 locus. *Mol. Cell. Biol.*, **11,** 3180.
93. Tschudi, C. and Ullu, E. (1988) Polygene transcripts are precursors to calmodulin mRNAs in trypanosomes. *EMBO J.*, **7,** 455.

94. Swindle, J., Ajioka, J., Eisen, H., Sanwal, B., Jacquemot, C., Browder, Z., and Buck, G. (1988) The genomic organization and transcription of the ubiquitin genes of *Trypanosoma cruzi*. *EMBO J.*, **7**, 1121.
95. Chung, S. H. and Swindle, J. (1990) Linkage of the calmodulin and ubiquitin loci in *Trypanosoma cruzi*. *Nucleic Acids Res.*, **18**, 4561.
96. Ajioka, J. and Swindle, J. (1993) The calmodulin-ubiquitin associated genes of *Trypanosoma cruzi*: their identification and transcription. *Mol. Biochem. Parasitol.*, **57**, 127.
97. Tschudi, C., Young, A. S., Ruben, L., Patton, C. L., and Richards, F. F. (1985) Calmodulin genes in trypanosomes are tandemly repeated and produce multiple mRNAs with a common 5′ leader sequence. *Proc. Natl. Acad. Sci. USA.*, **82**, 3998.
98. Hughes, D. E., Shonekan, O. A., and Simpson, L. (1989) Structure, genomic organization and transcription of the bifunctional dihydrofolate reductase-thymidylate synthase gene from *Crithidia fasciculata*. *Mol. Biochem. Parasitol.*, **34**, 155.
99. Murphy, W. J., Watkins, K. P., and Agabian, N. (1986) Identification of a novel Y branch structure as an intermediate in trypanosome mRNA processing: evidence for *trans*-splicing. *Cell*, **47**, 517.
100. Sutton, R. E. and Boothroyd, J. C. (1986) Evidence for *trans*-splicing in trypanosomes. *Cell*, **47**, 527.
101. Boothroyd, J. C. and Cross, G. A. M. (1982) Transcripts coding for surface glycoprotein in *Typanosoma brucei* have a short identical exon at their 5′ end. *Gene*, **20**, 281.
102. Van der Ploeg, L. T., Liu, A. Y. C., Michels, P. A. M., Delange, T., Majumder, H. K., Weber, H., Veenemen, G. H., and Van Bloom, J. (1982) RNA splicing is required to make the messenger RNA for the varient surface antigen in trypanosomes. *Nucleic Acids Res.*, **10**, 3591.
103. Ullu, E., Matthews, K. R., and Tschudi, C. (1993) Temporal order of RNA-processing reactions in trypanosomes: rapid *trans*-splicing precedes polyadenylation of newly synthesized tubulin transcripts. *Mol. Cell. Biol.*, **13**, 720.
104. Le Bowitz, J. H., Smith, H. Q., Rusche, L., and Beverley, S. M. (1993) Coupling of poly(A) site selection and *trans*-splicing in *Leishmania*. *Genes Dev.*, **7**, 996.
105. Bellofatto, V. and Cross, G. A. (1988) Characterization of RNA transcripts from the alpha tubulin gene cluster of *Leptomonas seymouri*. *Nucleic Acids Res.*, **16**, 3455.
106. Matthews, K. R., Tschudi, C., and Ullu, E. (1994) A common pyrimidine-rich motif governs *trans*-splicing and polyadenylation of tubulin polycistronic pre-mRNA in trypanosomes. *Genes Dev.*, **8**, 491.
107. Huang, J. and Van der Ploeg, L. H. (1991) Requirement of a polypyrimidine tract for *trans*-splicing in trypanosomes: discriminating the PARP promoter from the immediately adjacent 3′ splice acceptor site. *EMBO J.*, **10**, 3877.
108. Patzelt, E., Perry, K. L., and Agabian, N. (1989) Mapping of branch sites in *trans*-spliced pre-mRNAs of *Trypanosoma brucei*. *Mol. Cell. Biol.*, **9**, 4291.
109. ten Asbroek, A. L., Ouellette, M., and Borst, P. (1990) Targeted insertion of the neomycin phosphotransferase gene into the tubulin gene cluster of *Trypanosoma brucei*. *Nature*, **348**, 174.
110. Carruthers, V. B., Van der Ploeg, L. H., and Cross, G. A. (1993) DNA-mediated transformation of bloodstream-form *Trypanosoma brucei*. *Nucleic Acids Res.*, **21**, 2537.
111. Clayton, C. E., Fueri, J. P., Itzhaki, J. E., Bellofatto, V., Sherman, D. R., Wisdom, G. S., Vijayasarathy, S., and Mowatt, M. R. (1990) Transcription of the procyclic acidic repetitive protein genes of *Trypanosoma brucei*. *Mol. Cell. Biol.*, **10**, 3036.
112. Jefferies, D., Tebabi, P., Le Ray, D., and Pays, E. (1993) The *ble* resistance gene as a

new selectable marker for *Trypanosoma brucei*: fly transmission of stable procyclic transformants to produce antibiotic resistant bloodstream forms. *Nucleic Acids Res.*, **21,** 191.

113. ten Asbroek, A. L., Mol, C. A., Kieft, R., and Borst, P. (1993) Stable transformation of *Trypanosoma brucei. Mol. Biochem. Parasitol.*, **59,** 133.
114. Lee, M. G. and Van der Ploeg, L. H. (1990) Homologous recombination and stable transfection in the parasitic protozoan *Trypanosoma brucei. Science*, **250,** 1583.
115. Tomas, A. M. and Kelly, J. M. (1994) Transformation as an approach to functional analysis of the major cysteine protease of *Trypanosoma cruzi. Biochem. Soc. Trans.*, **22,** 90S.
116. Hariharan, S., Ajioka, J., and Swindle, J. (1993) Stable transformation of *Trypanosoma cruzi*: inactivation of the PUB12.5 polyubiquitin gene by targeted gene disruption. *Mol. Biochem. Parasitol.*, **57,** 15.
117. Kelly, J. M., Ward, H. M., Miles, M. A., and Kendall, G. (1992) A shuttle vector which facilitates the expression of transfected genes in *Trypanosoma cruzi* and *Leishmania. Nucleic Acids Res.*, **20,** 3963.
118. Gillespie, R. D., Ajioka, J., and Swindle, J. (1993) Using simultaneous, tandem gene replacements to study expression of the multicopy ubiquitin-fusion (FUS) gene family of *Trypanosoma cruzi. Mol. Biochem. Parasitol.*, **60,** 281.
119. Chung, S. H., Gillespie, R. D., and Swindle, J. (1994) Analyzing expression of the calmodulin and ubiquitin-fusion genes of *Trypanosoma cruzi* using simultaneous, independent dual gene replacements. *Mol. Biochem. Parasitol.*, **63,** 95.
120. Ryan, K. A., Dasgupta, S., and Beverley, S. M. (1993) Shuttle cosmid vectors for the trypanosomatid parasite *Leishmania. Gene*, **131,** 145.
121. Cruz, A., Coburn, C. M., and Beverley, S. M. (1991) Double targeted gene replacement for creating null mutants. *Proc. Natl. Acad. Sci. USA*, **88,** 7170.
122. Kapler, G. M., Coburn, C. M., and Beverley, S. M. (1990) Stable transfection of the human parasite *Leishmania major* delineates a 30-kilobase region sufficient for extrachromosomal replication and expression. *Mol. Cell. Biol.*, **10,** 1084.
123. Hughes, D., Simpson, L., Kayne, P. S., and Neckelmann, N. (1984) Autonomous replication sequences in the maxicircle kinetoplast DNA of *Leishmania tarentolae. Mol. Biochem. Parasitol.*, **13,** 263.
124. Wong, A. K., de Lafaille, M. A., and Wirth, D. F. (1994) Identification of a *cis*-acting gene regulatory element from the *lemdr1* locus of *Leishmania enriettii. J. Biol. Chem.*, **269,** 26497.
125. Zomerdijk, J. C., Kieft, R., and Borst, P. (1993) Insertion of the promoter for a variant surface glycoprotein gene expression site in an RNA polymerase II transcription unit of procyclic *Trypanosoma brucei. Mol. Biochem. Parasitol.*, **57,** 295.
126. Zomerdijk, J. C., Ouellette, M., ten Asbroek, A. L., Kieft, R., Bommer, A. M., Clayton, C. E., and Borst, P. (1990) The promoter for a variant surface glycoprotein gene expression site in *Trypanosoma brucei. EMBO J.*, **9,** 2791.
127. Curotto de Lafaille, M. A., Laban, A., and Wirth, D. F. (1992) Gene expression in *Leishmania*: analysis of essential 5' DNA sequences. *Proc. Natl. Acad. Sci. USA*, **89,** 2703.
128. Rudenko, G., Chung, H. M., Pham, V. P., and Van der Ploeg, L. H. (1991) RNA polymerase I can mediate expression of CAT and neo protein-coding genes in *Trypanosoma brucei. EMBO J.*, **10,** 3387.
129. Schneider, A., McNally, K. P., and Agabian, N. (1993) Splicing and 3'-processing of the tyrosine tRNA of *Trypanosoma brucei. J. Biol. Chem.*, **268,** 21868.
130. Sommer, J. M., Peterson, G., Keller, G. A., Parsons, M., and Wang, C. C. (1993) The C-

terminal tripeptide of glycosomal phosphoglycerate kinase is both necessary and sufficient for import into the glycosomes of *Trypanosoma brucei*. *FEBS Lett.*, **316,** 53.
131. Bello, A. R., Nare, B., Freedman, D., Hardy, L., and Beverley, S. M. (1994) PTR1: a reductase mediating salvage of oxidized pteridines and methotrexate resistance in the protozoan parasite *Leishmania major*. *Proc. Natl. Acad. Sci. USA*, **91,** 11442.
132. Ryan, K. A., Garraway, L. A., Descoteaux, A., Turco, S. J., and Beverley, S. M. (1993) Isolation of virulence genes directing surface glycosyl-phosphatidylinositol synthesis by functional complementation of *Leishmania*. *Proc. Natl. Acad. Sci. USA*, **90,** 8609.
133. Patnaik, P. K., Fang, X., and Cross, G. A. (1994) The region encompassing the procyclic acidic repetitive protein (PARP) gene promoter plays a role in plasmid DNA replication in *Trypanosoma brucei*. *Nucleic Acids Res.*, **22,** 4111.
134. Metzenberg, S. and Agabian, N. (1994) Mitochondrial minicircle DNA supports plasmid replication and maintenance in nuclei of *Trypanosoma brucei*. *Proc. Natl. Acad. Sci. USA*, **91,** 5962.
135. Bellofatto, V., Torres-Munoz, J. E., and Cross, G. A. (1991) Stable transformation of *Leptomonas seymouri* by circular extrachromosomal elements. *Proc. Natl. Acad. Sci. USA*, **88,** 6711.
136. Laban, A., Tobin, J. F., Curotto de Lafaille, M. A., and Wirth, D. F. (1990) Stable expression of the bacterial neoR gene in *Leishmania enriettii*. *Nature*, **343,** 572.
137. Cooper, R., de Jesus, A. R., and Cross, G. A. (1993) Deletion of an immunodominant *Trypanosoma cruzi* surface glycoprotein disrupts flagellum-cell adhesion. *J. Cell. Biol.*, **122,** 149.
138. Curotto de Lafaille, M. A. and Wirth, D. F. (1992) Creation of *Null/+* mutants of the alpha-tubulin gene in *Leishmania enriettii* by gene cluster deletion. *J. Biol. Chem.*, **267,** 23839.
139. Tobin, J. F., Laban, A., and Wirth, D. F. (1991) Homologous recombination in *Leishmania enriettii*. *Proc. Natl. Acad. Sci. USA*, **88,** 864.
140. Cruz, A. and Beverley, S. M. (1990) Gene replacement in parasitic protozoa. *Nature*, **348,** 171.
141. Bellofatto, V. and Cross, G. A. (1989) Expression of a bacterial gene in a trypanosomatid protozoan. *Science*, **244,** 1167.
142. Laban, A. and Wirth, D. F. (1989) Transfection of *Leishmania enriettii* and expression of chloramphenicol acetyltransferase gene. *Proc. Natl. Acad. Sci. USA*, **86,** 9119.
143. Lu, H. Y. and Buck, G. A. (1991) Expression of an exogenous gene in *Trypanosoma cruzi* epimastigotes. *Mol. Biochem. Parasitol.*, **44,** 109.
144. Le Bowitz, J. H., Coburn, C. M., and Beverley, S. M. (1991) Simultaneous transient expression assays of the trypanosomatid parasite *Leishmania* using beta-galactosidase and beta-glucuronidase as reporter enzymes. *Gene*, **103,** 119.
145. Hug, M., Carruthers, V. B., Hartmann, C., Sherman, D. S., Cross, G. A., and Clayton, C. (1993) A possible role for the 3'-untranslated region in developmental regulation in *Trypanosoma brucei*. *Mol. Biochem. Parasitol.*, **61,** 87.
146. Kaye, P. M., Coburn, C., McCrossan, M., and Beverley, S. M. (1993) Antigens targeted to the *Leishmania* phagolysosome are processed for CD4+ T cell recognition. *Eur. J. Immunol.*, **23,** 2311.
147. Coburn, C. M., Otteman, K. M., McNeely, T., Turco, S. J., and Beverley, S. M. (1991) Stable DNA transfection of a wide range of trypanosomatids. *Mol. Biochem. Parasitol.*, **46,** 169.
148. Lopez, J. A., Le Bowitz, J. H., Beverley, S. M., Rammensee, H. G., and Overath, P. (1993) *Leishmania mexicana* promastigotes induce cytotoxic T lymphocytes *in vivo* that do not recognize infected macrophages. *Eur. J. Immunol.*, **23,** 217.

149. Le Bowitz, J. H., Coburn, C. M., McMahon-Pratt, D., and Beverley, S. M. (1990) Development of a stable *Leishmania* expression vector and application to the study of parasite surface antigen genes. *Proc. Natl. Acad. Sci. USA*, **87,** 9736.
150. Zomerdijk, J. C., Kieft, R., Shiels, P. G., and Borst, P. (1991) Alpha-amanitin-resistant transcription units in trypanosomes: a comparison of promoter sequences for a VSG gene expression site and for the ribosomal RNA genes. *Nucleic Acids Res*, **19,** 5153.
151. Sutton, R. E. and Boothroyd, J. C. (1988) Trypanosome *trans*-splicing utilizes 2'-5' branches and a corresponding debranching activity. *EMBO J.*, **7,** 1431.
152. Agabian, N. (1990) *Trans* splicing of nuclear pre-mRNAs. *Cell*, **61,** 1157.
153. Beverley, S. M. and Clayton, C. E. (1993) Transfection of *Leishmania* and *Trypanosoma brucei* by electroporation. *Methods Mol. Biol.*, Vol. 21, p. 333, Humana Press Inc., Totowa, NJ.
154. Gibson, W. C., White, T. C., Laird, P. W., and Borst, P. (1987) Stable introduction of exogenous DNA into *Trypanosoma brucei* [erratum appears in *Ethiop. Med. J.* (1988), **26** 59]. *EMBO J.*, **6,** 2457.
155. Freedman, D. J. and Beverley, S. M. (1993) Two more independent selectable markers for stable transfection of *Leishmania*. *Mol. Biochem. Parasitol.*, **62,** 37.
156. Lee, M. G. and van der Ploeg, L. H. (1991) The hygromycin B-resistance-encoding gene as a selectable marker for stable transformation of *Trypanosoma brucei*. *Gene*, **105,** 255.
157. Otsu, K., Donelson, J. E., and Kirchhoff, L. V. (1993) Interruption of a *Trypanosoma cruzi* gene encoding a protein containing 14-amino acid repeats by targeted insertion of the neomycin phosphotransferase gene. *Mol. Biochem. Parasitol.*, **57,** 317.
158. de Jesus, A. R., Cooper, R., Espinosa, M., Gomes, J. E., Garcia, E. S., Paul, S., and Cross, G. A. (1993) Gene deletion suggests a role for *Trypanosoma cruzi* surface glycoprotein GP72 in the insect and mammalian stages of the life cycle. *J. Cell. Sci.*, **106,** 1023.
159. Rudenko, G., Blundell, P. A., Taylor, M. C., Kieft, R., and Borst, P. (1994) VSG gene expression site control in insect form *Trypanosoma brucei*. *EMBO J.*, **13,** 5470.
160. Swindle, J., Gillespie, R. D., Hariharan, S., Chung, S. H., and Ajioka, J. (1995) Gene replacements and genetics in *Trypanosoma cruzi*. In *Molecular Approaches to Parasitology*. (ed. J. C. Boothroyd and R. Komuniecki), Vol. 12, p. 26, Wiley-Liss, New York, NY.
161. Chung, S. H. (1993) in *Molecular genetic analysis of the calmodulin gene family of Trypanosoma cruzi*. Ph.D. thesis, University of Tennessee, Memphis.
162. Blackburn, E. H. and Challoner, P. B. (1984) Identification of a telomeric DNA sequence in *Trypanosoma brucei*. *Cell*, **36,** 447.
163. Van der Ploeg, L. H., Liu, A. Y., and Borst, P. (1984) Structure of the growing telomeres of trypanosomes. *Cell*, **36,** 459.
164. Kelly, J. M., Das, P., and Tomas, A. M. (1994) An approach to functional complementation by introduction of large DNA fragments into *Trypanosoma cruzi* and *Leishmania donovani* using a cosmid shuttle vector. *Mol. Biochem. Parasitol.*, **65,** 51.
165. Wirtz, E. and Clayton, C. (1995) Inducible gene expression in trypanosomes mediated by a prokaryotic repressor. *Science*, **268,** 1766.

3 | The three genomes of *Plasmodium*

JEAN E. FEAGIN and MICHAEL LANZER

1. Introduction

The Greek physician Hippocrates, who lived in the fifth century BC, was the first to describe the characteristic symptoms of a malaria infection: intermittent and often relapsing fever attacks that are accompanied by drenching sweats and followed by shaking chills. Centuries later, the medical almanac of 1888 referred to malaria as one of the great scourges of humanity, wreaking havoc in human civilizations throughout history and claiming more lives than any other infectious disease. Today, malaria is still a major health problem, causing an estimated 300 million cases of illness and killing 1–2 million people—mostly infants—every year. About 2.2 billion people (41% of the world's population) are now in danger of contracting malaria. The prospect for the future is bleak. The number of malaria cases is expected to rise due to the emergence and spread of drug-resistant malaria parasites and pesticide-resistant malaria-transmitting mosquitoes. Furthermore, a widely effective anti-malarial vaccine is not expected in the near future although several candidate vaccines are now in clinical trials.

Human malaria can be caused by four different but phylogenetically related protozoa of the genus *Plasmodium*, namely *P. vivax*, *P. ovale*, *P. malaria*, and *P. falciparum*. The last parasite is the most virulent and is responsible for the high infant mortality associated with malaria. The first aetiological agent of malaria was discovered by the French physician Laveran in 1880. However, it was not until 1922 that the entire life cycle and the route of transmission became known (Fig. 1).

Molecular and genetic studies on the biology and pathogenicity of these organisms have been complicated by the difficulties in maintaining the various parasite stages under laboratory conditions; only erythrocytic stages of *P. falciparum* can be cultured continuously *in vitro* (1). These developmental stages are obligate intracellular parasites and reside within a parasitophorous vacuole that forms during invasion of the erythrocyte. These properties have hampered the development of a transfection system for erythrocytic stages, precluding functional approaches to study of the parasite's genes. As an alternative, a structural approach has been employed by investigating chromosome organization and dynamics, revealing several interesting and peculiar features of the plasmodial genome, such as a high degree

36 | THE THREE GENOMES OF PLASMODIUM

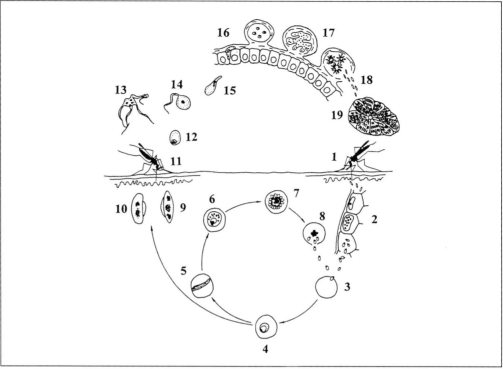

Fig. 1 Life cycle of *Plasmodium falciparum*. Infection is initiated by the bite of an infected mosquito (1). Sporozoites from the salivary glands migrate to the host liver and multiply in hepatocytes (2). These eventually burst, releasing thousands of merozoites which can infect erythrocytes (3). The 48 hour developmental cycle in the erythrocyte includes the ring (4), early trophozoite (5), late trophozoite (6), and schizont (7) stages. When schizogony is complete, the erythrocyte bursts and releases new merozoites (8) capable of reinitiating the erythrocytic cycle. In some cases, parasites in the newly infected erythrocytes will instead develop into female (9) and male (10) gametocytes. If taken up in a bloodmeal by a mosquito (11), these will develop into female (12) and male (13) gametes which fuse to form a zygote (14), the only diploid stage of the parasite life cycle. Following meiosis, the motile ookinete (15) migrates through the gut wall and develops into an oocyst on the outer surface (16). Sporogony (17) occurs in this location and when the sporozoites are mature, they burst the cyst and migrate (18) to the salivary glands (19), ready to initiate new infections.

of genetic variability that extends from allelic variation of single genes to polymorphisms of entire chromosomes. Furthermore, *Plasmodium* unexpectedly possesses not one but two organellar DNAs, one mitochondrial and the other of putative plastid origin. In this review, we will summarize current knowledge of genome organization in malaria parasites, focusing primarily on *P. falciparum*.

2. Nuclear genomic organization

As is common in protozoa, plasmodial chromosomes do not condense during mitosis and are not visible by light microscopy. Chromosome numbers were first estimated by counting the number of kinetochores (attachment sites of the spindle apparatus) in electron micrographs (2). Detailed information regarding chromosome number, size, and ploidy became available only with the advent of pulsed field gel electrophoresis (PFGE) (3, 4). These studies, and genetic crossing experiments, have shown that malaria parasites are haploid during most of the life cycle (3–5). A diploid stage exists only briefly in the mosquito midgut when a male and a female gamete fuse to form a zygote (Fig. 1). Immediately afterwards, meiotic reduction division occurs, giving rise to haploid daughter cells.

The haploid nuclear genome encompasses 30 Mb (6) and is organized as 14 chromosomes (7, 8). The size of the chromosomes depends on the species, as does the genomic A+T content (9–11). In *P. falciparum*, chromosomes range in size from 0.6Mb to 3.5Mb (7,8). The A+T content averages 80% and can exceed 90% in non-coding regions where long, consecutive runs of A and T nucleotides are present (6, 12), similar to the microsatellite DNA of other eukaryotes. The presence of repetitive sequences combined with a high A+T content poses a substantial obstacle to cloning *P. falciparum* DNA in prokaryotic vectors and hosts, as rearrangement and deletion of cloned DNA are frequently observed. Of the estimated 7000 *P. falciparum* genes, about 600 have been cloned and at least partially sequenced so far (13). Some, such as genes encoding antigens such as the major merozoite surface antigen (MSA-1), are unique to *P. falciparum* whereas others share homologies to known eukaryotic genes, e.g. housekeeping genes (13). Antigens frequently contain arrays of repetitive amino acid sequences that show species and even strain-dependent variations, as exemplified by MSA-1 (14). Chromosomal locations have been assigned to more than 100 genetic markers (15).

2.1 Polymorphisms at chromosome ends

Interestingly, homologous chromosomes from geographically different *P. falciparum* isolates display size variations of up to 20% (3, 4; Fig. 2). Such extensive polymorphism seems to be a hallmark of all plasmodial species since other human (*P. vivax*) and rodent (*P. chabaudi* and *P. berghei*) malaria parasites display a similar degree of chromosomal size variability, both in parasites cultured *in vitro* and in field isolates (10, 11, 16, 17). In other organisms, chromosomal polymorphisms are usually indicative of programmed DNA rearrangements that are involved in the control of gene expression, as exemplified by antigenic variation in African trypanosomes (see Chapter 6) or developmental chromosome diminution in the free-living ciliate *Paramecium* (18).

To determine the structural and molecular basis of *P. falciparum* karyotypic diversity, large fragments of genomic DNA have been stably cloned as yeast artificial chromosomes (YACs) (19, 20). This has allowed the physical cloning of entire chromosomes, thus far chromosomes 2 and 4 (21, 22). Contig maps developed for these chromosomes have demonstrated that *P. falciparum* chromosomes are compartmentalized into conserved central domains and polymorphic ends (21, 22; Figs. 2 and 3) Only 100kb on either end of a chromosome are highly variable (21; Fig. 3). Despite their general conservation, however, internal chromosome domains are

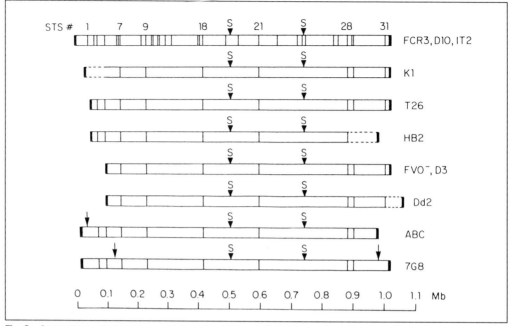

Fig. 2 Comparison of chromosome 2 from 11 different *P. falciparum* isolates. Genetic markers are indicated by vertical lines. The conserved chromosome domain extends from marker 7 to 28. The broken lines indicates translocations. The arrows indicate polymorphisms of major subpopulations within initially clonal lines. The positions of conserved *Sma*I restriction site(s) are indicated. STS, genetic marker defined by a sequence tag site.

also subject to DNA rearrangements, although less so than subtelomeric regions (see below).

2.2 Repetitive elements in subtelomeric domains

The confinement of polymorphisms to chromosome ends suggests that the structural organization of subtelomeric regions differs from the central domains. Several restriction sites, such as an *Apa*I site 10–15kb distal from the telomeres, are conserved among chromosome ends (14, 23). The terminal *Apa*I fragment contains 800–2000bp of telomere repeat sequences: GGGTTTA or, less frequently, GGGTTCA (24). Subcloning of DNA fragments from subtelomeric regions has revealed numerous repetitive sequence elements whose order is common in all wild-type chromosomes (25, 26). By contrast, the repetitiveness of individual elements is variable (16, 27). Contractions and expansions of subtelomeric sequences occur frequently in all plasmodial species, probably resulting from unequal crossing-over or slippage of the DNA polymerase during mitotic replication (28, 29). None of the subtelomeric elements are transcribed, as shown by northern analysis and transcription mapping of telomeric YAC clones (21, 26). Therefore, deletions of

Fig. 3 Organization of P. falciparum chromosomes. Wild-type chromosome ends are composed of repetitive, non-transcribed sequence elements, designated as the subtelomeric region (hatched boxes). The adjacent domains frequently contain antigen-encoding genes whereas housekeeping genes reside in central chromosomal domains (shaded box). Polymorphisms are usually in the subtelomeric repetitive region but can extend into the transcriptionally active, antigen-encoding domain. The central chromosome domains are conserved.

subtelomeric regions do not necessarily result in irreversible loss of genetic information, since these sequences are present on other chromosomes and could be retrieved by unequal crossing-over or gene conversion. Indeed, subtelomeric elements can be acquired during mitosis by chromosome ends that previously lacked this sequence (30).

In yeast, subtelomeric regions serve both as origins of replication and as buffers to prevent the silencing of telomere-proximal genes (31). Similar functions may be possible in P. falciparum but are not yet supported by experimental data. However, a role in chromosome pairing is likely (see below).

Unlike subtelomeric regions, central chromosomal domains are transcribed (21, 22). Thus, P. falciparum chromosomes are functionally compartmentalized into silent subtelomeric regions composed of repetitive sequence elements and central transcriptionally active domains containing the parasite's genes (Fig. 3).

2.3 Chromatin structure and breakage sites

Besides contractions and expansions of subtelomeric domains, other types of polymorphism such as translocations and truncations are evident (21). Chromosomal truncations occur spontaneously during mitotic propagation and are frequently observed in parasites cultured *in vitro* (32–38). They are initiated by internal double-strand breakages. While the terminal segment is lost, the centromere-containing fragment is mitotically stabilized by the *de novo* addition of telomere repeat sequences. It is likely that all 14 P. falciparum chromosomes are subject to breakage and healing events, though experimental proof is only available for chromosomes 1, 2, 8, 9, and 10 (32–38).

Interestingly, chromosomal truncations are not confined to silent subtelomeric repeats but can also extend into the transcribed domains of a chromosome (32–38). A hot spot for mitotic chromosome breakages has been mapped to the KAHRP gene on chromosome 2, 90kb from the telomere (39; Fig. 4). Surprisingly, the breakpoints are phased with a periodicity of 155 ± 4bp. This phasing is identical to the average P. falciparum nucleosome repeat unit, which consists of the internucleosomal linker (8 ± 3bp) and the core DNA fragment (148 ± 5bp) (39). Comparison of nucleosomal organization with chromosome breakage sites revealed that double-

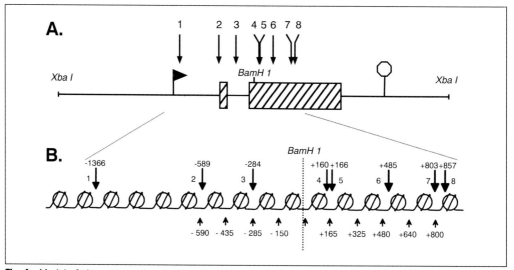

Fig. 4 Model of chromosome breakage in *P. falciparum*. A: Schematic drawing of the *KAHRP* gene locus. The coding region is indicated by rectangles, the transcription start site by an arrow and the termination site by a hexagon. Breakpoints are designated 1–8 and are indicated with arrows. Cleavage sites for *Bam*HI and *Xba*I are indicated. B: The primary chromatin structure of the KAHRP gene, as determined by digestion with micrococcal nuclease, with circles symbolizing nucleosome core particles. Breakpoints are indicated by arrows above the array of nucleosomes, micrococcal nuclease hypersensitive sites below. All sites are given in reference to the *Bam*HI restriction site.

stranded breakages preferentially occur in nucleosome linkers (Fig. 4), specifically those within coding regions (39).

The specificity for breaks within coding regions could result from differences in the chromatin structure between silent and transcribed sequences. Indeed, transcriptional activity disrupts the nucleosome structure and depletes histone H1, which extends into the internucleosomal linker (40). Although we know little about the composition of *P. falciparum* chromatin, initial data demonstrate the presence of the histones H1, H2A, H2B, H3, and H4 (41, 42) in plasmodial nucleosomes. Thus, chromosome breakages in *P. falciparum* may be initiated by an endonucleolytic activity that cleaves double-stranded DNA within H1-depleted nucleosome linkers. This enzymatic activity may possess a cleavage preference 3' to a CA dinucleotide since this dinucleotide is conserved among most chromosome breakpoints identified in *P. falciparum* so far (33–35, 37, 39). Alternatively, the CA dinucleotide could function in healing broken chromosomes. For example, following breakage, the protruding DNA could be degraded to the first accessible CA dinucleotide within a linker region. Since *P. falciparum* telomere repeat units are occasionally composed of GGGTTCA (24), a CA dinucleotide may be a necessary and perhaps sufficient signal for addition of telomere repeat sequences by the *P. falciparum* telomerase. This model is consistent with the observation that other eukaryotic telomerases possess a 3' to 5' exonuclease activity and are capable of elongating non-telomeric primers (43–45).

3. Biological ramifications of chromosomal polymorphism

P. falciparum-infected erythrocytes possess parasite-determined antigens on their surface that are potential targets for the host's immune response. By varying the antigenic profile, the parasite could divert the immune attack. Indeed, *P. falciparum* is capable of changing its antigen repertoire at a rate of 2% per generation during mitotic propagation (46, 47a). An example of parasite proteins expressed on the erythrocyte surface is the set of PfEMP1 (erythrocyte membrane protein 1) proteins. They are encoded by a large, diverse gene family and have been implicated in both antigenic variation and cytoadherence to endothelial cells (47a–d). PfEMP1 is responsible, at least in part, for the sequestration of erythrocytes in capillaries, contributing to oxygen deprivation of surrounding tissues and cerebral malaria.

3.1 Chromosomal rearrangements and antigenic variation

Each *P. falciparum* parasite contains a repertoire of several hundred PfEMP1 gene copies. The molecular mechanisms that control the expression of these genes are still obscure. The involvement of chromosomal rearrangements is an intriguing possibility but is not yet supported by experimental data. However, chromosomal rearrangements can affect the expression of other antigen-encoding genes such as KAHRP on chromosome 2, RESA on chromosome 1, HRPII on chromosome 8, and Pf11–1 on chromosome 10 (32–37). The expression of these genes can be abrogated by truncations that extend into the coding regions, resulting in mutants with altered antigenic phenotypes (32–37). However these mutations tend to be deleterious, as exemplified by KAHRP⁻ mutants. The KAHRP gene product is a component of electron-dense structures (knobs) on the surface of parasitized erythrocytes and is implicated in cytoadherence (48, 49). Chromosomal breakage events disrupting the KAHRP gene result in mutants that are knobless and cytoadherence-deficient (50). These mutants possess a growth advantage in cell culture, but are cleared by the host's spleen *in vivo* (51). Truncations of other chromosomes are also detrimental to parasite fitness and virulence, as demonstrated by a chromosome 9 truncation that abrogates gametogenesis (38).

As a result of chromosomal truncations, telomere repeats are relocated. In yeast, telomeres exert a positional effect by silencing proximal genes (44). However, this does not seem to be a mechanism of gene regulation in *P. falciparum* since chromosome 2 truncations, in which telomeres become juxtaposed with coding sequences, have no obvious effects on the developmental expression of these genes (52). While no direct function can be attributed to chromosomal truncations, such polymorphisms may be the result of a hyperactive recombinatory machinery that confers genomic flexibility during both mitotic and meiotic propagation. Such flexibility may be needed to respond to environmental changes. Indeed gross chromosomal rearrangements are observed when parasites are exposed to antimalarial drugs (see Chapter 10).

3.2 Sexual recombination and genetic diversity

Sexual recombination plays an important role not only in assorting the parasite's genes but also in creating genetic diversity, as demonstrated by several crossing experiments. While most crosses have been performed with animal malaria parasites (5, 53, 54), two crosses were conducted with *P. falciparum* (55, 56). These showed that genetic markers are inherited in a mendelian fashion with crossing-over and random assortment of sister chromatids. In one case, recombination maps to the MSA-1 gene, resulting in a phenotype serologically distinct from the parental alleles (57). This protein may function in erythrocyte invasion and antibodies directed against it are protective (58), so changing MSA-1 antigenicity could alter parasite fitness. Genetic crossing experiments have also been instrumental in elucidating mechanisms of drug resistance in malaria parasites (see Chapter 10).

4. Subtelomeric organization and chromosome pairing

Pairing between homologous chromosomes preferentially initiates at the telomere during meiosis, as shown in other species (59). From the telomeres, branch migration can progress into telomere-proximal domains, subjecting these regions to a high degree of genetic exchange. Interestingly, many *P. falciparum* antigens map to those domains immediately adjacent to the subtelomeric repeats (14, 15). Telomeres can also mediate association of heterologous chromosomes during both mitosis and meiosis (60, 61). It is tempting to speculate that the organization of *P. falciparum* subtelomeric regions as long arrays of repetitive sequences has evolved to facilitate pairing between heterologous as well as homologous chromosomes.

The limited existing genetic data indeed suggest a high crossing over frequency close to chromosome ends. Cross-fertilized progeny often display non-parental restriction fragment length polmorphism (RFLP) patterns in subtelomeric regions but not in internal chromosome domains (62). A location near the subtelomeric repeats could expose antigen-encoding genes to a high frequency of meiotic recombination, enhancing the emergence of new antigenic phenotypes. In this context it is worth mentioning that several antigens are encoded by gene families (14, 37). The individual members are located in subtelomere-proximal regions of different chromosomes. These gene families may have arisen from ancestral genes by ectopic recombination, that is, pairing of non-homologous chromosomes followed by duplicative transfer of telomere-proximal domains from one chromosome to another. By contrast, housekeeping genes reside in central chromosomal domains and are therefore less subject to recombination, leading to greater conservation.

Although the molecular mechanisms of translocations are only partially understood, initial evidence implicates meiotic recombination in this process. A 200kb segment in the *P. falciparum* clone HB3 contains several genes, including RESA-2 and Pf332 (63). This segment is present on two chromosomes: chromosome 11, its original location; and chromosome 13 as a result of a duplicative translocation. When the HB3 strain was crossed with the wild-type clone Dd2, progeny emerged that contained the chromosome 13 segment from HB3 back on chromosome 11 (63). This suggests that, during meiosis, the HB3 chromosome 13 and the Dd2 chromosome 11 have paired, resulting in the exchange of terminal chromosome fragments. Thus, heterologous chromosomes can pair during meiosis, facilitating illegitimate recombination that may then result in the exchange and possibly duplication of subtelomere-proximal genes. It is here that a hyperactive recombinatory machinery could bestow an advantage on the parasite. Once a gene family has emerged, the individual members can diverge independently, expanding the parasite's antigenic profiles.

5. The extrachromosomal DNAs

While the nuclear genome contains virtually all of the genes of *Plasmodium*, there are two extrachromosomal DNAs which have generated considerable interest (also recently reviewed in Ref. 64). One is a 6kb, tandemly reiterated element which appears to be the *Plasmodium* mitochondrial DNA. The second, a 35kb circular DNA, was described in *P. lophurae* 20 years ago and was thought to be mitochondrial DNA (65). More recent analysis of its structure and genetic content has suggested an intriguing alternative: the 35kb DNA looks strikingly like a remnant plastid genome. Both DNAs have unexpected features and potential as chemotherapeutic targets.

5.1 The mitochondrial genome

The *Plasmodium* mitochondrial genome consists of direct tandem repeats of a 6kb element (66, 67). Estimates indicate 15–20 copies in *P. gallinaceum* (66) and 150 copies in *P. yoelii* (67). The 6kb element has a lower A+T content than the nuclear DNA (68% versus >80%) and does not cross-hybridize with nuclear chromosomes (66, 68). While its mitochondrial location has not been unambiguously documented, polymerase chain reaction (PCR) analysis has shown that the 6kb element predominantly partitions with the fraction containing mitochondrial enzyme activities (69).

The 6kb element (Fig. 5) has been sequenced from several species (70–73), and is extremely well conserved, being 89% identical between *P. falciparum* and *P. yoelii* (72). There are three open reading frames (ORFs), with homology to apocytochrome *b* (CYb) and cytochrome *c* oxidase subunits I and III (COI and COIII). These are canonical mitochondrial genes and attest to the mitochondrial identity of the 6kb element (74). Specific amino acids known to be involved in the function of these proteins are conserved in the *P. falciparum* genes (74) and structural features of CYb may explain the sensitivity of *P. falciparum* to 8-aminoquinolines and hydroxynapthoquinones (73). Both the COI and COIII genes lack encoded AUGs near the 5′ end of their respective ORFs but analysis of the 5′ end of COI RNA has shown that it matches the genomic sequence (70, 74, 75). There is thus no evidence that RNA editing creates an initiation codon, as is the case in some trypanosomatid mitochon-

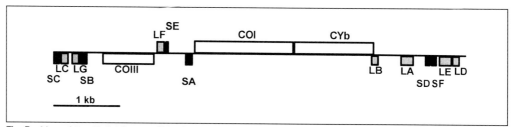

Fig. 5 Map of the *P. falciparum* 6kb element. Genes are shown above or below the line to indicate the direction of transcription (left to right above the line). Protein coding genes are white, and regions homologous to large subunit (LA–LG) and small subunit (SA–SF) rRNA sequences are shown in grey and black, respectively. Abbreviations are as in the text.

drial mRNAs (see Chapter 8). It is more likely that unusual initiation codons are employed, as seen in other mitochondrial genomes.

Ribosomal RNA genes are invariably components of mitochondrial genomes. The 6kb element has regions of sequence homologous to rRNA sequences but these are found as small (30–180 nt) scattered patches (70–72, 75). Fragmented rRNAs, while rare, have been described in other lower eukaryotes (76), but the *Plasmodium* fragments have a cumulative size much smaller than expected. As yet, there is no direct evidence that these unusual rRNAs function in organelle protein synthesis but several lines of indirect evidence support this possibility (72):

1. Abundant small transcripts map to the predicted rRNA fragments.
2. The fragments correspond to those rRNA regions which are most strongly conserved between diverse species.
3. The nucleotide sequence of the rRNA fragments is conserved between *Plasmodium* species as strongly as the amino acid sequence of the mitochondrial protein coding genes, arguing that evolutionary pressure prevents their divergence.
4. Sites implicated in function, such as tRNA binding sites, tend to be conserved.
5. The protein products of the mitochondrial genes are present (see below), arguing that the small rRNA fragments assemble in *trans* in functional ribosomes to translate the mitochondrial mRNAs.

5.2 Mitochondrial expression and function

The mitochondrial mRNAs have transcripts which are similar in size to the corresponding ORFs, suggesting that any untranslated regions are short (75, 77, 78). Indeed, the 5′ untranslated regions examined are less than 20 nt (74, 75) and the COI and CYb ORFs are separated by only 20–27 nt, depending on species (70, 73–75). Poly(A) tails have not yet been directly demonstrated but these transcripts are retained on oligo(dT) columns (67, 78). The mature mitochondrial rRNAs are quite small and they do not appear to become covalently joined into larger rRNAs (72, 78). Larger size transcripts up to 5kb have been reported, although the sizes vary between species (67, 73, 75, 77). These might be precursor transcripts but this possibility requires further analysis. Promoter location also remains undetermined but recent analyses indicate that there is considerable co-transcription of flanking genes, suggesting one or a very few sites of transcription initiation per DNA strand (Y. Ji, B. L. Mericle, and J. E. Feagin, unpublished).

There has been little direct analysis of mitochondrial protein products. However, all three are crucial components of the electron transport chain and inhibitors which target these proteins have antimalarial effects (79, 80). For example, the effects of cytochrome oxidase inhibitors provide evidence for existence of the COI and COIII polypeptides, which are both required for its catalytic activity; cytochrome oxidase activity has also been detected (81). In addition, spectral assays of isolated *Plasmodium* mitochondria show a peak for type *b* cytochrome (82).

Plasmodium species have complex life cycles and many genes are expressed in a stage-specific fashion. The mitochondrial genes are no exception (78). Transcripts of the protein coding genes are hard to detect in the ring stage and peak in late trophozoites and schizonts. In contrast, the rRNAs appear constitutively expressed until the late schizont stage when their abundance increases, possibly coincident with the formation of new merozoites. The difference in expression pattern between the protein coding genes and the rRNA fragments indicates that the abundance changes cannot be mediated solely by transcription rate (since at least some of these appear to be co-transcribed) but also involve transcript stability (78).

The role of the *Plasmodium* mitochondrion is not well understood. In *Plasmodium* species that parasitize mammals, energy generation apparently varies between life cycle stages, with reliance on glycolysis in erythrocytic stages, while mitochondrial respiration is thought to be more important in insect stages (83). Despite this, the *P. falciparum* mitochondrial genes are clearly transcribed during erythrocytic stages and compounds inhibiting mitochondrial function have antimalarial effects. What, then, is the role of the mitochondrion in erythrocytic stages of mammalian *Plasmodium* species?

While the answer to the foregoing question is not certain—indeed, there probably is not a single role but several—one possibility is a role in pyrimidine biosynthesis (84). This process includes conversion of dihydroorotate to orotate by dihydroorotate dehydrogenase, a mitochondrial enzyme which passes electrons to the electron transport chain. *Plasmodium* species make pyrimidines *de novo*, using this enzyme (84–86). The observation that the peak of pyrimidine biosynthesis (87), the timing of maximum DNA synthesis (RNA synthesis peaks more broadly over the trophozoite and schizont stages) (88), and the greatest abundance of transcripts for mitochondrially encoded components of the electron transport chain (78) all coincide at the late trophozoite/early schizont portion of the erythrocytic cycle is suggestive that there is, in fact, a mitochondrial role in pyrimidine biosynthesis.

5.3 The putative plastid genome

The 35kb circular DNA is even more A+T-rich than the nuclear genome and this has hampered efforts to clone and sequence its DNA. The most prominent organ-

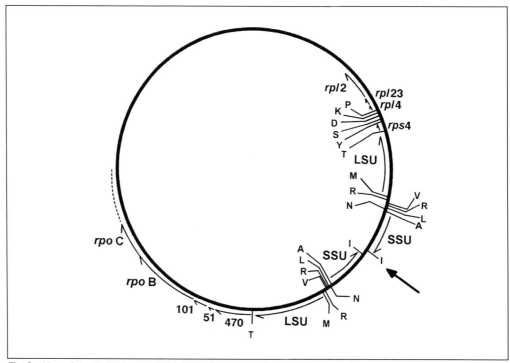

Fig. 6 Map of the *P. falciparum* 35kb DNA. Genes are shown as arrows inside or outside the circle depending on transcription direction (clockwise outside). A dotted line shows the expected extent of the incompletely sequenced *rpoC* gene. Exact duplication of the tRNAs within the inverted repeat (the axis of symmetry between the SSU rRNA genes is marked by an arrow) is assumed but has not been directly demonstrated. The sequence of the 'blank region' between *rpoC* and *rpl2* has not yet been completed. Abbreviations are as in the text with the following additions: tRNAs are identified with the one letter code, unidentified ORFs are identified with numbers, and the *rpl* and *rps* genes code for large and small subunit ribosomal proteins, respectively.

izational feature of the molecule is a large inverted repeat, composed of rRNA and tRNA genes (Fig. 6). Such structures are highly characteristic of chloroplast genomes and have fueled further investigations of the provenance of the 35kb DNA (discussed below). The intracellular location of this extrachromosomal DNA remains a mystery. It does not co-localize with the 6kb element in subcellular fractionations (69) and thus is presumably not in the mitochondrion. *Plasmodium* species contain an organelle of unknown function, the spherical body, which has been suggested as a possible site since it has multiple bounding membranes similar to plastids (89).

Like the 6kb element, the 35kb DNA encodes organelle-type rRNAs. However, they are not fragmented and have a potential secondary structure similar to that of *E. coli* rRNAs (90, 91). Each arm of the inverted repeat contains a large (LSU) and small (SSU) rRNA gene, transcribed in divergent directions and separated at their 5' ends by a cluster of tRNA genes (92; Fig. 6). Additional tRNA genes are located at the 3' ends of the rRNAs (90, 91) and in another cluster (93), yielding a total of 14 different tRNA genes. The protein coding genes include those for ribosomal proteins (93) and the RNA polymerase subunits *rpo*B and *rpo*C (94, 95), plus three ORFs. With the exception of one ORF of unknown function, all genes identified to date are involved with organelle transcription or translation. As data on the 35kb DNA have increased, the support for the hypothesis of its plastid origin has also grown (reviewed in Refs 64, 96, 97)). The following all point to such a scenario:

1. Plastid genomes commonly encode rRNAs and tRNAs in an inverted repeat but this is rare for mitochondrial genomes.

2. Mitochondria do not encode their own RNA polymerases but plastid genomes commonly encode subunits of a eubacterial-like polymerase, as found in the 35kb DNA.

3. The *rpo*B (95) and *rpo*C (98) genes are more similar to chloroplast than bacterial counterparts.

4. ORF470 from the 35kb DNA has >50% amino acid identity to a partially sequenced, unidentified ORF in the plastid genome of a red alga (99).

5. Plastid genomes in non-photosynthetic plants are greatly reduced, having lost many genes but retaining the inverted repeat and genes for transcription and translation (100). They thus strongly resemble the 35kb DNA.

5.4 Expression and function of the 35kb DNA

As a potential remnant plastid genome in a clearly non-photosynthetic organism, the role of the 35kb DNA is unclear. A first question is whether the genes are functional; to date, it is clear that they are at least transcribed. RNase protection assays have shown that both rRNA genes are co-transcribed with flanking tRNAs, followed by processing (90, 91). The abundance of these rRNAs varies during the erythrocytic cycle, being least in early trophozoites and greatest in late schizonts (78). Transcripts for the *rpo*B and *rpo*C genes are not very abundant at any time in the erythrocytic cycle and are virtually undetectable in the ring stage (78). The large size of these transcripts (multiple transcripts >7kb) suggested that they might be polycistronic (94), and this has been confirmed by PCR analysis (78, 95).

The remaining unknown regions of the 35kb DNA presumably harbour other protein coding genes whose identity may clarify the cellular function(s) of this element. Initially, it was thought that the parasite's sensitivity to rifampicin (101), an inhibitor of eubacterial RNA polymerases, indicated an important role for gene products from this genome (94). Recent analyses suggest, however, that the principal inhibitory effects of this drug are directed at other targets (102). Still, the retention and expression of the 35kb DNA in *Plasmodium* species and other apicomplexans (93) suggest that the parasites derive some benefit from it. The possibility that a function encoded by the 35kb DNA might be a target for chemotherapeutic intervention is an exciting prospect.

5.5 Extrachromosomal DNA inheritance

The two extrachromosomal DNAs are extremely well conserved between *P. falciparum* stocks (68, 103) with few of the genetic polymorphisms necessary for analysis of inheritance. Nevertheless, using single nucleotide differences between strains, both DNAs have now been shown to be uniparentally inherited (68, 103). Inheritance is probably through the female gamete, since the organelle DNAs are difficult to detect in male gametes (68, 103). There also appears to be dominance of specific stocks in the crosses examined but, as these have been relatively few, this observation may reflect stock differences rather than a specific genetic phenomenon.

6. Conclusions

The imminent threat malaria poses to humankind requires a comprehensive effort to understand the biochemical and genetic properties of its causative agent. The ability to clone entire chromosomes will facilitate the physical characterization of the *P. falciparum* genome. This and the recent successes in transient (104, 105) and stable (106) transfection of erythrocytic and sexual stages of *Plasmodium* species, added to other molecular genetic tools, will allow detailed investigation of the pathogenicity of malaria parasites, including antigenic variation, drug resistance, host cell invasion, cytoadherence, and general cellular functions. The potential for development of new therapeutic strategies, identification of novel drug targets, and increased understanding of basic biochemical processes is enormous.

References

1. Trager, W. and Jensen, J. B. (1976) Human malaria parasites in continuous culture. *Science*, **193**, 673.
2. Slomianny, C. and Prensier, G. (1986) Application of the serial sectioning and tridimensional reconstruction techniques to the morphological study of the *Plasmodium falciparum* mitochondrion. *J. Parasitol.*, **72**, 595.
3. Kemp, D. J., Corcoran, L. M., Coppel, R. L., Stahl, H.-D., Bianco, A. E., Brown, G. V., and Anders, R. F. (1985) Size variation in chromosomes from independent cultured isolates of *Plasmodium falciparum. Nature*, **315**, 347.
4. Van der Ploeg, L. H. T., Smits, M., Ponnudurai, T., Meuwissen, J. H. E. T., and Langsley, G. (1985) Chromosome-sized DNA molecules of *Plasmodium falciparum. Science*, **229**, 658.
5. Walliker, D., Carter, R., and Sanderson, A. (1975) Genetic studies on *Plasmodium chabaudi*: recombination between enzyme markers. *Parasitology*, **70**, 19.
6. Goman, M., Langsley, G., Hyde, J. E., Yankovsky, N. K., Zolg, J. W., and Scaife, J. G. (1982) The establishment of genomic DNA libraries for the human malaria parasite *Plasmodium falciparum* and identification of individual clones by hybridisation. *Mol. Biochem. Parasitol.*, **5**, 391.
7. Wellems, T. E., Walliker, D., Smith, C. L., do Rosario, V. E., Maloy, W. L., Howard, R. J., Carter, R., and McCutchan, T. F. (1987) A histidine-rich protein gene marks a linkage group favored strongly in a genetic cross of Plasmodium falciparum. *Cell*, **49**, 633.

8. Kemp, D. J., Thompson, J. K., Walliker, D., and Corcoran, L. M. (1987) Molecular karyotype of *Plasmodium falciparum*: Conserved linkage groups and expendable histidine-rich protein genes. *Proc. Natl. Acad. Sci. USA*, **84,** 7672.
9. Langsley, G., Sibili, L., Mattei, D., Falanga, P., and Mercereau-Puijalon, O. (1987) Karyotype comparison between *P. chabaudi* and *P. falciparum*: analysis of a *P. chabaudi* cDNA containing sequences highly repetitive in *P. falciparum*. *Nucleic Acids Res.*, **15,** 2203.
10. Sharkey, A., Langsely, G., Patarapotikul, J., Mercereau-Puijalon, O., McLean, A. P., and Walliker, D. (1988) Chromosome size variation in the malaria parasite of rodents, *Plasmodium chabaudi*. *Mol. Biochem. Parasitol.*, **22,** 47.
11. Sheppard, M., Thompson, J. K., Anders, R. F., Kemp, D. J., and Lew, A. M. (1989) Molecular karyotyping of the rodent malarias *Plasmodium chabaudi, Plasmodium berghei* and *Plasmodium vinckei*. *Mol. Biochem. Parasitol.*, **34,** 45.
12. Pollack, Y., Katzen, A. L., Spira, D. T., and Golenser, J. (1982) The genome of *Plasmodium falciparum*. I: DNA base composition. *Nucleic Acids Res.*, **10,** 539.
13. Reddy, G. R., Chakrabarti, D., Schuster, S. M., Ferl, R. J., Almira, E. C., and Dame, J. B. (1993) Gene sequence tags from *Plasmodium falciparum* genomic DNA fragments prepared by the genase action of mung bean nuclease. *Proc. Natl. Acad. Sci. USA*, **90,** 9867.
14. Kemp, D. J., Cowman, A. F., and Walliker, D. (1990) Genetic diversity in *Plasmodium falciparum*. *Adv. Parasitol.*, **29,** 75.
15. Walker-Jonah, A., Dolan, S. A., Gwadz, R. W., Panton, L. J., and Wellems, T. E. (1992) A RFLP map of the *Plasmodium falciparum* genome, recombination rates and favored linkage groups in a genetic cross. *Mol. Biochem. Parasitol.*, **51,** 313.
16. Corcoran, L. M., Forsyth, K. P., Bianco, A. E., Brown, G. V., and Kemp, D. J. (1986) Chromosome size polymorphisms in Plasmodium falciparum can involve deletions and are frequent in natural parasite populations. *Cell*, **44,** 87.
17. Biggs, B. A., Kemp, D. J., and Brown, G. V. (1989) Subtelomeric chromosome deletions in field isolates of *Plasmodium falciparum* and their relationship to loss of cytoadherence *in vitro*. *Proc. Natl. Acad. Sci. USA*, **86,** 2428.
18. Blackburn, E. H. and Karrer, K. M. (1986) Genomic reorganization in ciliated protozoans. *Annu. Rev. Genet.*, **20,** 501.
19. Triglia, T. and Kemp, D. J. (1991) Large fragments of *Plasmodium falciparum* DNA can be stable when cloned in yeast artificial chromosomes. *Mol. Biochem. Parasitol.*, **44,** 207.
20. deBruin, D., Lanzer, M., and Ravetch, J. V. (1992) Characterization of yeast artificial chromosomes from *Plasmodium falciparum*: Construction of a stable, representative library and cloning of telomeric DNA fragments. *Genomics*, **14,** 332.
21. Lanzer, M., de Bruin, D., and Ravetch, J. V. (1993) Transcriptional differences in polymorphic and conserved domains of a complete *P. falciparum* chromosome. *Nature*, **361,** 654.
22. Rubio, J. P., Triglia, T., Kemp, D. J., de Bruin, D., Ravetch, J. V., and Cowman, A. F. (1994) A YAC contig map of *Plasmodium falciparum* chromosome 4: Transcriptional analysis and characterization of a DNA amplification between two recently separated isolates. *Genomics*, **21,** 447.
23. Scotti, R., Pace, T., and Ponzi, M. (1993) A 40-kilobase subtelomeric region is common to most *Plasmodium falciparum* 3D7 chromosomes. *Mol. Biochem. Parasitol.*, **58,** 1.
24. Vernick, K. D. and McCutchan, T. F. (1988) Sequence and structure of a *Plasmodium falciparum* telomere. *Mol. Biochem. Parasitol.*, **28,** 85.
25. Patarapotikul, J. and Langsley, G. (1988) Chromosome size polymorphism in *Plasmodium falciparum* can involve deletions of the subtelomeric pPFrep20 sequence. *Nucleic Acids Res.*, **16,** 4331.

26. de Bruin, D., Lanzer, M., and Ravetch, J. V. (1994) The polymorphic subtelomeric regions of *Plasmodium falciparum* chromosomes contain arrays of repetitive sequence elements. *Proc. Natl. Acad. Sci. USA*, **91,** 619.
27. Corcoran, L. M., Thompson, J. K., Walliker, D., and Kemp, D. J. (1988) Homologous recombination within subtelomeric repeat sequences generates chromosome size polymorphism in P. falciparum. *Cell*, **53,** 807.
28. Ponzi, M., Janse, C. J., Dore, E., Scotti, R., Pace, T., Reterink, T. J. F., Van der Berg, F. M., and Mons, B. (1990) Generation of chromosome size polymorphism during in vivo mitotic multiplication of *Plasmodium berghei* involves both loss and addition of subtelomeric repeat sequences. *Mol. Biochem. Parasitol.*, **41,** 73.
29. Dore, E., Pace, T., Ponzi, M., Picci, L., and Frontali, C. (1990) Organization of subtelomeric repeats in *Plasmodium bergehi*. *Mol. Cell. Biol.*, **10,** 2423.
30. Janse, C. J., Ramesar, J., and Mons, B. (1992) Chromosome translocation in *Plasmodium berghei*. *Nucleic Acids Res.*, **20,** 581.
31. Biessmann, H. and Mason, J. M. (1992) Genetics and molecular biology of telomeres. *Adv. Genet.*, **30,** 185.
32. Pologe, L. G. and Ravetch, J. V. (1986) A chromosomal rearrangement in a *P. falciparum* histidine-rich protein gene is associated with the knobless phenotype. *Nature*, **322,** 474.
33. Pologe, L. G. and Ravetch, J. V. (1988) Large deletions result from breakage and healing of P. falciparum chromosomes. *Cell*, **55,** 869.
34. Pologe, L. G., de Bruin, D., and Ravetch, J. V. (1990) A and T homopolymeric stretches mediate a DNA inversion in *Plasmodium falciparum* which results in loss of gene expression. *Mol. Cell. Biol.*, **10,** 3243.
35. Scherf, A. and Mattei, D. (1992) Cloning and characterization of chromosome breakpoints of *Plasmodium falciparum*: breakage and new telomere formation occurs frequently randomly in subtelomeric genes. *Nucleic Acids Res.*, **20,** 1491.
36. Shirley, M. W., Biggs, B. A., Forsyth, K. P., Brown, H. J., Thompson, J. K., Brown, G. V., and Kemp, D. J. (1990) Chromosome 9 from independent clones and isolates of *Plasmodium falciparum* undergoes subtelomeric deletions with similar breakpoints in vitro. *Mol. Biochem. Parasitol.*, **40,** 137.
37. Scherf, A., Carter, R., Petersen, C., Alano, P., Nelson, R., Aikawa, M., Mattei, D., Pereira da Silva, L., and Leech, J. (1992) Gene inactivation of Pf11–1 of *Plasmodium falciparum* by chromosome breakage and healing: identification of a gametocyte-specific protein with a potential role in gametocytogenesis. *EMBO J.*, **11,** 2293.
38. Day, K. P., Karamalis, F., Thompson, J., Barnes, D. A., Peterson, C., Brown, H., Brown, G. V., and Kemp, D. J. (1993) Genes necessary for expression of a virulence determinant and for transmission of *Plasmodium falciparum* are located on a 0.3-megabase region of chromosome 9. *Proc. Natl. Acad. Sci. USA*, **90,** 8292.
39. Lanzer, M., Wertheimer, S. P., de Bruin, D., and Ravetch, J. V. (1994) Chromatin structure determines the sites of chromosome breakages in *Plasmodium falciparum*. *Nucleic Acids Res.*, **22,** 3099.
40. Clark, D. J. and Felsenfeld, G. (1992) A nucleosome core is transferred out of the path of a transcribing polymerase. *Cell*, **71,** 11.
41. Cary, C., Lamont, D., Dalton, J. P., and Doerig, C. (1994) *Plasmodium falciparum* chromatin: nucleosomal organisation and histone-like proteins. *Parasitol. Res.*, **80,** 255.
42. Creedon, K. A., Kaslow, D. C., Rathod, P. K., and Wellems, T. E. (1992) Identification of a *Plasmodium falciparum* histone 2A gene. *Mol. Biochem. Parasitol.*, **54,** 113.

43. Collins, K. and Greider, C. W. (1993) *Tetrahymena* telomerase catalyzes nucleolytic cleavage and nonprocessive elongation. *Genes Dev.*, **7b,** 1364.
44. Harrington, L. A. and Greider, C. W. (1991) Telomerase primer specificity and chromosome healing. *Nature*, **353,** 451.
45. Morin, G. B. (1991) Recognition of a chromosome truncation site associated with a-thalassemia by human telomerase. *Nature*, **353,** 454.
46. Roberts, D. J., Craig, A. G., Berendt, A. R., Pinches, R., Nash, G., Marsh, K., and Newbold, C. I. (1992) Rapid switching to multiple antigenic and adhesive phenotypes in malaria. *Nature*, **357,** 689.
47a. Biggs, B.-A., Gooze, L., Wycherley, K., Wollish, W., Southwell, B., Leech, J. H., and Brown, G. V. (1991) Antigenic variation in *Plasmodium falciparum*. *Proc. Natl. Acad. Sci. USA*, **88,** 9171.
47b. Baruch, D. I., Pasloske, B. L., Singh, H. B., Bi, X., Ma, X. C., Feldman, M., Taraschi, T. F., and Howard, R. J. (1995) Cloning the P. falciparum gene encoding PfEMP1, a malarial variant antigen and adherence receptor on the surface of parasitized human erythrocytes. *Cell*, **82,** 77.
47c. Su, X.-Z., Heatwole, V. M., Wertheimer, S. P., Guinet, F., Herrfeldt, J. A., Peterson, D. S., Ravetch, J. A., and Wellems, T. E. (1995) The large diverse gene family *var* encodes proteins involved in cytoadherence and antigenic variation of Plasmodium falciparum-infected erythrocytes. *Cell*, **82,** 89.
47d. Smith, J. D., Chitnis, C. E., Craig, A. G., Roberts, D. J., Hudson-Taylor, D. E., Peterson, D. S., Pinches, R., Newbold, C. I., and Miller, L. H. (1995) Switches in expression of Plasmodium falciparum *var* genes correlate with changes in antigenic and cytoadherent phenotypes of infected erythrocytes. *Cell*, **82,** 101.
48. Aikawa, M., Rabbage, J. R., and Wellde, B. T. (1972) Junctional apparatus in erythrocytes infected with malaria parasites. *Zeitschrift Zellforschung Mikroscopische Anat.*, **124,** 722.
49. Raventos-Suarez, C., Kaul, D. K., Macaluso, F., and Nagel, R. L. (1985) Membrane knobs are required for the microcirculatory obstruction induced by *Plasmodium falciparum*-infetced erythrocytes. *Proc. Natl. Acad. Sci. USA*, **82,** 3829.
50. Udeinya, I. J., Schmidt, J. A., Aikawa, M. A., Miller, L. H., and Green, I. (1981) Falciparum malaria infected erythrocytes specifically bind to cultured human endothelial cells. *Science*, **213,** 555.
51. Barnwell, J. W., Howard, R. J., and Miller, L. H. (1983) Influence of the spleen on the expression of surface antigens on parasitized erythrocytes. *Ciba Found. Symp.* **94,** 117.
52. Lanzer, M., de Bruin, D., Wertheimer, S. P., and Ravetch, J. V. (1994) Transcriptional and nucleosomal characterization of a subtelomeric gene cluster flanking a site of chromosomal rearrangements in *Plasmodium falciparum*. *Nucleic Acids Res.*, **22,** 4176.
53. Walliker, D. (1989) Genetic recombination in malaria parasites. *Exp. Parasitol.*, **69,** 303.
54. Greenberg, J. and Trembley, H. L. (1954) The apparent transfer of pyrimethamine resistance from the BI strain of *Plasmodium gallinaceum* to the M strain. *J. Parasitol.*, **40,** 667.
55. Walliker, D., Quakyi, I. A., Wellems, T. E., McCutchan, T. F., Szarfman, A., London, W. T., Corcoran, L. M., Burkot, T. R., and Carter, R. (1987) Genetic analysis of the human malaria parasite *Plasmodium falciparum*. *Science*, **236,** 1661.
56. Wellems, T. E., Panton, L. J., Gluzman, I. Y., do Rosario, V. E., Gwadz, R. W., Walker-Jonah, A., and Krogstad, D. J. (1990) Chloroquine resistance not linked to *mdr*-like genes in a *Plasmodium falciparum* cross. *Nature*, **345,** 253.

57. Kerr, P. J., Ranford-Cartwright, L. C., and Walliker, D. (1994) Proof of intragenic recombination in *Plasmodium falciparum*. *Mol. Biochem. Parasitol.*, **66**, 241.
58. Holder, A. A. and Blackman, M. J. (1994) What is the function of MSP-I on the malaria merozoite? *Parasitol. Today*, **10**, 182.
59. Zickler, D., Moreau, P. J., Huynh, A. D., and Slezac, A. M. (1992) Correlation between pairing initiation sites, recombination modules and meiotic recombination in *Sordasia macrospora*. *Genetics*, **132**, 135.
60. Loidl, J., Schertan, H., and Kaback, D. B. (1994) Physical association between nonhomologous chromosomes precedes distributive disjunction in yeast. *Proc. Natl. Acad. Sci. USA*, **91**, 331.
61. Dancis, B. M. and Holmquist, G. P. (1979) Telomere replication and fusion in eukaryotes. *J. Theor. Biol.*, **78**, 211.
62. Vernick, K. D., Walliker, D., and McCutchan, T. F. (1988) Genetic hypervariability of telomere-related sequences is associated with meiosis in *Plasmodium falciparum*. *Nucleic Acids Res.*, **16**, 6973.
63. Hinterberg, K., Mattei, D., Wellems, T. E., and Scherf, A. (1994) Interchromosomal exchange of a large subtelomeric segment in a *Plasmodium falciparum* cross. *EMBO J.*, **13**, 4174.
64. Feagin, J. E. (1994) The extrachromosomal DNAs of apicomplexan parasites. *Annu. Rev. Microbiol.*, **48**, 81.
65. Kilejian, A. (1975) Circular mitochondrial DNA from the avian malarial parasite *Plasmodium lophurae*. *Biochim. Biophys. Acta*, **390**, 276.
66. Joseph, J. T., Aldritt, S. M., Unnasch, T., Puijalon, O., and Wirth, D. F. (1989) Characterization of a conserved extrachromosomal element isolated from the avian malarial parasite *Plasmodium gallinaceum*. *Mol. Cell. Biol.*, **9**, 3621.
67. Vaidya, A. B. and Arasu, P. (1987) Tandemly arranged gene clusters of malarial parasites that are highly conserved and transcribed. *Mol. Biochem. Parasitol.*, **22**, 249.
68. Vaidya, A. B., Morrisey, J., Plowe, C. V., Kaslow, D. C., and Wellems, T. E. (1993) Unidirectional dominance of cytoplasmic inheritance in two genetic crosses of *Plasmodium falciparum*. *Mol. Cell. Biol.*, **13**, 7349.
69. Wilson, R. J. M., Fry, M., Gardner, M. J., Feagin, J. E., and Williamson, D. H. (1992) Subcellular fractionation of the two organelle DNAs of malaria parasites. *Curr. Genet.*, **21**, 405.
70. Joseph, J. T. (1990) *The mitochondrial microgenome of malaria parasites*. Ph.D. thesis, Harvard Medical School, Boston.
71. Vaidya, A. B., Akella, R., and Suplick, K. (1989) Sequences similar to genes for two mitochondrial proteins and portions of ribosomal RNA in tandemly arrayed 6-kilobase-pair DNA of a malarial parasite. *Mol. Biochem. Parasitol.*, **35**, 97.
72. Feagin, J. E., Werner, E., Gardner, M. J., Williamson, D. H., and Wilson, R. J. M. (1992) Homologies between the contiguous and fragmented rRNAs of the two *Plasmodium falciparum* extrachromosomal DNAs are limited to core sequences. *Nucleic Acids Res.*, **20**, 879.
73. Vaidya, A. B., Lashgari, M. S., Pologe, L. G., and Morrisey, J. (1993) Structural features of *Plasmodium* cytochrome *b* that may underlie susceptibility to 8-aminoquinolines and hydroxynaphthoquinones. *Mol. Biochem. Parasitol.*, **58**, 33.
74. Feagin, J. E. (1992) The 6 kb element of *Plasmodium falciparum* encodes mitochondrial cytochrome genes. *Mol. Biochem. Parasitol.*, **52**, 145.
75. Suplick, K., Morrisey, J., and Vaidya, A. B. (1990) Complex transcription from the extra-

chromosomal DNA encoding mitochondrial functions of *Plasmodium yoelii*. *Mol. Cell. Biol.*, **10,** 6381.

76. Gray, M. W. and Schnare, M. N. (1990) Evolution of the modular structure of rRNA. In *The Ribosome: structure, function, and evolution* (ed. W. E. Hill, A. Dahlberg, R. A. Garrett, P. B. Moore, D. Schlessinger, and J. R. Warner), p. 589. American Society for Microbiology, Washington, D. C.
77. Aldritt, S. M., Joseph, J. T., and Wirth, D. F. (1989) Sequence identification of cytochrome *b* in *Plasmodium gallinaceum*. *Mol. Cell. Biol.*, **9,** 3614.
78. Feagin, J. E. and Drew, M. E. (1995) *Plasmodium falciparum*: Alterations in organelle transcript abundance during the erythrocytic cycle. *Exp. Parasitol.* **80,** 430.
79. Divo, A. A., Geary, T. G., and Jensen, J. B. (1985) Oxygen- and time-dependent effects of antibiotics and selected mitochondrial inhibitors on *Plasmodium falciparum* in culture. *Antimicrob. Agents Chemother.*, **27,** 21.
80. Ginsburg, H., Divo, A. A., Geary, T. G., Boltand, M. T., and Jensen, J. B. (1986) Effects of mitochondrial inhibitors on intraerythrocytic *Plasmodium falciparum* in in vitro cultures. *J. Protozool.*, **33,** 121.
81. Scheibel, L. W. and Pflaum, W. K. (1970) Cytochrome oxidase activity in platelet-free preparations of *Plasmodium falciparum*. *J. Parasitol.*, **56,** 1054.
82. Fry, M. and Beesley, J. E. (1991) Mitochondria of mammalian *Plasmodium* spp. *Parasitology*, **102,** 17.
83. Scheibel, L. W. (1988) Plasmodial metabolism and related organellar function during various stages of the life-cycle: carbohydrates. In *Malaria: principles and practice of malariology* (ed. W. H. Wernsdorfer, and I. McGregor), p. 171. Churchill Livingstone, Edinburgh.
84. Gutteridge, W. E., Dave, D., and Richards, W. H. G. (1979) Conversion of dihydroorotate to orotate in parasitic protozoa. *Biochim. Biophys. Acta*, **582,** 390.
85. Krungkrai, J., Cerami, A., and Henderson, G. B. (1991) Purification and characterization of dihydroorotate dehydrogenase from the rodent malaria parasite *Plasmodium berghei*. *Biochemistry*, **30,** 1934.
86. Scheibel, L. W. and Sherman, I. W. (1988) Plasmodial metabolism and related organellar function during various stages of the life cycle: proteins, lipids, nucleic acids and vitamins. In *Malaria: principles and practice of malariology* (ed. W. H. Wernsdorfer and I. McGregor), p. 219. Churchill Livingstone, Edinburgh.
87. Seymour, K. K., Lyons, S. D., Phillips, L., Rieckmann, K. H., and Christopherson, R. I. (1994) Cytotoxic effects of inhibitors of *de novo* pyrimidine biosynthesis upon *Plasmodium falciparum*. *Biochemistry*, **33,** 5268.
88. Gritzmacher, C. A. and Reese, R. T. (1984) Protein and nucleic acid synthesis during synchronized growth of *Plasmodium falciparum*. *J. Bacteriol.*, **160,** 1165.
89. Kilejian, A. (1991) Spherical bodies. *Parasitol. Today*, **7,** 309.
90. Gardner, M. J., Feagin, J. E., Moore, D. J., Spencer, D. F., Gray, M. W., Williamson, D. H., and Wilson, R. J. M. (1991) Organization and expression of small subunit ribosomal RNA genes encoded by a 35-kb circular DNA in *Plasmodium falciparum*. *Mol. Biochem. Parasitol.*, **48,** 77.
91. Gardner, M. J., Feagin, J. E., Moore, D. J., Rangachari, K., Williamson, D. H., and Wilson, R. J. M. (1993) Sequence and organization of large subunit rRNA genes from the extrachromosomal 35 kb circular DNA of the malaria parasite *Plasmodium falciparum*. *Nucleic Acids Res.*, **21,** 1067.
92. Gardner, M., Preiser, P., Rangachari, K., Moore, D., Feagin, J. E., Williamson, D. H., and

Wilson, R. J. M. (1994) Nine duplicated tRNA genes on the plastid-like DNA of the malaria parasite *Plasmodium falciparum*. *Gene*, **144**, 307.

93. Wilson, I., Gardner, M., Rangachari, K., and Williamson, D. (1993) Extrachromosomal DNA in the Apicomplexa. In *Toxoplasmosis* (ed. J. E. Smith), p. 51. Springer-Verlag NATO:ASI Series, Heidelberg.

94. Gardner, M. J., Williamson, D. H., and Wilson, R. J. M. (1991) A circular DNA in malaria parasites encodes an RNA polymerase like that of prokaryotes and chloroplasts. *Mol. Biochem. Parasitol.*, **44**, 115.

95. Gardner, M. J., Goldman, N., Barnett, P., Moore, P. W., Rangachari, K., Strath, M., Whyte, A., Williamson, D. H., and Wilson, R. J. M. (1994) Phylogenetic analysis of the *rpo*B gene from the plastid-like DNA of *Plasmodium falciparum*. *Mol. Biochem. Parasitol.*, **66**, 221.

96. Wilson, R. J. M., Gardner, M. J., Williamson, D. H., and Feagin, J. E. (1991) Have malaria parasites three genomes? *Parasitol. Today*, **7**, 134.

97. Feagin, J. E. (1995) Exploring the organelle genomes of malaria parasites. In *Biology of parasitism: modern approaches* (ed. J. C. Boothroyd and R. Komuniecki), p. 136. Wiley, New York.

98. Howe, C. J. (1992) Plastid origin of an extrachromosomal DNA molecule from *Plasmodium*, the causative agent of malaria. *J. Theor. Biol.*, **158**, 199.

99. Williamson, D. H., Gardner, M. J., Preiser, P., Moore, D. J., Rangachari, K., and Wilson, R. J. M. (1994) The evolutionary origin of the 35 kb circular DNA of *Plasmodium falciparum*: New evidence supports a possible rhodophyte ancestry. *Mol. Gen. Genet.*, **243**, 249.

100. Morden, C. W., Wolfe, K. H., dePamphilis, C. W., and Palmer, J. D. (1991) Plastid translation and transcription genes in a non-photosynthetic plant: intact, missing and pseudo genes. *EMBO J.*, **10**, 3281.

101. Geary, T. G. and Jensen, J. B. (1983) Effects of antibiotics on *Plasmodium falciparum in vitro. Am. J. Trop. Med. Hyg.*, **32**, 221.

102. Strath, M., Scott-Finnigan, T., Gardner, M., Williamson, D., and Wilson, I. (1993) Antimalarial activity of rifampicin *in vitro* and in rodent models. *Trans. R. Soc. Trop. Med. Hyg.*, **87**, 211.

103. Creasey, A. M., Ranford-Cartwright, L. C., Moore, D. J., Williamson, D. H., Wilson, R. J. M., Walliker, D., and Carter, R. (1993) Uniparental inheritance of the *mitochondrial* gene cytochrome b in *Plasmodium falciparum*. *Curr. Genet.*, **23**, 360.

104. Goonewardene, R., Daily, J., Kaslow, D., Sullivan, T. J., Duffy, P., Carter, R., Mendis, K., and Wirth, D. (1993) Transfection of the malaria parasite and expression of firefly luciferase. *Proc. Natl. Acad. Sci. USA*, **90**, 5234.

105. Wu, Y., Sifri, C. D., Lei, H.-H., Su, X.-Z., and Wellems, T. E. (1995) Transfection of *Plasmodium falciparum* within human red blood cells. *Proc. Natl. Acad. Sci. USA*, **92**, 973.

106. Van Dijk, M. R., Waters, A. P., and Janse, C. J. (1995) Stable transfection of malaria parasite blood stages. *Science*, **268**, 1358.

4 | *Toxoplasma* as a model genetic system

L. DAVID SIBLEY, DAN K. HOWE, KIEW-LIAN WAN,
SHAHID KHAN, MARTIN A. ASLETT and JAMES W. AJIOKA

1. Introduction

1.1 Advantages of *Toxoplasma* as a model genetic system

Several important biological attributes make *Toxoplasma* well suited for genetic studies. First, the parasite is easily maintained in the laboratory, having the ability to grow productively in virtually any vertebrate cell line. Second, it is possible to produce mutants and to propagate clones indefinitely. Phenotypic analysis of mutants is aided by the fact that replicative stages are haploid. Third, it is possible to conduct classical genetic studies and rudimentary genetic and physical linkage maps are available. Finally, molecular genetic studies are supported by DNA transformation using a variety of selectable markers. These properties make *Toxoplasma* unique among the world of parasites where biological and technological limitations have generally limited application of genetics.

1.2 Life cycle of *Toxoplasma*

The protozoan parasite *Toxoplasma gondii* infects a wide range of vertebrate hosts including humans (1). *Toxoplasma* is an obligate intracellular parasite that propagates mitotically as a haploid cell during most of its life cycle (Fig. 1). During the acute infection, the parasites replicate rapidly as tachyzoites that undergo successive rounds of invasion, replication, and host cell lysis. Virtually all nucleated cells are susceptible to invasion by tachyzoites (2). A fraction of the tachyzoites differentiate into slow-growing forms called bradyzoites that give rise to intracellular cysts (3). The eventual rupture of some mature cysts releases bradyzoites that in turn infect new host cells producing a chronic, long-term infection (4). Despite the presence of parasite cysts in the tissues, they cause little cellular response and chronic infections persist for the life of the host without significant pathology (3, 4).

Sexual reproduction is only known to occur in members of the cat family (5). Following a mitotic replication phase within cat intestinal epithelial cells, the parasites differentiate into male and female gametocytes. There does not appear to be a fixed

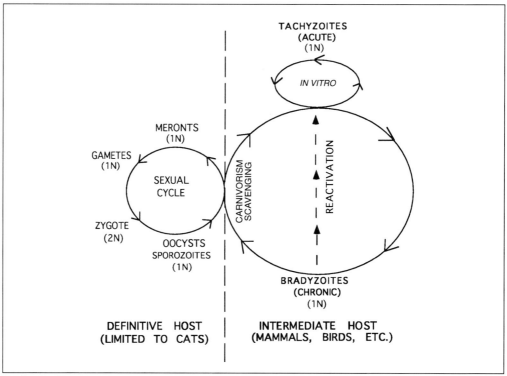

Fig. 1 The life cycle of *Toxoplasma gondii* involves both sexual and asexual replication cycles that are interconnected. Members of the cat family (*Felidae*) are the only known definitive host, but a wide variety of vertebrates, including humans, serve as intermediate hosts. In the intestinal epithelium of the cat, *Toxoplasma* undergoes several rounds of asexual division (schizogony) before differentiating into micro- and macrogametocytes. Fusion of gametes results in the formation of a zygote which develops a thick-walled oocyst that is shed in the cat faeces. Sporulation occurs in the external environment and the sporozoites within mature oocysts are then infectious. After ingestion, the parasite grows as a rapidly proliferating tachyzoite stage. Differentiation into slow-growing bradyzoites that reside within cysts is responsible for the chronic phase of infection which lasts for the life of the host. The asexual phase of the life cycle can be perpetuated in successive intermediate hosts by carnivorous feeding or scavenging, without the need for the sexual phase.

mating type, as a single organism is capable of undergoing the complete life cycle in cats (6, 7). All the replicative stages of *Toxoplasma* are haploid, including gametocytes, and the zygote is the only diploid stage (8). Following the fusion of gametes, the zygote develops into an oocyst that is shed in the faeces where it undergoes meiosis yielding eight haploid progeny. Oocysts are extremely resistant to environmental conditions and their dissemination probably accounts for the high prevalence of infection in herbivorous animals (9).

1.3 Human infection and pathogenesis

The wide-spread nature of *Toxoplasma* infections is partly due to the ease of parasite transmission through the food chain. Humans become infected by eating undercooked meat that contains cysts, or by ingestion of oocysts (9, 10). Serological studies indicate that the prevalence of infection by *Toxoplasma* ranges from 20% to more than 80% in different populations of adult humans (1). *Toxoplasma* infection typically causes only mild or inapparent symptoms in healthy adults (10), but can result in severe disease in immunocompromised individuals. The most severe cases of toxoplasmosis occur in congenital infections (11, 12), in organ transplant recipients (13), and in AIDS patients (14).

2. Genomic organization

2.1 The genomes

The haploid nuclear genome size of *Toxoplasma* has been estimated at 7.8×10^7 bp based on cytofluorometric measurements (8). Separation of *Toxoplasma* chromosomes by pulsed-field gel electrophoresis (PFGE) and construction of a preliminary genetic linkage map indicates that there are at least 11 distinct chromosomes, ranging in size from 2Mb to greater than 10Mb (15, 16). In addition to the nuclear genome, two organellar genomes have been described in the related Apicomplexan parasite *Plasmodium*: a linear mitochondrial genome (17, 18) and a 35kb circular plastid-like genome (19, 20; see Chapter 3). *Toxoplasma* probably contains similar genomes as indicated by cross-hybridization (17) and electron microscopic studies (21); however, their analysis is complicated by the presence of numerous pseudogenes derived from the sequences of cytochrome oxidase subunit 1 and apocytochrome *b* that have been integrated throughout the nuclear genome (22).

2.2 The genes

To date approximately 30 genes have been cloned and characterized from *Toxoplasma* including a variety of housekeeping genes, genes encoding surface antigens and secretory proteins, and ribosomal RNA genes (GenBank release date March 1995). The nucleotide content of coding regions is approximately 50–55% GC, and there is no strong codon bias (23). The majority of genes are single copy and there is no evidence for polycistronic transcripts or for *trans*-splicing. These features make cloning and expression of *Toxoplasma* genes relatively straightforward particularly in comparison with the AT-rich *P. falciparum* genome.

Many protein-coding genes do not contain introns while others are interrupted by one or several small introns that are cis-spliced using conventional intron–exon sites. The exceptions to this pattern are the genes encoding dihydrofolate reductase (DHFR) (24), uracil phosphoribosyl transferase (D. Roos, unpublished; GenBank number U10246), and hypoxanthine xanthine guanosine phosphoribosyl transferase (D. Roos, unpublished; GenBank number U09219), all of which are interrupted by numerous introns. The reason for this unusual pattern is not known, as other housekeeping genes such as those encoding the tubulins (25), NTPases (26), and actin (J. D. Dobrowolski and L. D. Sibley, GenBank number 410429) are not interrupted by frequent introns.

2.3 Repetitive DNAs

Several repetitive DNAs have been characterized in *Toxoplasma* and used for detection and strain identification. The first of these is a 2kb element called B1 that is tandemly repeated 35 times on chromosome *IX* (27). Although the B1 gene is not polymorphic, it is a useful target for detection of *Toxoplasma* using the polymerase chain reaction (PCR) for amplification due to its high sensitivity and specificity (27–29). Three different families of dispersed repetitive elements have been described; their extreme polymorphism makes them useful for strain identification (30–33).

The nuclear rDNA genes are tandemly repeated in *Toxoplasma* (34) occurring in a single cluster of approximately 100 copies located on chromosome *IX*. The 8.2kb repeat unit contains the small subunit, two short transcribed spacers flanking the 5.8S RNA, the large subunit, the 5S RNA and an untranscribed intergenic region (34, 35). The rDNA locus has also been used for sensitive PCR based detection (36) and for phylogenetic comparisons (35, 37–39).

2.4 Gene expression

The 5' untranslated regions of *Toxoplasma* mRNAs are 100–500bp in length. The upstream regions do not contain the usual TATA or CAAT boxes typical of genes transcribed by RNA polymerase II (pol II) in higher eukaryotes (40). It is not clear if *Toxoplasma* uses a divergent sequence element for binding a TATA-binding protein homologue or if the site of transcription initiation is determined by a different mechanism. Indeed, there are precedents even among genes of higher eukaryotes for pol II transcription occurring in the absence of TATA sequences (41). Messenger RNAs in *Toxoplasma* are typically polyadenylated, although transcriptional terminators and polyadenylation signals have not been characterized. Translation initiates at an ATG where the -3 position is typically adenosine and the +1 position is less constrained than in higher eukaryotes. Several genes are translated from the second in-frame ATG (42), rather than the first, indicating the importance of the surrounding residues.

The development of DNA transfection techniques (see below) has enabled direct analysis of transcriptional elements in *Toxoplasma* based on the activity of chloramphenicol acetyl transferase (CAT) encoded by the bacterial *cat* gene (42). The first element described consists of a series of six 27bp repeats that lie within 200bp upstream of the *SAG1* gene encoding the surface antigen p30, also known as SAG1. This repeat functions in part as an enhancer; decreases in the number of repeats from the 5' end greatly reduce CAT activity (43). These repeats also function as a site selector to position RNA polymerase for initiation of transcription; successively moving the repeats upstream shifts the transcription initiation site accordingly (43). Independently, a conserved heptanucleotide (A/TGAGACG) has been identified as a common sequence motif upstream of the *GRA* and *TUB1* genes (C. Mercier, S. Lefebvre-Van Hende, and M. F. Cesbron, unpublished). A similar heptanucleotide sequence occurs in the core of the 27bp *SAG1* repeat. The heptanucleotide element can be found in both forward and reverse orientations which typically lie within 100–300bp upstream of the transcription initiation site. The importance of this element has been confirmed by deletions or point mutations within the core sequence that greatly reduce transcription (C. Mercier, S. Lefebvre-Van Hende, and M. F. Cesbron, unpublished).

2.5 Stage-specific gene expression

A number of stage-specific antigens have been characterized in *Toxoplasma*. Two surface antigens, SAG1 and SAG2, are abundantly expressed on the tachyzoite but are down regulated during the conversion to bradyzoites in which they are not detected (44). Conversely, a number of specific antigens are only expressed in bradyzoites (45–49). The interconversion between tachyzoite and bradyzoite forms is of central importance to understanding the reactivation of chronic infection that occurs in immunocompromised patients. The recent development of *in vitro* methods to induce stage conversion (50–51) should greatly aid our understanding of developmental gene regulation in *Toxoplasma*.

3. Population genetic structure

3.1 Strain-specific antigenic markers

Toxoplasma strains are remarkably similar in their antigenic make-up, host range, and morphology, allowing isolates from widely different hosts and widely different geographical regions to be grouped into a single species. While the different life cycle stages express distinct surface antigens (44, 46, 52, 53), the antigens of any one stage are highly conserved between strains (54, 55). Minor antigenic differences do occur and these have typically been recognized by western blot analysis using sera from chronically infected individuals (56, 57). Although the majority of monoclonal antibodies raised against *Toxoplasma* react with all strains of the parasite, several strain-specific monoclonal antibodies have also been described (58, 59).

The antigenic similarity of *Toxoplasma* strains is supported by sequence analyses of the major surface antigens (SAGs) of the tachyzoite. Only two different forms of the proteins SAG1 and SAG2 have been described from a large number of independently isolated strains (60–62). Thus, unlike the situation in *Plasmodium* (63), antigenic diversity in *Toxoplasma* is low. While these results could indicate sequence conservation due to functional constraints, as discussed below, a similar low degree of polymorphism is observed at most genetic loci.

3.2 Polymorphic isoenzyme and DNA markers

Screening for polymorphic isoenzymes or DNA markers also supports a low degree of genetic diversity among *Toxoplasma* strains. Only six polymorphic isoenzymes have been identified from a total of 18 enzymes and the majority of these exhibit only two alleles (64, 65). Likewise, the frequency of restriction fragment length polymorphisms (RFLPs) detected by single-copy DNA markers is relatively low in *Toxoplasma*; an average of more than 12 separate restriction digests are needed to reveal a single polymorphism for a single copy gene (16). One advantage of this low degree of genetic diversity is that markers developed in one strain have a high probability of being useful for all strains. With the development of PCR, it is possible to amplify specific genetic loci for RFLP analysis from very small numbers of cells. In addition to the obvious practical advantages, this allows for direct analysis of strains without the need for extensive growth and possible alteration outside the original host. Strain-specific DNA polymorphisms in *Toxoplasma* have also been identified using PCR to amplify DNA fragments based on the random amplified polymorphic DNA (RAPD) technique (66, 67).

3.3 The clonal population structure of *Toxoplasma*

Despite the presence of a sexual phase in the life cycle, *Toxoplasma* strains show much lower genetic variation than expected for a panmictic population when analysed by RFLP (31) or isoenzyme markers (64, 65). This clonality is manifested by the repeated isolation of the same genotype from independent samples and by the absence of many possible recombinant genotypes. *Toxoplasma* strains isolated from humans and animals can be readily grouped into three distinct clonal lineages, each of which is globally distributed (68, 69). The ability of *Toxoplasma* to be transmitted horizontally through the food chain by carnivorous feeding or scavenging has probably contributed to the spread of clonal lineages (Fig. 1). Nevertheless, the three *Toxoplasma* lineages are extremely similar overall and do not constitute separate species as they occasionally undergo genetic recombination in the wild (68).

Clonal population structures have also been described for a number of other parasitic protozoa including *Leishmania* spp. and *Trypanosoma cruzi* (70, 71; see Chapter 2). One consequence of such clonality is that specific clones are likely to exhibit distinct biological traits related to pathogenicity. In the case of *Toxoplasma*, there are clear correlations between the genotype of the parasite and pathogenicity in mice, which may also be important in human disease (68). Type I strains are readily distinguished by their unique allele at the *SAG1* locus, which is correlated with high parasitaemia and acute virulence in mice (31, 62, 68). Type II strains are prone to reactivation of chronic infection in mice and are the most common cause of toxoplasmosis in AIDS patients (68). It should be possible to map and identify genes involved in pathogenesis caused by *Toxoplasma* using the genetic approaches described below combined with available animal models for toxoplasmosis.

4. Classical genetic analyses

4.1 Tools for *in vitro* analysis

Although *Toxoplasma* is an obligate intracellular parasite, the ability of tachyzoites to propagate *in vitro* in virtually any cell line provides a convenient experimental system. Parasite viability can be quantified by the formation of plaques on cell monolayers (72) and selective uptake of uracil (unlike the host cell, the parasite efficiently salvages uracil but does not salvage thymidine) (73). Individual clones can be obtained by limiting dilution and these can be propagated indefinitely or easily cryopreserved. It is possible to produce *Toxoplasma* mutants using UV irradiation or a variety of chemical mutagens (74) and to isolate temperature-sensitive mutants (75). Despite the difficulty of having to conduct crosses in cats, there are several features of the life cycle that favour classical genetic approaches. First, because oocyst maturation (and meiosis) occurs outside the cat, only a single round of meiosis results from experimental infection. Second, all the progeny resulting from one meiosis are contained within a single oocyst which should allow tetrad analysis. Third, a large number of oocysts are obtained from a single infected cat and they can be stored for several years prior to analysis.

4.2 Use of drug resistance markers in genetic crosses

The first genetic experiments in *Toxoplasma* involved the production of drug resistance mutants that were then crossed by co-infection in cats (76, 77). These studies, conducted with a single cloned line called CTG (isolated from a naturally infected cat), demonstrated that *Toxoplasma* does not have a predetermined mating type but is capable of both self-mating and cross-fertilization. Segregation of loci was detected between two unlinked drug resistance markers by monitoring the frequency of progeny with doubly resistant or doubly sensitive phenotypes. Because of the expense and difficulty in producing a large number of mutants with unique markers, it is not feasible to develop a linkage map from such crosses.

4.3 RFLP linkage studies

In contrast to phenotypic traits like drug-resistance markers, RFLPs are easily assayed. There are two additional advantages of RFLP markers for genetic studies. First, a large number of independent loci can be mapped in a single cross. (Traditional phenotypic markers typically require a series of separate crosses, each of which can analyse only a small number of markers simultaneously.) RFLP linkage maps have proven useful in systems where classical genetics were already well established (78, 79) and in systems where detailed genetic maps were not previously available (80, 81). Second, RFLPs can be used to identify genetic correlates of phenotypes in systems where experimental genetic studies are not feasible, such as in the human genome (82, 83).

The genetic map of *Toxoplasma* currently consists of 64 RFLP markers defining 77 loci in 11 linkage groups (16). As classically defined, the genome consists of a total of 147 map units (centimorgans; cM), where the current lower limit of resolution is 5.25cM, where 1cM is 300–500kb (16). This relatively low rate of recombination allows direct mapping of a locus to a specific chromosome using a small number of markers. For example, a locus that mediates resistance to adenine arabinoside maps to chromosome *V*, while resistance to sinefungin maps to chromosome *IX* (16). One limitation imposed by the low recombination rate in *Toxoplasma* is that finer mapping of genes will require the analysis of a large number of recombinant progeny in order to identify informative cross-overs. While much of this preliminary analysis can be conducted by PCR-RFLP testing of clones, thus obviating the need to grow large amounts of parasites, new RFLP markers will be required for finer positional mapping.

5. Genome mapping

A useful definition of genome mapping is the synthesis of genetic linkage and physical mapping information, i.e. the positioning of genetic information in a physical context (Fig. 2) (84, 85). This approach not only provides a multifaceted tool to investigate the genetic basis of interesting phenotypes but will help elucidate many

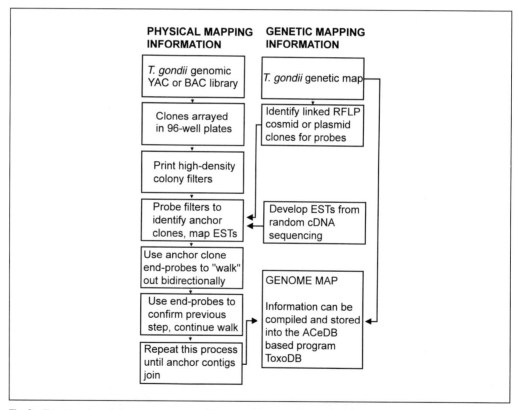

Fig. 2 Flow diagram of genome map assembly. A combination of genetic linkage studies and physical mapping strategies are being used to generate an integrated genome map. In addition to providing specific information about chromosome structure and function, this map will be useful for positional cloning of genes and for identification of genes that regulate important biological phenotypes involved in pathogenesis.

aspects of chromosome structure and function. The main use of a genome map is to facilitate the identification and cloning of genetic loci. Positional cloning, or finding a gene by linkage to a known marker, requires that the cloned genomic fragments used for physical mapping approach the size encompassed by one cM. Thus, even if an interesting phenotype is genetically mapped, physically defining a candidate locus for positional cloning is not realistic with the present low resolution of the *Toxoplasma* genetic map. Therefore the application of genetics in this species depends on the development of a higher resolution genetic map and the generation of a genome map.

5.1 Physical mapping, construction of contigs and mapping of cDNAs

Physical mapping of eukaryotic genomes has been made possible with the advent of 'big DNA' technology. The key components are the ability to separate large fragments of DNA (50kb to several Mb) by PFGE (86, 87) and to clone these fragments using cosmids (88), yeast artificial chromosomes (YACs) (89), and bacterial artificial chromosomes (BACs) (90). Since the genome can be cloned and analysed in bigger pieces, the number of clones that need to be sorted and aligned is *at least two orders of magnitude* less than the number required when using conventional cloning vectors like bacteriophage lambda.

There are several methods for constructing physical maps based on overlapping clones or contigs, each with their own advantages and disadvantages. They can be divided into two basic categories: analysis of DNA fingerprints following cleavage with restriction endonucleases and identification of sequence–tagged sites (STSs) (91). The first method has been used to construct the *Caenorhabditis elegans* physical map (79, 92); its main drawback is the need for sophisticated digital scanning equipment and computer analysis.

In the STS approach, contigs are built by successively identifying overlapping clones by virtue of their hybridization to the ends of anchor clones. The process is streamlined by determining the STS content of individual clones by PCR. This method requires a large number of high-density clone filters but lends itself to automated robotics. STS-based mapping has been used to generate contigs of several human chromosomes and to identify the basis of several genetic diseases including cystic fibrosis (93, 94).

The generation and mapping of expressed sequence tags (ESTs) has become an important complementary 'genetic' approach to RFLP mapping. With the advent of technical developments such as PCR amplification (95) and semi-automated DNA sequencing, rapid sequencing of cDNAs has become the fastest growing part of DNA sequence databases (92, 93, 96–98). Once EST mapping reaches saturation, as is nearly the case for some of the chromosomes of *C. elegans* (91, 97), positional cloning of specific genes is reduced to testing sets of ESTs (or the corresponding genomic clones).

5.2 Physical mapping of parasite genomes

Because of the difficulty in performing classical or molecular genetics in parasite systems, contig mapping may prove a useful alternative approach for identification and cloning of genes. A contig of chromosome 2 in *Plasmodium* has been generated from a series of overlapping YAC clones and used to generate a transcript map of this chromosome (99) (see Chapter 3). A similar approach is being used to map genes on chromosome 7 that are linked to chloroquine resistance in *Plasmodium* (T. Wellems, personal communication). To facilitate the construction of a genome map in *Toxoplasma*, a BAC library consisting of more than 6100 clones with an average insert size of 60kb has been constructed from the RH strain and is being used to generate contigs by STS mapping (S. Khan and J. Ajioka, unpublished).

In the case of protozoan parasites where transmission genetics are difficult or impossible, using EST information provides an alternative for developing genetic maps. The current EST database for *Toxoplasma* (RH strain) consists of 237 independent randomly selected cDNA sequences (GenBank accession numbers T62239–T62475; K.-L. Wan and J. Ajioka, submitted). Analysis of these cDNAs by database searches (100) yields results consistent with other protozoan EST projects (101; J. Ajioka, unpublished data on *Leishmania major* ESTs): highly expressed or conserved genes, such as certain cytoskeletal and ribosomal proteins, are readily identified, while the majority of sequences do not show significant homologies. Despite their unknown function, these genes will provide useful physical markers for anchoring an integrated genomic map.

5.3 Database development

Information storage and access is the most crucial part of genome analysis. The accumulation of data has accelerated to the point that the dissemination of information via traditional peer-reviewed publications is no longer feasible. To address this problem, Richard Durbin (MRC, LMB, Cambridge, UK) and Jean Thierry-Mieg (CNRS, Montpellier, France) developed a database program for the *C. elegans* genome project, ACeDB (a *C. elegans* database) that can be reconfigured to suit any particular genome project (102).

A modified version of ACeDB was chosen as the framework for *Toxoplasma* (as well as for *Leishmania* and *Trypanosoma b. brucei*) due to its flexibility, widespread use and adaptability to future needs. The *Toxoplasma* database (ToxoDB, M. Aslett and J. Ajioka, unpublished) currently runs on a UNIX–based Sun SPARC workstation, although a PC version is being refined for general release (H. Cobb and M. Aslett, unpublished). As of April 1995, ToxoDB holds genetic maps for 11 chromosomes, 77 loci assigned to the 11 chromosomes, all *Toxoplasma* DNA sequences in the EMBL database, 237 ESTs and approximately 1800 references. The example shown in Fig. 3 illustrates the use of ToxoDB to find information on one of the DNA markers located on chromosome *VIII* of the *Toxoplasma* RFLP linkage map. The user is able to display graphic representations of genetic maps and retrieve information

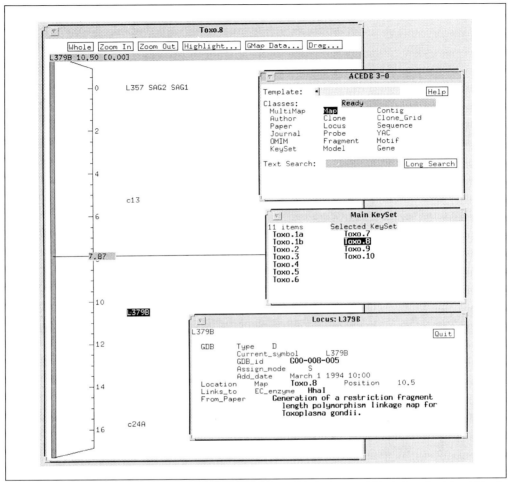

Fig. 3 Illustration of the database for *Toxoplasma gondii*, ToxoDB. In the top right window, the 'class' Map is selected in the opening menu ACeDB (ToxoDB). This opens the central window, Main Keyset, listing the 11 *Toxoplasma* chromosomes. Toxo.8 is selected from the list which opens the left window containing a graphic depiction of markers on chromosome *VIII*. Selection of the marker L379B opens the final window (Locus: L379B) that displays information on this marker, including the restriction enzyme used to show the polymorphism at this site and the paper which describes this locus. The boldface fields can be selected for further information.

on specific genomic clones, cDNA clones, gene mutants, and relevant literature citations.

6. Molecular genetics

Progress in molecular genetics of *Toxoplasma* has been particularly rapid in the past several years. While it is not possible to document all of the recent developments here, a brief review of the available techniques will be provided. Readers are encouraged to consult two other recent reviews for additional details on the specific methodology employed (103, 104).

6.1 DNA transfection

One of the initial barriers to the development of DNA transfection techniques is the extreme fastidiousness of many intracellular parasites. Fortunately, it is possible to obtain large numbers of extracellular *Toxoplasma* parasites and to maintain their viability in medium for several hours. This flexibility makes it straightforward to electroporate DNA into extracellular parasites followed by recovery and continued culture. Two different reporter genes have been used for transient expression assays in *Toxoplasma*; initially, the *cat* gene, which encodes chloramphenicol acetyl transferase (42) and more recently, the *lacZ* gene, which encodes β-galactosidase (F. Seeber and J. C. Boothroyd, submitted; M. Messina and L. D. Sibley, unpublished) (Table 1). Proper expression requires that the reporter gene is driven by the 5' flanking region from a *Toxoplasma* gene. Transient expression levels increase steadily for the first 24 hours and maintain detectable levels for several days after electroporation (42). In addition to being useful as reporters, these markers are being used as heterologous probes to study protein trafficking and secretion in *Toxoplasma* (D. Soldati, unpublished; M. Messina and L. D. Sibley, unpublished).

6.2 Stable transformation

One initial limitation to developing stable DNA transformation in *Toxoplasma* was the inability to use existing selectable markers. Selection strategies based on resistance to aminoglycosides like G418 (neomycin) are limited by toxicity to the host cells at levels of drug necessary to inhibit *Toxoplasma*. The antibiotics used for most prokaryotic expression vectors, including ampicillin, tetracycline, and kanamycin, are of limited use in selecting against the growth of *Toxoplasma*. Thus, development of stable DNA transformation required the identification of novel selectable markers and the construction of vectors for expression of these heterologous genes in *Toxoplasma*. Currently, four independent dominant selectable markers are being used for stable DNA transformation in *Toxoplasma* (Table 1). Three of these rely on antibiotic resistance selection (105, 106a, 106b; K. Kim, J. Kampmeier, and J. C. Boothroyd, unpublished), while the fourth is based on rescue from tryptophan auxotrophy (107) (Table 1). The transformation efficiency is roughly 10^{-4} for the three heterologous markers (*ble*, *cat*, *trp*), while that achieved with DHFR mutants encoding pyrimethamine resistance (Pyr^r) is much higher. In general, stable transformants contain the marker gene integrated into the genome, primarily at non-homologous sites. By varying the selection strategies, it is possible to obtain transformants that contain either single or multiple copies of the marker. Vectors based on the above markers are currently being used for expression of heterologous genes, for expression of epitope-tagged forms of endogenous proteins (C. Mercier and L. D. Sibley, unpublished) and for targeted gene knockouts in *Toxoplasma*.

Table 1 Transient and stable DNA transformation markers in *Toxoplasma*

Transient expression

Gene	Activity	Substrates	Ref.
cat	chloramphenicol acetyl transferase	^{14}C acetate	42
lacZ	β-galactosidase	ONPG[a], MUG[b]	Unpublished[c]

Selectable markers

Gene	Selection		Ref.
ble	phleomycin		Unpublished[d]
cat	chloramphenicol		103
dhfr-Pyr^r	pyrimethamine		104, 110
trp	tryptophan−/indole+		105

[a] o-nitrophenyl-β-D-galactopyranoside
[b] 4-methylumbelliferyl β-D-galactoside
[c] M. Messina and L. D. Sibley, unpublished; F. Seeber and J. C. Boothroyd, submitted
[d] 106a; K. Kim, J. Kampmeier, and J. C. Boothroyd, unpublished

6.3 Targeted gene disruption

DNA transformation is critical to reverse genetics, the evaluation of gene function by phenotypic analysis of gene knockouts. Reverse genetics generally relies on homologous recombination which occurs in all organisms but with different frequencies than random integration at non-homologous sites. In yeast (108) and other protozoan parasites like *Leishmania* (109) and *Trypanosoma* (110, 111; see Chapter 2), homologous recombination is the preferred means of integration. Despite the low rate of homologous recombination in *Toxoplasma*, targeting specific genetic loci for recombination has been accomplished in two cases. Transfection of the *cat* gene flanked by *ROP1* sequences was used to generate a knockout by a double cross-over event (105). Transfection with *DHFR*-Pyr^r has been used for either random or targeted integration, depending on the length of flanking homologous sequences (112).

The high frequency of stable transformation with *DHFR*-Pyr^r has been exploited to create banks of mutants by random insertional mutagenesis (104). Following transformation with *DHFR*-Pyr^r, parasites are first selected with pyrimethamine to enrich for transformants and then selected with a second drug to select for mutants. Mutants resistant to 5-fluorouridine, 6-thio-guanosine and adenine arabinoside have been obtained by this method and the corresponding genes have been isolated by cloning out the genomic regions that flank the site of original insertion (104). This promises to be an important technique for cloning of non-essential genes where a direct selection scheme can be identified. Additionally, isolation of the target genes should allow development of markers for negative selection which may prove useful in selecting for homologous integration, using a combination of positive and negative selection (104).

7. Future developments

All the essential tools for both forward and reverse genetics are currently available in *Toxoplasma*. The rapid advances in molecular genetics provide an array of complementary tools for isolating genes and directly testing their functions. Continued developments to integrate genetic and physical markers into a genome map will facilitate the identification of genes that are not easily obtained by direct transformation procedures. Among the more important areas for future applications of genetics are identification of virulence and pathogenesis genes, identification of genes related to invasion and intracellular survival, and determination of the molecular basis of stage-specific differentiation.

References

1. Dubey, J. P. and Beattie, C. P. (1988) *Toxoplasmosis of animals and man*. p. 220 CRC Press, Boca Raton.
2. Werk, R. (1985) How does *Toxoplasma gondii* enter host cells? *Rev. Infect. Dis.*, **7**, 449.
3. Remington, J. S. and Cavanaugh, E. N. (1965) Isolation of the encysted form of *Toxoplasma gondii* from human skeletal muscle and brain. *N. Engl. J. Med.*, **273**, 1308.
4. Frenkel, J. K. and Escajadillo, A. (1987) Cyst rupture as a pathogenic mechanism of toxoplasmic encephalitis. *Am. J. Trop. Med. Hyg.*, **36**, 517.
5. Dubey, J. P. and Frenkel, J. F. (1972) Cyst-induced toxoplasmosis in cats. *J. Protozool.*, **19**, 155.
6. Cornelissen, A. W. C. A. and Overdulve, J. P. (1985) Sex determination and sex differentiation in coccidia: gametogony and oocyst production after monoclonal infection of cats with free-living and intermediate host stages of *Isospora (Toxoplasma) gondii*. *Parasitol.*, **90**, 35.
7. Pfefferkorn, E. R., Pfefferkorn, L. C., and Colby, E. D. (1977) Development of gametes and oocysts in cats fed cysts derived from cloned trophozoites of *Toxoplasma gondii*. *J. Parasitol.*, **63**, 158.
8. Cornelissen, A. W. C. A., Overdulve, J. P., and Van der Ploeg, M. (1984) Determination of nuclear DNA of five Eucoccidian parasites, *Isospora (Toxoplasma) gondii, Sarcocystis cruzi, Eimeria tenella, E. Acervulina*, and *Plasmodium beghei*, with special reference to gametogenesis and meiosis in *I. (T.) gondii*. *Parasitology*, **88**, 531.
9. Frenkel, J. K. and Ruiz, A. (1980) Human toxoplasmosis and cat contact in Costa Rica. *Am. J. Trop. Med. Hyg.*, **29**, 1167.
10. Frenkel, J. K. (1988) Pathophysiology of toxoplasmosis. *Parasitol. Today*, **4**, 273.
11. Wong, S. and Remington, J. S. (1994) Toxoplasmosis in pregnancy. *Clin. Infect. Dis.*, **18**, 853.
12. Desmonts, G. and Couvreur, J. (1974) Congenital toxoplasmosis: a prospective study of 378 pregnancies. *N. Engl. J. Med.*, **290**, 1110.
13. Israelski, D. M. and Remington, J. S. (1993) Toxoplasmosis in the non-AIDS immunocompromised host. *Curr. Clin. Top. Infect. Dis.*, **13**, 322.
14. Luft, B. J. and Remington, J. S. (1992) Toxoplasmic encephalitis in AIDS. *Clin. Infect. Dis.*, **15**, 211.
15. Sibley, L. D. and Boothroyd, J. C. (1992) Construction of a molecular karyotype for *Toxoplasma gondii*. *Mol. Biochem. Parasitol.*, **51**, 291.

16. Sibley, L. D., LeBlanc, A. J., Pfefferkorn, E. R., and Boothroyd, J. C. (1992) Generation of a restriction fragment length polymorphism linkage map for *Toxoplasma gondii*. *Genetics*, **132,** 1003.
17. Joseph, J. T., Aldritt, S. M., Unnasch, T., Puijalon, O., and Wirth, D. F. (1989) Characterization of a conserved extrachromosomal element isolated from the avian malarial parasite *Plasmodim gallinaceum*. *Mol. Cell. Biol.*, **9,** 3621.
18. Vaidya, A. B., Akella, R., and Suplick, K. (1989) Sequences similar to genes for two mitochondrial proteins and portions of ribosomal RNA in tandemly arrayed 6-kilobase-pair DNA of a malarial parasite. *Mol. Biochem. Parasitol.*, **35,** 97.
19. Feagin, J. E., Werner, E., Gardner, M. J., Williamson, D. H., and Wilson, R. J. M. (1992) Homologies between the contiguous and fragmented rRNAs of the two *Plasmodium falciparum* extrachromosomal DNAs are limited to core sequences. *Nucleic Acids Res.*, **20,** 879.
20. Gardner, M. J., Feagin, J. E., Moore, D. J., Spencer, D. F., Gray, M. W., Williamson, D. H., and Wilson, R. J. M. (1991) Organisation and expression of small subunit ribosomal RNA genes encoded by a 35-kilobase circular DNA in *Plasmodium falciparum* . *Mol. Biochem. Parasitol.*, **48,** 77.
21. Borst, P., Overdulve, J. P., Weijer, P. J., Fase-Fowler, F., and Van der Berg, M. (1984) DNA circles with cruciforms from *Isospora* (*Toxoplasma*) *gondii*. *Biochem. Biophys. Acta*, **781,** 100.
22. Ossorio, P. N., Sibley, L. D., and Boothroyd, J. C. (1991) Mitochondrial-like DNA sequences flanked by direct and inverted repeats in the nuclear genome of *Toxoplasma gondii*. *J. Mol.. Biol.*, **222,** 525.
23. Johnson, A. (1990) Comparison of dinucleotide frequency and codon usage in *Toxoplasma* and *Plasmodium*: evolutionary implications. *J. Mol. Evol.*, **30,** 383.
24. Roos, D. S. (1993) Primary structure of the dyhydrofolate reductase–thymidylate synthase gene from *Toxoplasma gondii*. *J. Biochem.*, **268,** 6269.
25. Nagel, S. D. and Boothroyd, J. C. (1988) The alpha- and beta-tubulins of *Toxoplasma gondii* are encoded by single copy genes containing multiple introns. *Mol. Biochem. Parasitol.*, **29,** 261.
26. Asai, T., Miura, S., Sibley, L. D., Okabayashi, H., and Takeuchi, T. (1995) Biochemical and molecular characterization of nucleoside triphosphate hydrolase isozymes from the parasitic protozoan *Toxoplasma gondii*. *J. Biol. Chem.*, **270,** 11391.
27. Burg, L. J., Grover, C. M., Pouletty, P., and Boothroyd, J. C. (1989) Direct and sensitive detection of a pathogenic protozoan, *Toxoplasma gondii* by polymerase chain reaction. *J. Clin. Microbiology*, **27,** 1787.
28. Grover, C. M., Thulliez, P., Remington, J. S., and Boothroyd, J. C. (1990) Rapid prenatal diagnosis of congenital *Toxoplasma* infection by using polymerase chain reaction and amniotic fluid. *J. Clin. Microbiology*, **28,** 2297.
29. Hohlfeld, P., Daffos, F., Costa, J., Thulliez, P., Forestier, F., and Vidaud, M. (1994) Prenatal diagnosis of congenital toxoplasmosis with a ploymerase chain reaction test on amniotic fluid. *New Engl. J. Med.*, **331,** 695.
30. Cristina, N., Liaud, M. F., Santoro, F., Oury, B., and Ambroise-Thomas, P. (1991) A family of repeated DNA sequences in *Toxoplasma gondii*: cloning, sequence analysis, and use in strain characterization. *Exp. Parasitol.*, **73,** 73.
31. Sibley, L. D. and Boothroyd, J. C. (1992) Virulent strains of *Toxoplasma gondii* comprise a single clonal lineage. *Nature*, **359,** 82.
32. Blanco, J. C., Angel, S. O., Maero, E., Pszenny, V., Serpente, P., and Garberi, J. C. (1992)

Cloning of repetitive DNA sequences from *Toxoplasma gondii* and their usefulness for parasite detection. *Am. J. Trop. Med. Hyg.*, **46**, 350.

33. Howe, D. K. and Sibley, L. D. (1994) *Toxoplasma gondii*: analysis of different laboratory stocks of RH strain reveals genetic heterogeneity. *Exp. Parasitol.*, **78**, 242.
34. Guay, J. M., Huot, A., Gagnon, S., Tremblay, A., and Levesque, R. C. (1992) Physical and genetic mapping of cloned ribosomal DNA from *Toxoplasma gondii*: primary and secondary structure of the 5S gene. *Gene*, **114**, 165.
35. Gagnon, S., Levesque, R. C., Sogin, M. L., and Gajadhar, A. A. (1993) Molecular cloning, complete sequence of the small subunit ribosomal RNA coding region and phylogeny of *Toxoplasma gondii*. *Mol. Biochem. Parasitol.*, **60**, 145.
36. Guay, J., Dubois, D., Morency, M., Gagnon, S., Mercier, J., and Levesque, R. C. (1993) Detection of the pathogenic parasite *Toxoplasma gondii* by specific amplification of ribosomal sequences using comultiplex polymerase chain reaction. *J. Clin. Microbiology*, **31**, 203.
37. Johnson, A. M., Murray, P. J., Illana, S., and Baverstock, P. J. (1987) Rapid nucleotide sequence analysis of the small subunit ribosomal RNA of *Toxoplasma gondii*: evolutionary implications for the Apicomplexa. *Mol. Biochem. Parasitol.*, **25**, 239.
38. Brindley, P. J., Gazzinelli, R. T., Denkers, E. Y., Davis, S. W., Dubey, J. P., Belfort Jr., R., Martins, M. C., Silveira, C., Jamra, L., and Waters, A. P. (1993) Differentitiation of *Toxoplasma gondii* from closely related coccidia by riboprint analysis and a surface antigen gene polymerase chain reaction. *Am. J. Trop. Med. Hyg.*, **48**, 447.
39. Luton, K., Gleeson, M., and Johnson, A. M. rRNA gene sequence heterogeneity among *Toxoplasma gondii* strains. *Parasitol. Res.*, **81**, 310.
40. Peterson, M. G., Tanese, N., Pugh, B. F., and Tjian, R. (1990) Functional domains and upstream activation properties of cloned human TATA binding protein. *Science*, **248**, 1625.
41. Tjian, R. and Maniatis, T. (1994) Transcriptional activation: a complex puzzle with few easy pieces. *Cell*, **77**, 5.
42. Soldati, D. and Boothroyd, J. C. (1993) Transient transfection and expression in the obligate intracellular parasite, *Toxoplasma gondii*. *Science*, **260**, 349.
43. Soldati, D. and Boothroyd, J. C. (1995) A selector of transcription initiation in the protozoan parasite *Toxoplasma gondii*. *Mol. Cell. Biol.*, **15**, 87.
44. Kasper, L. H. and Ware, P. L. (1985) Recognition and characterization of stage-specific oocyst/sporozoite antigens of *Toxoplasma gondii* by human antisera. *J. Clin. Invest.*, **75**, 1570.
45. Bohne, W., Heesemann, J., and Gross, U. (1993) Induction of bradyzoite-specific *Toxoplasma gondii* antigens in gamma interferon-treated mouse macrophages. *Infect. Immun.*, **61**, 1141.
46. Omata, Y., Igarashi, M., Ramos, M. I., and Nakabayashi, T. (1989) *Toxoplasma gondii*: antigenic differences between endozoites and tachyzoites defined by monoclonal antibodies. *Exp. Parasitol.*, **75**, 189.
47. Soete, M., Fortier, B., Camus, D., and Dubremetz, J. F. (1993) *Toxoplasma gondii*: kinetics of bradyzoite–tachyzoite interconversion *in vitro*. *Exp. Parasitol.*, **76**, 259.
48. Tomavo, S., Fortier, B., Soete, M., Ansel, C., Camus, D., and Dubremetz, J. F. (1991) Characterization of bradyzoite-specific antigens of *Toxoplasma gondii*. *Infect. Immun.*, **59**, 3750.
49. Parmley, S. F., Yang, S., Harth, G., Sibley, L. D., Sucharczuk, A., and Remington, J. S. (1994) Molecular charcterization of a 65-kilodalton *Toxoplasma gondii* antigen expressed abundantly in the matrix of tissue cysts. *Mol. Biochem. Parasitol.*, **66**, 283.

50. Bohne, W., Heesemann, J., and Gross, U. (1994) Reduced replication of *Toxoplasma gondii* is necessary for induction of bradyzoite-specific antigens: a possible role for nitric oxide in triggering stage conversion. *Infect. Immun.*, **62**, 1761.
51. Soete, M., Camus, D., and Dubremetz, J. F. (1994) Experimental induction of bradyzoite-specific antigen expression and cyst formation by the RH strain of *Toxoplasma gondii in vitro. Exp. Parasitol.*, **78**, 361.
52. Kasper, L. H., Bradley, M. S., and Pfefferkorn, E. R. (1984) Identification of stage-specific sporozoite antigens of *Toxoplasma gondii* by monoclonal antibodies. *J. Immunol.*, **132**, 443.
53. Lunde, M. N. and Jacobs, L. (1983) Antigenic differences between endozoites and cystozoites of *Toxoplasma gondii. J. Parasitology*, **65**, 806.
54. Handman, E., Goding, J. W., and Remington, J. S. (1980) Detection and characterization of membrane antigens of *Toxoplasma gondii. J. Immunol.*, **124**, 2578.
55. Couvreur, G., Sadak, A., Fortier, B., and Dubremetz, J. F. (1988) Surface antigens of *Toxoplasma gondii. Parasitology*, **97**, 1.
56. Ware, P. L. and Kasper, L. H. (1987) Strain-specific antigens of *Toxoplasma gondii. Infect. Immun.*, **55**, 778.
57. Weiss, L. M., Udem, S. A., Tanowitz, H., and Wittner, M. (1987) Western blot analysis of the antibody response of patients with AIDS and toxopalsmic encephalitis: antigenic diversity among *Toxoplasma* strains. *J. Infect. Dis.*, **157**, 7.
58. Bohne, W., Gross, U., and Heesemann, J. (1993) Differentiation between mouse-virulent and -avirulent strains of *Toxoplasma gondii* by a monoclonal antibody recognizing a 27 kilodalton antigen. *J. Clin. Microbiology*, **31**, 1641.
59. Gross, U., Muller, W. A., Knapp, S., and Heesemann, J. (1991) Identification of a virulence-associated antigen of *Toxoplasma gondii* by use of a mouse monoclonal antibody. *Infect. Immun.*, **59**, 4511.
60. Buelow, R. and Boothroyd, J. C. (1991) Protection of mice from fatal *Toxoplasma* infection by immunization with p30 antigen in liposomes. *J. Immunol.*, **147**, 3496.
61. Parmley, S. F., Gross, U., Sucharczuk, A., Windeck, T., Sgarlato, G. D., and Remington, J. S. (1994) Two alleles of the gene encoding surface antigen P22 in 25 strains of *Toxoplasma gondii. J. Parasitol.*, **80**, 293.
62. Rinder, H., Thomschke, A., Dardé, M. L., and Loscher, T. (1995) Specific DNA polymorphisms discriminate between virulence and non-virulence to mice in nine *Toxoplasma gondii* strains. *Mol. Biochem. Parasitol.*, **69**, 123.
63. Miller, L. H., Good, M. F., and Milon, G. (1994) Malaria pathogenesis. *Science*, **264**, 1878.
64. Dardé, M. L., Bouteille, B., and Pestre-Alexandre, M. (1988) Isoenzyme characterization of seven strains of *Toxoplasma gondii* by isoelectric focusing in polyacrylamide gels. *Am. J. Trop. Med. Hyg.*, **39**, 551.
65. Dardé, M. L., Bouteille, B., and Pestre-Alexander, M. (1992) Isoenzyme analysis of 35 *Toxoplasma gondii* isolates: biological and epidemiological implications. *J. Parasitol.*, **78**, 786.
66. Williams, J. G. K., Kubelik, A. R., Livak, K. J., Rafalski, J. A., and Tingey, S. V. (1990) DNA polymorphisms amplified by arbitrary primers are useful as genetic markers. *Nucleic. Acids Res.*, **18**, 6531.
67. Guo, Z. G. and Johnson, A. M. (1995) Genetic comparison of *Neospora caninum* with *Toxoplasma* and *Sarcocystis* by random amplified polymorphic DNA-polymerase chain reaction. *Parasitol. Res.* **81**, 365.

68. Howe, D. K. and Sibley, L. D. (1995) *Toxoplasma gondii* is comprised of three clonal lineages; correlation of parasite genotype with human toxoplasmosis. *J. Infect. Dis.*, (in press).
69. Cristina, N., Dardé, M. L., Boudin, C., Tavernier, G., Pestre-Alexandre, M., and Ambroise-Thomas, P. (1995) A DNA fingerprinting method for individual characterization of *Toxoplasma gondii* strains: combination with isoenzymatic characters for determination of linkage groups. *Parasitol. Res.*, **81**, 32.
70. Tibayrenc, M., Kjellberg, F., Araud, J., Oury, B., Breniere, S. F., Dardé, M. L., and Ayala, F. J. (1991) Are eukaryotic microorganisms clonal or sexual? A population genetics vantage. *Proc. Natl. Acad. Sci. USA*, **88**, 5129.
71. Tibayrenc, M., Kjellberg, F., and Ayala, F. J. (1990) A clonal theory of parasitic protozoa: the population structures of *Entamoeba, Giardia, Leishmania, Naegleria, Plasmodium, Trichamonas*, and *Trypanosoma* and their medical and taxonomic consequences. *Proc. Natl. Acad. Sci., USA*, **87**, 2414.
72. Chaparas, S. D. and Schlesinger, R. W. (1959) Plaque assay of *Toxoplasma* on monolayers of chick embryo fibroplasts. *Proc. Soc. Exp. Biol. Med.*, **102**, 431.
73. Pfefferkorn, E. R. and Pfefferkorn, L. C. (1977) Specific labeling of intracellular *Toxoplasma gondii* with uracil. *J. Protozool.*, **24**, 449.
74. Pfefferkorn, E. R. and Pfefferkorn, L. C. (1979) Quantitative studies of the mutagenesis of *Toxoplasma gondii*. *J. Parasitol.*, **65**, 363.
75. Pfefferkorn, E. R. and Pfefferkorn, L. C. (1976) *Toxoplasma gondii*: isolation and preliminary characterization of temperature sensitive mutants. *Exp. Parasitol.*, **39**, 365.
76. Pfefferkorn, L. C. and Pfefferkorn, E. R. (1980) *Toxoplasma gondii*: genetic recombination between drug resistant mutants. *Exp. Parasitol.*, **50**, 305.
77. Pfefferkorn, E. R. and Kasper, L. H. (1983) *Toxoplasma gondii*: Genetic crosses reveal phenotypic suppression of hydroxyurea resistance by fluorodeoxyuridine resistance. *Exp. Parasitol.*, **55**, 207.
78. Chang, C., Bowman, J. L., DeJohn, A. W., Lander, E. S., and Meyerowitz, E. M. (1988) Restriction fragment polymorphism linkage map for *Arabidopsis thaliana*. *Proc. Natl. Acad. Sci. USA*, **85**, 6856.
79. Coulson, A., Sulston, J., Brenner, S., and Karn, J. (1986) Toward a physical map of the genome of the nematode *Caenorhabditis elegans*. *Proc. Natl. Acad. Sci. USA*, **83**, 7821.
80. Walker-Jonah, A., Dolan, S. A., Gwadz, R. W., Panton, L. J., and Wellems, T. E. (1992) An RFLP map of the *Plasmodium falciparum* genome, recombination rates and favored linkage groups in a genetic cross. *Mol. Biochem. Parasitol.*, **51**, 313.
81. Tzeng, T. H., Lyngholm, L. K., Ford, C. F., and Bronson, C. R. (1992) A restriction fragment length polymorphism map and electrophoretic karyotype of the fungal maize pathogen *Cochliobolus heterostrophus*. *Genetics*, **130**, 81.
82. Botstein, D., White, R. L., Skolnick, M., and Davis, R. W. (1980) Construction of a genetic linkage map in man using restriction fragment length polymorphisms. *Am. J. Human Genetics*, **32**, 314.
83. Donis-Keller, H., Green, P., Helms, C., Cartinhour, S., Weiffenbach, B., Stephens, K., Keith, T. P., Bowden, D. W., Smith, D. R., and Lander, E. S. (1987) A genetic linkage map of the human genome. *Cell*, **51**, 319.
84. Collins, F. S. (1992) Positional cloning: Let's not call it reverse anymore. *Nature Genetics*, **1**, 3.
85. Ajioka, J.W., Smoller, D. A., Jones, R. W., Carulli, J. P., Vellek, A. E. C., Garza, D., Link, A. J., Duncan, I. W., and Hartl, D. L. (1991) *Drosophila* genome project: one-hit coverage in yeast artifical chromosomes. *Chromosoma*, **100**, 495.

86. Chu, G., Voolrath, D., and David, R. W. (1986) Separation of large DNA molecules by contour-clamped homogeneous electric fields. *Science*, **234,** 1582.
87. Vollrath, D. and Davis, R. W. (1987) Resolution of DNA molecules greater than 5 megabases by contour clamped homogenous electric fields. *Nucleic Acids Res.*, **15,** 7865.
88. Ish-Horowicz, D. and Burke, J. F. (1981) Rapid and effective cosmid cloning. *Nucleic Acids Res.*, **9,** 2989.
89. Burke, D. T., Carle, G. F., and Olson, M. V. (1987) Cloning large segments of exogenous DNA into yeast by means of artificial chromosome vectors. *Science*, **236,** 806.
90. Shizuya, H., Birren, B., Kim, U.-J., Mancino, V., Slepak, T., Tachiri, Y., and Simon, M. (1992) Cloning and stable maintenance of 300-kilobasepair fragments of human DNA in *Escherichia coli* using a F-factor-based vector. *Proc. Natl. Acad. Sci., USA*, **89,** 8794.
91. Olsen, M. V., Hood, L., Cantor, C., and Botstein, D. (1989) A common language for physical mapping of the human genome. *Science*, **245,** 1434.
92. Sulston, J., Du, Z., Thomas, K., Wilson, R., Hillier, L., Staden, R., Halloran, N., Green, P., Thierry-Mieg, J., and Qiu, L. (1992) The *Caenorhabditis elegans* genome sequencing project: a beginning. *Nature*, **356,** 37.
93. Green, E. D. and Olson, M. V. (1990) Chromosomal region of the cystic fibrosis gene in yeast artificial chromosomes: a model for human genome mapping. *Science*, **250,** 94.
94. Foote, S., Vollrath, D., Hilton, A., and Page, D. C. (1992) The human Y chromosome: Overlapping DNA clones spanning the euchromatic region. *Science*, **258,** 60.
95. Saiki, R. K., Gelfand, D. H., Stoffel, S., Scharf, S. J., Higuchi, R., Horn, G. T., Mullis, K. B., Erlich, H. A. (1988) Primer-directed enzymatic amplification of DNA with a thermostable DNA ploymerase. *Science*, **239,** 487.
96. Waterson, R., Martin, C., Craxton, M., Huynh, C., Coulson, A., Hillier, A., Durbin, R., Green, P., Shownkeen, R., Halloran, N. (1992) A survey of expressed genes in *Caenorhabditis elegans*. *Nature Genetics*, **1,** 114.
97. Adams, M. D., Kelly, J. M., Gocayne, J. D., Dubnick, M., Polymeropolous, M. H., Xiao, H., Merril, C. R., Wu, A., Olde, B., and Moreno, R. F. (1991) Complementary DNA sequencing: expressed sequence tags and human genome project. *Science*, **252,** 1651.
98. McCombie, W. R., Adams, M. D., Kelley, J. M., FitzGerald, M. G., Utterback, T. R., Khan, M., Dubnick, M., Kerlavage, A. R., Venter, J. C., and Fields, C. R. (1992) *Caenorhabditis elegans* expressed sequence tags identify gene families and potential disease gene homologues. *Nature Genetics*, **1,** 124.
99. Lanzer, M., De Bruin, D., and Ravetch, J. V. (1993) Transcriptional differences in polymorphic and conserved domains of a complete cloned *P. falciparum* chromosome. *Nature*, **361,** 654.
100. Boguski, M. S., Lowe, T. M. J., and Tolstoshev, C. M. (1993) dBEST-database for 'expressed sequence tags'. *Nature Genetics* **4,** 332.
101. Durbin, R. and Thierry-Mieg, J. (1991) A *C. elegans* database, documentation code and data. Available from anonymous ftp servers at lirmm.lirmm.fr,mrc-lmb.cam.ac.uk or ncbi.nlm.nih.gov.
102. Chakrabarti, D., Reddy, G. R., Dame, J. B., Almira, E. C., Lapis, P. J., Ferl, R. J., Yang, T. P., Rowe, T. C., and Schuster, S. M. (1994) Analysis of expressed sequence tags from *Plasmodium falciparum* . *Mol. Biochem. Parasitol.*, **66,** 97.
103. Boothroyd, J. C., Black, M., Kim, K., Pfefferkorn, E. R., Seeber, F., Sibley, L. D., and Soldati, D. (1995) Forward and reverse genetics in the study of the obligate, intracelluar parasite *Toxoplasma gondii*. *Methods in Molecular Genetics*. Vol. 6, Microbial Gene Techniques (ed K. Adolph), pp. 3–29. Academic Press, New York.

104. Roos, D. S., Donald, R. G. K., Morrissette, N. S., and Moulton, A. L. (1994) Molecular tools for genetic dissection of the protozoan parasite *Toxoplasma gondii*. *Methods Cell Biol.*, **45,** 27.
105. Kim, K. and Boothroyd, J. C. (1993) Gene replacement in *Toxoplasma gondii* with chloramphenicol acetyltransferase as selectable marker. *Science*, **262,** 911.
106a. Donald, R. G. K. and Roos, D. S. (1993) Stable molecular transformation of *Toxoplasma gondii*: A selectable DHFR-TS marker based on drug resistance mutations in malaria. *Proc. Natl. Acad. Sci. USA*, **90,** 11703.
106b. Messina, M., Niesman, I. R., Mercier, C. and Sibley, L. D. (1995) Stable DNA transformation of *Toxoplasma gondii* using phleomycin selection. *Gene*, in press.
107. Sibley, L. D., Messina, M., and Niesman, I. R. (1994) Stable transformation in the obligate intracellular parasite *Toxoplasma gondii* by complementation of tryptophan auxotrophy. *Proc. Natl. Acad. Sci. USA*, **91,** 5508.
108. Scherer, S. and Davis, R. W. (1979) Replacement of chromosome segments with altered DNA sequences constructed *in vitro*. *Proc. Natl. Acad. Sci. USA*, **76,** 4951.
109. Cruz, A. and Beverley, S. M. (1990) Gene replacement in parasitic protozoa. *Nature*, **348,** 171.
110. Lee, M. G. S. and Van der Ploeg, L. H. T. (1990) Homologous recombination and stable transfection in the parasitic protozoan *Trypanosoma brucei*. *Science*, **250,** 1583.
111. Asbroek, A. L. M. A., Ouellette, M., and Borst, P. (1990) Targeted insertion of the neomycin phosphotransferase gene into the tubulin gene cluster of *Tyrpanosoma brucei*. *Nature*, **348,** 174.
112. Donald, R. G. K. and Roos, D. S. (1994) Homologous recombination and gene replacement at the dihydrofolate reductase-thymidilate synthase locus in *Toxoplasma gondii*. *Mol. Biochem. Parasitol.*, **63,** 243.

5 | Kinetoplast DNA: structure and replication

PAUL T. ENGLUND, D. LYS GUILBRIDE, KUO-YUAN HWA,
CATHARINE E. JOHNSON, CONGJUN LI, LAURA J. ROCCO and
AL F. TORRI

1. Introduction

Trypanosomatid protozoa are parasites which are responsible for important human diseases in tropical areas of the world. Examples are *Trypanosoma brucei*, an African trypanosome, *T. cruzi*, a South American trypanosome, and *Leishmania*. These parasites are among the most primitive of the eukaryotes and, because of their ancient lineage, it is not surprising that they have unusual biological properties. They are among the first eukaryotes to have mitochondria and one of their most remarkable features is their mitochondrial DNA, known as kinetoplast DNA (kDNA). Unlike any other DNA in nature, kDNA is organized into a network containing several thousand topologically interlocked DNA circles. The circles are of two types. There are several thousand small minicircles and a few dozen larger maxicircles. Figure 1 shows an electron micrograph of part of a kDNA network from the trypanosomatid *Crithidia fasciculata*. Each cell contains a single network within the matrix of its single mitochondrion (for reviews, see Refs. 1–5).

It is still not clear why trypanosomatids are the only eukaryotes which organize their mitochondrial DNA in a network. However, there is now considerable information about the genetic function of maxicircles and minicircles. The maxicircles resemble conventional mitochondrial DNAs, such as those in mammals or yeast, in that they encode ribosomal RNAs and a small number of proteins. Most maxicircle gene products are involved in mitochondrial energy transduction and examples include subunits of cytochrome oxidase and NADH dehydrogenase. However, the expression of maxicircle protein coding genes is unexpectedly complex. Their transcripts undergo RNA editing, a process in which uridine residues are added to or deleted from precise positions in the transcript, to create the open reading frame (6). In some cases editing can be on a massive scale, with more than half of the nucleotides in a reading frame being added by this process (7). In contrast, the genetic function of minicircles was a mystery for many years, but recently it was discovered that they encode small guide RNAs which control the specificity of RNA editing (8; for a review on editing, see Chapter 8).

Fig. 1 Electron micrograph of part of a kDNA network from *C. fasciculata*. The small loops are 2.5kb minicircles

2. Structure of a kDNA network

2.1 The isolated network

Our laboratory has studied extensively the kDNA network from *C. fasciculata*. This parasite, which infects insects and has no mammalian host, has been exceedingly valuable for these studies primarily because it is convenient to culture in large quantities. A *C. fasciculata* network not undergoing replication has about 5000 minicircles, each about 2.5kb, and about 25 maxicircles, each about 37kb (9). As viewed by electron microscopy, the isolated network is a planar structure, organized like the chain mail of medieval armour. It is elliptically shaped, about 10μm by 15μm in size (10). The structure of the network is too complex to deduce by inspection of electron micrographs, but we do have some important information about how it is organized. First, each minicircle in a network is joined to its neighbours by single interlocks, an arrangement which allows strain free linkages (11). Second, although minicircles in a non-replicating network are covalently closed, they are topologically relaxed (11). This finding, to be discussed again below, is surprising in that other circular DNAs, in either prokaryotes or eukaryotes, are negatively supercoiled. Third, recent experiments, based on the quantification of the minicircle products from partial restriction enzyme digests of kDNA, indicate that each *C. fasciculata* minicircle is linked to an average of three neighbors (12). This value is termed the minicircle valence, and we shall discuss below the changes in valence which occur during network replication.

2.2 Organization of the network *in vivo?*

The isolated 10μm × 15μm *C. fasciculata* network is larger than the cell from which it is derived and this fact raises the important question of how it is condensed inside the mitochondrial matrix. Electron microscopy of thin sections, and confocal fluorescence microscopy of cells in which the kDNA is stained with acridine, indicate that the network is compacted into a disc about 1μm in diameter and 0.4μm thick (13, 14). Although the isolated network is a monolayer, it is also a monolayer when condensed into a disc inside the cell (14). In isolated kDNA, the minicircles lie flat in the plane of the network (Fig. 2A) whereas, inside the cell, the minicircles stretch out and stand perpendicular to the plane of the network (14–16 Fig. 2B). As predicted by Fig. 2B, the thickness of the disk is roughly half the circumference of a minicircle. It is not known what stabilizes the network in its condensed conformation. Proteins must be involved and candidates are a recently discovered family of highly basic DNA binding proteins which by immunofluorescence localize to the kinetoplast disc *in vivo* (17; D. Ray, personal communication). Also, each minicircle has a segment of bent helix which may facilitate packing of the network into a structure like that shown in Fig. 2B (16).

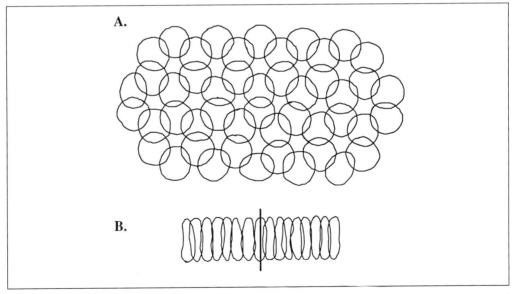

Fig. 2 The organization of kDNA. A: diagrammatic representation of an isolated network. B: the organization of a kDNA network *in vivo*. This diagram shows a section through the kinetoplast disc. The vertical line represents the disc's axis. Maxicircles are not shown on these models. See text for further discussion.

In cells with networks undergoing replication, the kinetoplast disk is flanked by two complexes of proteins involved in DNA replication, situated on opposite sides of the disk. These complexes, roughly 0.4μm in diameter, contain a topoisomerase II (18), a DNA polymerase (14), and probably other enzymes involved in DNA replication. They also contain minicircles which are probably replication intermediates (14). The role of these structures will be discussed below.

2.3 Conditions required for network formation

It is likely that a topoisomerase II will catenate minicircles into a network provided they are relaxed and aligned at high concentration in a monolayer, in the volume delimited by the 1μm × 0.4μm disc (12). The reason that they must be relaxed is that there is much more space available in the centre of a relaxed circle than one which is supercoiled, facilitating interpenetration and catenation. With other DNAs, negative supercoiling provides crucial energy for strand unwinding during processes such as replication or transcription. Therefore trypanosomatids must have sacrificed the advantages of supercoiling minicircles so that these molecules can be assembled into a network. We do not know what keeps the minicircles inside the volume of this small disc, but one candidate for constraining the minicircles at the periphery is the mitochondrial membrane. Many electron micrographs of thin sections show that the kinetoplast disc directly abuts the mitochondrial membrane (for example, see Ref. 13). It is likely that the minicircle valence, the number of neighbours linked to each minicircle, is determined by the concentration of minicircles when the topological bonds are formed. In *C. fasciculata*, the 5000 minicircles in the network have an average valence of three. If there were more minicircles in the same space, or if they were constrained in a smaller space, then the valence would be higher. Below we shall discuss the fact that the average minicircle valence actually increases as the minicircle copy number increases during kDNA replication.

3. Replication of the kDNA network

In contrast to replication of mitochondrial DNA in mammalian cells, which occurs throughout the entire cell cycle (19), kDNA replication occurs during a distinct S phase, approximately coincident with that of the nucleus (20–22). During S phase, there is replication of each individual minicircle and maxicircle. The progeny circles are ultimately distributed into two daughter networks which segregate into the daughter cells at the time of cell division.

3.1 Free minicircles

A crucial feature of the current model, shown in Fig. 3, is that minicircles do not replicate while linked to the network but only after they are released from it by a topoisomerase II (23). The network never completely decatenates, and only a few

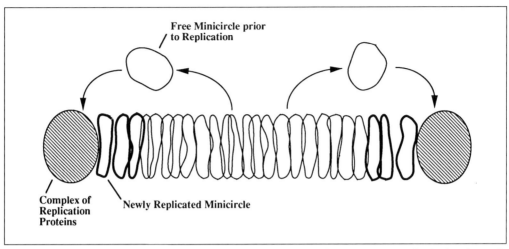

Fig. 3 Model for kDNA replication *in vivo*. The minicircles shown in bold are newly synthesized and contain nicks. Maxicircles are not shown on this model. See text for further discussion.

hundred minicircles are free at any time. The advantage of this scheme is that the free minicircles can replicate by a standard mechanism, like that of a plasmid or other circular DNA, without the topological problems which would arise if they replicated while linked to other minicircles in the network. The free minicircles, after release from the network, are thought to migrate (or be transported) to one of the two antipodal protein complexes where they undergo replication as θ-structures (14; see below). It is important that the minicircles are covalently closed prior to replication and nicked or gapped afterwards (we shall subsequently refer to both types of interruptions in the newly synthesized strand as nicks). Some of the nicks in each molecule persist until all the minicircles have finished replicating (24). It is hypothesized that nicking plays a 'book-keeping' function, allowing the cell to keep track of which minicircles have replicated and which have not. Therefore each minicircle is sure to replicate once and only once per generation. After replication in one of the two protein complexes, the newly synthesized nicked minicircles are attached to the network's periphery. Because two daughter minicircles are attached for every one removed, the minicircle copy number increases during the replication process; the network grows from about 5000 minicircles to ultimately about 10 000 before it splits in two to form two daughter networks (24).

3.2 Structure of a replicating network

Inspection of partly replicated networks, either by electron microscopy after network isolation (10), or by fluorescence *in situ* hybridization (14), reveals that at all stages of replication the nicked minicircles are distributed uniformly around the network periphery. This is shown clearly in electron micrographs of networks at different stages of replication (Fig. 4). The newly synthesized minicircles in these

KINETOPLAST DNA: STRUCTURE AND REPLICATION

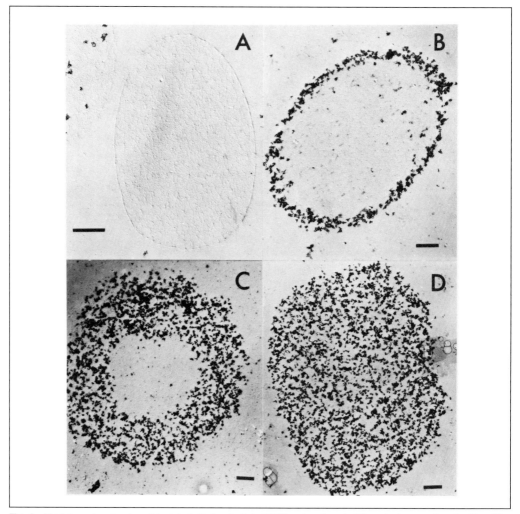

Fig. 4 Electron micrographs of isolated *C. fasciculata* kDNA networks at different stages of replication. The networks were radiolabeled with [³H]dNTPs by nick translation at endogenous nicks. They were then subjected to autoradiography and the silver grains indicate the location of endogenously nicked minicircles. A: pre-replication network. B: an early replicative form. C; a later replicative form. D; a double size fully replicated network in which all the minicircles are nicked. Scale bars, 2μm. (From Ref. 10, with permission.)

networks were labeled *in vitro* with [³H]dNTPs in a nick translation reaction at their endogenous nicks. Then the DNA was spread on a grid, coated with photographic emulsion and subjected to autoradiography (10). The silver grains mark the location of the radiolabeled nicked minicircles. Panel A shows a pre-replication network, in which all minicircles are covalently closed; therefore there are no silver grains. Panels B and C show networks undergoing replication. These structures resemble doughnuts because the silver grains form a ring around the network periphery. The clear space in the interior of these networks (the doughnut hole) contains covalently closed minicircles, not yet replicated, which were not labeled in the nick translation reaction. As replication proceeds, the doughnut hole shrinks in size. At the conclusion of replication, all of the minicircles in the double-size network are nicked (panel D) and the doughnut hole disappears.

3.3 Distribution of newly synthesized minicircles around the network periphery

If newly synthesized minicircles are attached to the network only adjacent to the two protein complexes, how do they become distributed uniformly around the periphery? This problem was first raised during attempts to interpret early autoradiographic studies, at the light microscope level, of isolated kDNA which had been pulse labeled *in vivo* with [³H]thymidine. After a very short pulse (1 minute), the silver grains were restricted to two peripheral sites on opposite sides of the network (25). These sites presumably had been associated with the two complexes of replication proteins. The silver grains marked minicircles which had replicated during the pulse and then attached to the network adjacent to the protein complexes. In contrast, in kDNA labeled *in vivo* for a longer period, the silver grains were localized around the entire periphery (26).

Recent studies have clarified the question of minicircle attachment. The experimental approach was to perform autoradiography of networks labeled *in vivo* with [³H]thymidine, exactly as done in the earlier studies. However, the labeled networks were visualized by electron microscopy to localize the newly synthesized minicircles at higher resolution (27). One example of an EM autoradiograph, a network labeled *in vivo* for 6 minutes, is shown in Fig. 5. As expected, the silver grains are located on the periphery and they are clearly in two zones. Surprisingly, the grains in each zone form a gradient in density and the gradient in the two zones is in the same direction. This density gradient is due to the rising specific radioactivity of the DNA precursor during the course of the labeling. As the exogenous [³H]thymidine mixes with the endogenous pool of DNA precursors, there is a gradual increase in specific radioactivity of the deoxynucleoside triphosphates. The presence of the two labeling zones is consistent with the minicircles being attached at two sites. However, the gradients of silver grains indicate that the minicircles are attached in an orderly and sequential manner around the network periphery. Since silver grains on opposite sides of the network have approximately the same density, the sites of attachment must always be on opposite sides of the disc.

3.4 The spinning kinetoplast model

The best explanation of these and related data is that there is a relative movement of the network and the two protein complexes (27). One possibility is that the disc is fixed and the complexes move around the circumference of the network. Another is that the two complexes are fixed and the kinetoplast disc actually rotates between

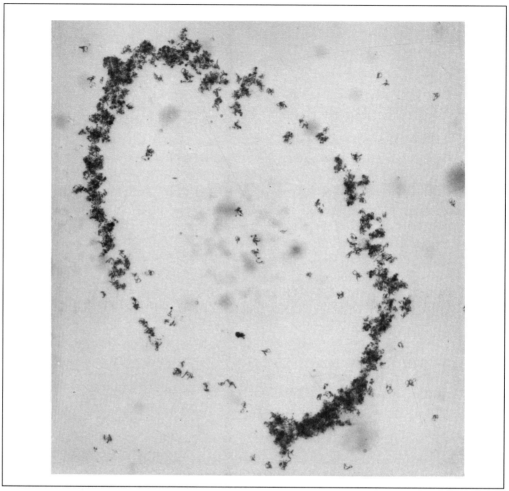

Fig. 5 Electron micrograph of a kDNA network labeled *in vivo* for 6 minutes with [^{3}H]thymidine. After isolation the network was subjected to autoradiography. (From Ref. 27, with permission.)

them (see Fig. 6). Although it is essential to obtain much more evidence on this point, we tentatively favour the second possibility (27). Since it takes about 6 minutes to radiolabel the entire network periphery using two complexes of replication proteins, the kinetoplast disk would make about one turn in 12 minutes. Presumably after each turn it continues to rotate in the same direction, allowing minicircle reattachment to occur in a spiral pattern. It probably takes about seven turns to replicate the entire network (27). Recent studies on *T. brucei* kDNA, however, suggested a fundamental difference in the kDNA replication mechanism from that of *C. fasciculata*. Although newly synthesized minicircles are attached to opposite sides of the network as they are in *C. fasciculata* (implying the existence of two protein complexes), the kinetoplast does not appear to rotate (28, 29).

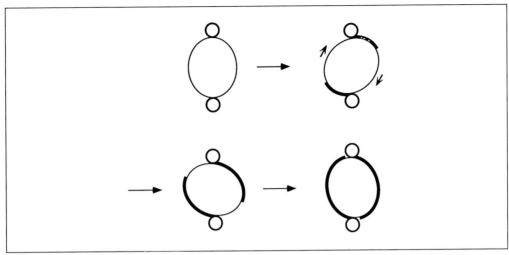

Fig. 6 Model showing the rotation of the *C. fasciculata* kinetoplast *in vivo*. The kinetoplast is depicted as an ellipse and the two complexes of replication proteins as small circles. The bold lines represent rows of newly replicated minicircles. Small arrows indicate the direction of rotation of kinetoplast between fixed replication complexes. See text for further discussion. (From Ref. 27, with permission.)

3.5 Changes in minicircle valence during replication

As kDNA replication proceeds throughout the S phase, minicircles are continuously released from the network's central zone and their progeny are attached around the periphery. Interestingly, in *C. fasciculata* the newly attached minicircles have a higher valence than those in a non-replicating network. The average valence may be as high as five or six, rather than three as is found for a non-replicating network (J. Chen, P. T. Englund, and N. R. Cozzarelli, manuscript in preparation). A probable reason for this difference is that the number of minicircles is increasing during replication, but the available space (perhaps limited by the mitochondrial membrane) is not. Therefore the minicircles must pack more tightly together. Finally, when replication is complete the network contains about 10,000 nicked minicircles. Because the network is still constrained in a limited space, all the minicircles are packed at a very high density. The minicircle valence is as high as five or six and the surface area of the isolated network has grown little from that of the pre-replication network (see Fig. 7).

3.6 Final stages of network replication

The network undergoes dramatic remodeling after replication of minicircles has been completed (Fig. 7). As the available space for the network increases, possibly due to enlargement of the mitochondrial membrane, continuing topoisomerase action results in a gradual drop in minicircle valence and a corresponding increase in network surface area (J. Chen, P. T. Englund, and N.R. Cozzarelli, manuscript in

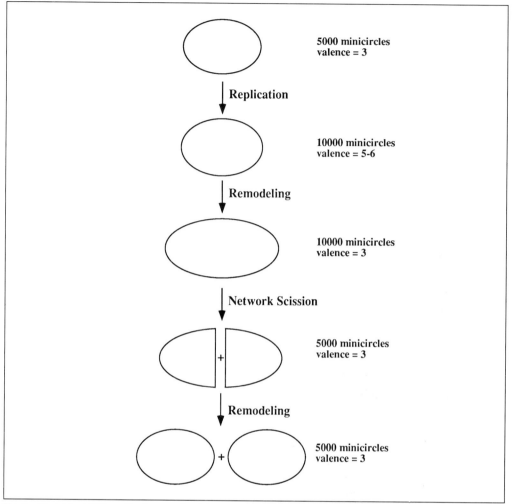

Fig. 7 Model for kDNA replication emphasizing changes in minicircle valence and remodeling of the network. See text for further discussion.

preparation). During this time there is a gradual repair of the minicircle nicks and gaps (10). Finally, when the minicircle valence reaches three and the network has a surface area double that of a pre-replication network, the structure splits in two. This scission undoubtedly occurs by topoisomerase action, unlinking neighbouring circles between the two nascent daughter networks. Electron microscopy of newly formed networks has revealed that they have a characteristic flat edge (10), as if the splitting of the double-size network occurs by a straight cut across its centre (as shown diagramatically in Fig. 7). Subsequent remodeling of the flat-edged network must result in the characteristic elliptical shape. Perhaps this final form of the network is shaped by the contour of the mitochondrial membrane.

4. A closer look at the replication of minicircles and maxicircles

4.1 Minicircle replication

Replication begins soon after the covalently closed minicircle is decatenated from the network and it is thought to occur in one of the two complexes of replication proteins. Minicircle replication occurs via θ-intermediates and is unidirectional (23, 30). Leading strand synthesis initiates complementary to the universal minicircle sequence, GGGGTTGGTGTA, a 12-mer found in a conserved region of minicircles from nearly all trypanosomatids that have been examined (31, 32). The leading strand has one or more ribonucleotides at its 5′ end, indicating that DNA synthesis is initiated by a primase (31, 32). A second sequence, ACGCCC, which is less well conserved, is located 70–90 nucleotides downstream from the 12-mer; the first Okazaki fragment initiates complementary to this 6-mer (33). In both *C. fasciculata* and *T. equiperdum*, the leading strand is synthesized continuously around the molecule and the lagging strand Okazaki fragments are roughly 100 nucleotides in size (33, 34). As mentioned above, covalent closure of minicircles occurs only after all have replicated and their progeny attached to the network.

4.2 Maxicircle replication

Recent electron microscopy of maxicircles released from *C. fasciculata* networks by restriction enzyme cleavage revealed many branched replication intermediates. Measurement of these molecules indicated that maxicircle replication initiates in a unique region and proceeds unidirectionally as a θ-structure (L. J. Rocco and P. T. Englund, submitted for publication). Surprisingly, these data contrast with an earlier report which suggested that maxicircles replicate by a rolling circle mechanism (35). A likely explanation for the early data is that replicating maxicircles are susceptible to cleavage by topoisomerase II during cell lysis in the presence of strong detergent (36). In *T. brucei* there are many topoisomerase II consensus binding sites near the replication origin (37) and the same could be true for *C. fasciculata*. Topoisomerase-induced cleavage of θ-structures could result in molecules which look like rolling circles.

The current electron microscopy studies indicate that the maxicircle replication initiates in the 'variable' region, a non-coding segment of the molecule which contains many repetitive sequences. Although not sequenced in *C. fasciculata*, this region in *T. brucei* contains two copies of the GGGGTTGGTGT sequence found at the minicircle replication origin (see previous section) (37, 38). If similar sequences exist in this region of *C. fasciculata* maxicircles and if they serve as replication origins, then the mechanism of maxicircle replication might be very similar to that of minicircles. Minicircles and maxicircles replicate simultaneously (35) and both replicate unidirectionally as θ-structures. If they initiate at similar sequences, the factors which control initiation would probably be the same. The major difference is that minicircles replicate free of the network whereas maxicircles replicate while still linked to it (35 and unpublished observations of L. J. Rocco). The significance of this difference is certainly important, but it is not yet understood. Solving this problem, and many others raised in this chapter, will provide exciting challenges for future years.

References

1. Shapiro, T. A. and Englund, P. T. (1995) The structure and replication of kinetoplast DNA. *Annu. Rev. Microbiol.*, **49**, 117.
2. Stuart, K. and Feagin, J. E. (1992) Mitochondrial DNA of kinetoplastids. *Int. Rev. Cytol.*, **141**, 65.
3. Simpson, L. (1987) The mitochondrial genome of kinetoplastid protozoa: Genomic organization, transcription, replication, and evolution. *Annu. Rev. Microbiol.*, **41**, 363.
4. Ray, D. S. (1987) Kinetoplast DNA minicircles: High-copy-number mitochondrial plasmids. *Plasmid*, **17**, 177.
5. Shlomai, J. (1994) The assembly of kinetoplast DNA. *Parasitol. Today*, **10**, 341.
6. Benne, R., Van den Burg, J., Brakenhoff, J. P., Sloof, P., Van Boom, J. H. and Tromp, M. C. (1986) Major transcript of the frameshifted cox II gene from trypanosome mitochondria contains four nucleotides that are not encoded in the DNA. *Cell*, **46**, 819.
7. Feagin, J. E., Abraham, J. M. and Stuart, K. (1988) Extensive editing of the cytochrome c oxidase III transcript in *Trypanosoma brucei*. *Cell*, **53**, 413.
8. Sturm, N. R. and Simpson, L. (1990) Kinetoplast DNA minicircles encode guide RNAs for editing of cytochrome oxidase subunit III mRNA. *Cell*, **61**, 879.
9. Marini, J. C., Miller, K. G., and Englund, P. T. (1980) Decatenation of kinetoplast DNA by topoisomerases. *J. Biol. Chem.*, **255**, 4976.
10. Pérez-Morga, D. and Englund, P. T. (1993) The structure of replicating kinetoplast DNA networks. *J. Cell Biol.*, **123**, 1069.
11. Rauch, C. A., Pérez-Morga, D., Cozzarelli, N. R., and Englund, P. T. (1993) The absence of supercoiling in kinetoplast DNA minicircles. *EMBO J.*, **12**, 403.
12. Chen, J., Rauch, C. A., White, J. H., Englund, P. T., and Cozzarelli, N. R. (1995) The topology of the kinetoplast DNA network. *Cell*, **80**, 61.
13. Renger, H. C. and Wolstenholme, D. R. (1972) The form and structure of kinetoplast DNA of *Crithidia*. *J. Cell Biol.*, **54**, 346.
14. Ferguson, M., Torri, A. F., Ward, D. C., and Englund, P. T. (1992) *In situ* hybridization to the *Crithidia fasciculata* kinetoplast reveals two antipodal sites involved in kinetoplast DNA replication. *Cell*, **70**, 621.
15. Delain, E. and Riou, G. (1969) Ultrastructure du DNA du kinétoplaste de *Trypanosoma cruzi* cultivé *in vitro*. *C. R. Acad. Sci. [D] (Paris)*, **268**, 1225.
16. Marini, J. C., Levene, S. D., Crothers, D. M., and Englund, P. T. (1983) A bent helix in kinetoplast DNA. *Cold Spring Harbor Symp. Quant. Biol.*, **47**, 279.
17. Xu, C. and Ray, D.S. (1993) Isolation of proteins associated with kinetoplast DNA networks *in vivo*. *Proc. Natl. Acad. Sci. USA*, **90**, 1786.
18. Melendy, T., Sheline, C., and Ray, D. S. (1988) Localization of a type II DNA topoisomerase to two sites at the periphery of the kinetoplast DNA of *Crithidia fasciculata*. *Cell*, **55**, 1083.
19. Clayton, D.A. (1991) Replication and transcription of vertebrate mitochondrial DNA. *Annu. Rev. Cell Biol.*, **7**, 453.

20. Cosgrove, W. B. and Skeen, M. J. (1970) The cell cycle in *Crithidia fasciculata*. Temporal relationships between synthesis of deoxyribonucleic acid in the nucleus and in the kinetoplast. *J. Protozool.*, **17**, 172.
21. Simpson, L. and Braly, P. (1970) Synchronization of *Leishmania tarentolae* by hydroxyurea. *J. Protozool.*, **17**, 511.
22. Woodward, R. and Gull, K. (1990) Timing of nuclear and kinetoplast DNA replication and early morphological events in the cell cycle of *Trypanosoma brucei*. *J. Cell Sci.*, **95**, 49.
23. Englund, P. T. (1979) Free minicircles of kinetoplast DNA in *Crithidia fasciculata*. *J. Biol. Chem.*, **254**, 4895.
24. Englund, P. T. (1978) The replication of kinetoplast DNA networks in *Crithidia fasciculata*. *Cell*, **14**, 157.
25. Simpson, A. M. and Simpson, L. (1976) Pulse-labeling of kinetoplast DNA: localization of 2 sites of synthesis within the networks and kinetics of labeling of closed minicircles. *J. Protozool.*, **23**, 583.
26. Simpson, L., Simpson, A. M., and Wesley, R. D. (1974) Replication of the kinetoplast DNA of *Leishmania tarentolae* and *Crithidia fasciculata*. *Biochim. Biophys. Acta*, **349**, 161.
27. Pérez-Morga, D. and Englund, P. T. (1993) The attachment of minicircles to kinetoplast DNA networks during replication. *Cell*, **74**, 703.
28. Ferguson, M. F., Torri, A. F., Pérez-Morga, D., Ward, D. C., and Englund, P. T. (1994) Kinetoplast DNA replication: Mechanistic differences between *Trypanosoma brucei* and *Crithidia fasciculata*. *J. Cell. Biol.*, **126**, 631.
29. Robinson, D. R. and Gull, K. (1994) The configuration of DNA replication sites within the *Trypanosoma brucei* kinetoplast. *J. Cell Biol.*, **126**, 641.
30. Ryan, K. A. and Englund, P. T. (1989) Synthesis and processing of kinetoplast DNA minicircles in *Trypanosoma equiperdum*. *Mol. Cell Biol.*, **9**, 3212.
31. Ntambi, J. M., Shapiro, T. A., Ryan, K. A., and Englund, P. T. (1986) Ribonucleotides associated with a gap in newly replicated kinetoplast DNA minicircles from *Trypanosoma equiperdum*. *J. Biol. Chem.*, **261**, 11890.
32. Birkenmeyer, L., Sugisaki, H. and Ray, D. S. (1987) Structural characterization of site-specific discontinuities associated with replication origins of minicircle DNA from *Crithidia fasciculata*. *J. Biol. Chem.*, **262**, 2384.
33. Ryan, K. A. and Englund, P. T. (1989) Replication of kinetoplast DNA in *Trypanosoma equiperdum*. Minicircle H strand fragments which map at specific locations. *J. Biol. Chem.*, **264**, 823.
34. Kitchin, P. A., Klein, V. A., Fein, B. I., and Englund, P. T. (1984) Gapped Minicircles. A novel replication intermediate of kinetoplast DNA. *J. Biol. Chem.*, **259**, 15532.
35. Hajduk, S. L., Klein, V. A. and Englund, P. T. (1984) Replication of kinetoplast DNA maxicircles. *Cell*, **36**, 483.
36. Schneider, E., Hsiang, Y.-H., and Liu, L. F. (1990) DNA topoisomerases as anticancer drug targets. *Adv. Pharmacol.*, **21**, 149.
37. Myler, P. J., Glick, D., Feagin, J. E., Morales, T. H., and Stuart, K. D. (1993) Structural organization of the maxicircle variable region of *Trypanosoma brucei*: identification of potential replication origins and topoisomerase II binding sites. *Nucleic Acids Res.*, **21**, 687.
38. Sloof, P., de Haan, A., Eier, W., van Iersel, M., Boel, E., Van Steeg, H., and Benne, R. (1992) The nucleotide sequence of the variable region in *Trypanosoma brucei* completes the sequence analysis of the maxicircle component of mitochondrial kinetoplast DNA. *Mol. Biochem. Parasitol.*, **56**, 289.

6 | Developmental regulation of gene expression in African trypanosomes

ETIENNE PAYS and LUC VANHAMME

1. Introduction

The parasitic way of life of many trypanosomatids requires frequent, drastic, and rapid adaptations to the different environments encountered during their development. In particular, *Trypanosoma* and *Leishmania* are cyclically transmitted from different insects to mammals, where some species proliferate intracellularly and others extracellularly. In their mammalian hosts, the parasites live at 37°C and have to protect themselves against lytic factors and cells. They are able not only to recognize and use specific components of the host, such as growth factors, cytokines and surface receptors, but also to produce molecules that influence the activity of mammalian cells, for instance to modify the phenotype of macrophages and lymphocytes. When taken up by insect vectors, they undergo a cold shock as well as a complete change of the extracellular milieu, and must interact with totally different cells. The adaptations of the parasite are achieved through extensive re-programming of gene expression, comparable to that occurring during cell differentiation in metazoans. In this chapter, we review the information available on the genetic mechanisms involved in these differentiation processes.

1.1 A model: *Trypanosoma brucei*

Several features of the biology of the African trypanosome *Trypanosoma brucei brucei* make it a suitable laboratory model to study the genetic basis of differentiation in parasitic protozoans.

First, trypanosomes of the *T. b. brucei* sub-species are not pathogenic to humans, as they are lysed by human serum. Apart from this characteristic, these trypanosomes are virtually indistinguishable from *T. b. rhodesiense* and *T. b. gambiense*, the causative agents of the human sleeping sickness disease.

Second, these parasites show a relatively simple life cycle with no intracellular development. Two of the major stages are readily available in the laboratory: the bloodstream form of the mammalian host; and the procyclic form that resides in the

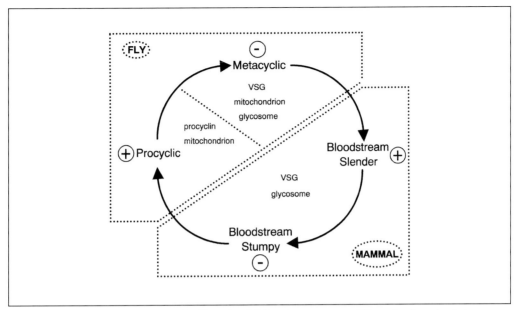

Fig. 1 Differentiation markers during the life cycle of African trypanosomes. This cycle alternates between proliferative (+) and quiescent (−) stages. The expression of the major surface glycoproteins, VSG and procyclin, are indicated, as are the primary sites of energy metabolism. Not all differentiation intermediates are mentioned.

Glossina vector (Fig. 1). In each host, periods of cell multiplication are terminated by differentiation into quiescent forms (the stumpy form in the bloodstream and the metacyclic form in the fly salivary glands). Only the latter forms appear to be competent for transformation into the next developmental stage, indicating that differentiation may be triggered only in a definite window of the cell cycle (1).

Third, each of these major stages is characterized by a typical, abundant, and easily detectable surface glycoprotein. The VSG (variant surface glycoprotein) is the major antigen of the bloodstream form. This protein appears at the end of the development in the fly, during the metacyclic stage, and persists throughout the development in the mammalian blood. It forms a dense coat of 10 000 000 tightly packed molecules covering the entire cell surface and acts as a protective barrier against the lytic elements of the blood. Moreover, continuous antigenic variation by expression of different VSGs allows the parasite to escape the immune reaction of the host. Procyclin, or procyclic acidic repetitive protein (PARP), is the major antigen of the procyclic form. It probably protects the parasite against lytic components from the fly. These two proteins are synthesized in a mutually exclusive manner and therefore represent excellent markers of differentiation. Their genes have been cloned and sequenced (2, 3).

Fourth, energy metabolism in *T. brucei* is also subject to a remarkable developmental regulation. While bloodstream forms rely on the oxidation of glucose in specialized organelles termed glycosomes, the procyclic forms develop a fully operational mitochondrion and cytochrome-mediated respiration. The genes for some of the glycosomal and mitochondrial proteins have been characterized (see Chapters 8 and 9).

Fifth, both bloodstream and procyclic forms can be cloned, stored frozen, and produced in large numbers *in vitro*. In addition, some of the life-cycle transitions can be reproduced *in vitro*. This is the case for the differentiation from the bloodstream to the procyclic form. Under well defined conditions, in which temperature appears to play a major role, this process occurs efficiently and synchronously (4).

Finally, electroporation allows the transfection of foreign DNA into trypanosomes. In *T. b. brucei*, circular DNA molecules do not integrate into the genome, but persist for a few days as extrachromosomal episomes (5). In contrast, free ends of DNA tend to recombine very efficiently with genomic sequences, provided that there are regions of homology between the two (6–8). Therefore transgenic trypanosomes can be generated easily by targeted homologous recombination. Thus the role of sequences involved in the control of gene expression can be monitored via the assay of reporter DNA constructs, either transiently or stably expressed (see Chapter 2).

Together these characteristics have allowed a detailed study of the developmental regulation of gene expression in this protozoan parasite.

1.2 Transcription units of the genes for the major stage-specific antigens of *T. brucei*

Like most genes in *T. brucei*, the VSG and procyclin genes are contained in polycistronic transcription units (Fig. 2).

While the trypanosome genome contains several hundred VSG genes, there are only six to 20 VSG units that can be expressed in the bloodstream form. These are often referred to as the VSG gene expression sites. Only one is transcriptionally active at any given time, leading to the synthesis of a coat made of a single type of VSG. All expression sites characterized to date are telomeric and share a similar size (45–60kb) and structure (9; *B VSG* map in Fig. 2). The VSG gene is the most distal from the promoter. It is surrounded by arrays of repeats which extend over several kilobases: downstream telomeric repeats including the hexamer TTAGGG, and upstream 76bp repeats which act as recombination hot-spots. A battery of expression site-associated genes (ESAGs) is located between the VSG gene and the transcription promoter. Several of these genes encode surface proteins. For example, the product of *ESAG 4* is a transmembrane receptor-like adenylate cyclase (10), while those of *ESAG 7* and *ESAG 6* associate into a heterodimeric receptor for the growth factor transferrin (11–13). It seems that *ESAG 8* encodes a nuclear regulatory factor able to interact both with nucleic acids and proteins (14, 15). Approximately half of the expression sites are provided with an additional adjacent region of 13kb at their 5' extremity (16, 17). This region starts with a second copy of the VSG promoter. It contains *ESAG 10,* a gene potentially encoding a transmembrane transporter-like

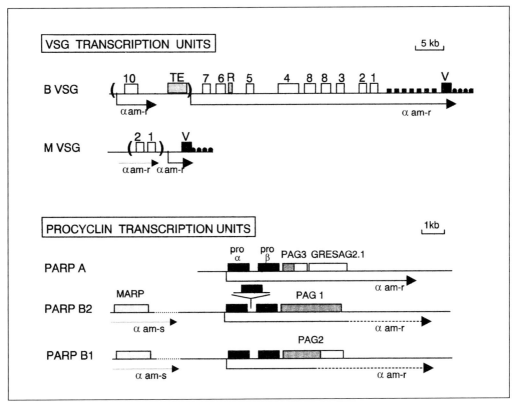

Fig. 2 The VSG and procyclin transcription units. In the different maps, the arrows define the regions transcribed, by either α-amanitin-resistant (α am-r) or α-amanitin-sensitive (α am-s) RNA polymerase; the promoters are located at the vertical bar at the beginning of the arrows. The bloodstream (B) and metacyclic (M) VSG transcription units are both telomeric. While the metacyclic units only contain the VSG gene, the bloodstream units also harbour a collection of expression site-associated genes (ESAGs), numbered in the map (see the text for details of the proteins encoded by these genes). A 13kb region (marked in parenthesis) is present in approximately half of the bloodstream expression sites. It contains another copy of the VSG promoter, *ESAG 10 and a* transposable element (TE). The small squares indicate the arrays of 76bp repeats; circles represent telomeric repeats. In the M VSG map, the bracketed region encompassing *ESAG 1* and *ESAG 2* is not always present and is transcribed independently of the VSG gene. R, *RIME* transposon; V, VSG gene. The different procyclin transcription units contain procyclin-associated genes (PAGs). *PAG 1* is related to *ESAG 7* and *ESAG 6*, *PAG 2* and *PAG 3* share a long sequence with *PAG 1* (shaded region), and *GRESAG 2.1* is related to *ESAG 2*. The size of the *PARP A* locus is around 8kb, while that of *PARP B* probably does not exceed 10kb. MARP, microtubule-associated gene; pro α and pro β = α and β procyclin genes.

protein (18), as well as a transposable element (19). While the ESAGs, together with the VSG gene, are only expressed in the bloodstream form, genes related to some ESAGs are transcribed independently of the VSG expression site (18, 20–23).

The VSG expression sites utilised at the metacyclic stage (*M VSG* map in Fig. 2) are located at telomeres of the largest chromosomes. Unlike the bloodstream sites, they seem to contain only the VSG gene, although some ESAGs may be present in the immediate vicinity. In the 2kb region between the gene and the promoter, there is no trace of the 76bp repeats that characterize the 5' environment of the VSG gene in the bloodstream expression sites (9, 24–27). These differences may prevent the metacyclic VSG genes from recombining with those of the bloodstream repertoire (27). The metacyclic expression sites are only active in the salivary glands of the fly, although one of these units has recently been found to be transcribed in the bloodstream form (26).

The procyclin genes are arranged as tandem copies contained within 8 to 10kb transcription units in two different, non-telomeric and diploid loci (21, 28, 29; *PARP A* and *PARP B* maps in Fig. 2). They are located immediately downstream from the promoter. A few procyclin-associated genes (PAGs) have been described but their function is unknown. Interestingly, two of these genes are related to ESAGs (21, 22). All of the procyclin units appear to be active simultaneously but only in the procyclic form (28–30).

1.3 Possible levels of control for gene expression

In trypanosomes, the genome is organized in polycistronic transcription units (see Chapter 2). With such a gene organization, any control at the level of transcription initiation would imply that all genes belonging to the same unit are regulated in the same way. This does not appear to be the case, except for the units of the genes for the major stage-specific antigens VSG and procyclin.

Data generated to date indicate that the majority of trypanosome genes are constitutively transcribed in very long polycistronic units under the control of elusive RNA polymerase II (pol II) promoters (31). Thus, unlike the situation in higher eukaryotes, gene expression in trypanosomes is not regulated at the levels of transcription initiation or elongation. By contrast, many contiguous genes belonging to the same transcription unit are differentially regulated during parasite development by stage-specific modulation of mRNA abundance. Examples of such regulation include the cluster of genes for phosphoglycerate kinase (32) and the sequences linked to that of a gene for a Ca^{2+}-ATPase of the endoplasmic reticulum (33; Fig. 3). Moreover, in a given developmental stage, the different genes of a transcription unit are generally not expressed at the same level. This is most dramatically illustrated in the case of the VSG transcription unit in the bloodstream form, where the VSG mRNA is several hundred times more abundant than those of some ESAGs (34). Furthermore, there are marked differences between the levels of each of the ESAG mRNAs (35; Fig. 3). Thus, developmental changes in specific protein levels are frequently regulated by a modulation of the final mRNA abundance, and this regulation is determined by post-transcriptional controls. Possible targets for the controlling factors include the RNA maturation processes, *trans*-splicing, and polyadenylation. In addition, some examples of translational and post-translational controls have been documented.

The expression of the VSG and procyclin genes exhibits particular characteristics. In contrast to the majority of genes, their developmental regulation is largely effected at the levels of transcription initiation, or elongation, or both (Fig. 3). Prob-

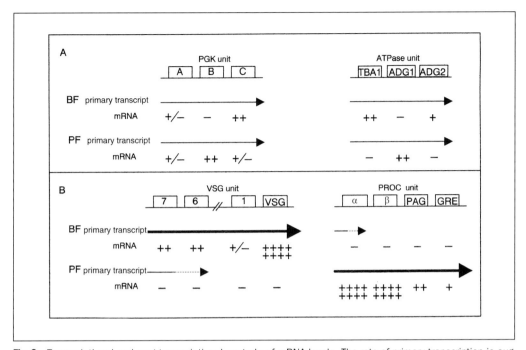

Fig. 3 Transcriptional and post-transcriptional controls of mRNA levels. The rate of primary transcription is suggested by the relative thickness of horizontal arrows, and the final abundance of the mRNAs is indicated by +. A: Segments from the transcription units of the genes for the different phosphoglycerate kinase (*PGK*) isoforms and a Ca^{2+}-ATPase of the endoplasmic reticulum (*TBA1*) are taken as examples of the expression of 'housekeeping' genes. The primary transcription of these genes is performed by pol II and does not appear to be regulated. The final amounts of individual mRNAs are determined during RNA processing. B: The VSG and procyclin units are transcribed by a ribosomal-like RNA polymerase. These units undergo opposite stage-specific controls acting at both transcriptional and post-transcriptional levels. BF, bloodstream form; PF, procyclic form.

ably because the genes for the major antigens must be expressed at very high levels, these units are transcribed by a pol I-like RNA polymerase (9, 36). Furthermore, in the case of the VSG units, additional mechanisms modulate gene expression to achieve antigenic variation in the bloodstream form. These mechanisms include frequent DNA rearrangements and reversible switches of the transcriptional competence of telomeres.

In the rest of this chapter, we detail the different steps, from transcription initiation to protein stability, whereby the expression of trypanosome genes is regulated.

2. Promoters and the control of transcription initiation

The polycistronic nature of the transcription units has rendered the quest for trypanosome promoters very difficult. The promoters for the ribosomal RNA genes

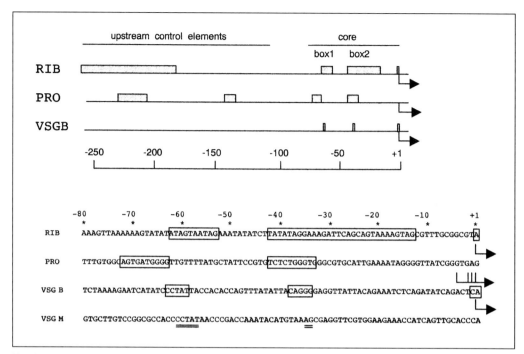

Fig. 4 Comparison of different promoters of *T. brucei*. The VSG, procyclin, and ribosomal promoters are aligned on the transcription start sites (arrows). The -70 to +1 region of the bloodstream VSG promoter (*VSG B*) is sufficient for maximal activity, while the corresponding regions of the ribosomal (RIB) and procyclin (PRO) promoters confer activities stimulated by the presence of upstream elements. The boxed regions in the line drawings have been defined as crucial for activity in transient transfection assays (RIB, Ref. 49a; PRO, Refs 45,46; VSG B, Ref. 49b). In the -80 to +1 promoter core (sequences below), these regions include box 1 and box 2, as well as the transcription start site. The bottom sequence (VSG M) is that of the *AnTat 11.17* metacyclic VSG promoter (49b). The sequence elements shared with the crucial regions of the bloodstream VSG promoter are double underlined.

and some small RNA genes, which recruit pol I and pol III, respectively (37, 38), do not appear to be stage-regulated and will not be discussed here.

2.1 The few promoters known

To date, there are no well-established examples of pol II promoters in trypanosomes. The activity of the putative actin promoter (39) is very weak in transient assays and remains to be demonstrated by expression from a transcriptionally silent region of the genome. Another proposed pol II promoter is that of the spliced leader RNA genes (40). However, the particular divalent cation dependence exhibited by the transcription of these genes is difficult to reconcile with this view (9, 41). Finally, in *Leishmania enriettii* a *cis*-acting DNA fragment able to confer strandedness of transcription of a multidrug resistance gene has been uncovered (42), but this sequence does not appear to be necessary and sufficient to organize transcription initiation.

The only two promoters identified for protein-encoding genes are those of the VSG and procyclin genes of *T. brucei* (28, 29, 43, 44). These promoters resemble ribosomal promoters, both structurally and functionally (Fig. 4). Despite a lack of sequence homology, their 80bp core region shares a similar structure, with three important regions around -60, -35, and +1, the spacing of which appears to be critical (9, 45–49b). Elements of the crucial boxes from the bloodstream VSG promoters are conserved in those for a metacyclic VSG gene (Fig. 4). While the core region is sufficient for full activity of the VSG promoters, upstream control elements located between -250 and -150 are necessary to stimulate the basal activity of the same region from the ribosomal and procyclin promoters. The important boxes of the core region are relatively interchangeable, as hybrid constructs containing boxes from the different promoters still show some activity (49a, 49b). Moreover, the second box of all three promoters seems to bind the same proteins. Interestingly, the specific binding of proteins to these promoters requires that the DNA is single-stranded (50). This observation suggests that the promoters must be denatured to be functional.

2.2 Regulation of promoter activity

The promoters for the majority of the trypanosome genes studied to date do not appear to be developmentally controlled, as primary transcription is generally constitutive (31, 51). As described above, the VSG and procyclin units represent exceptions, since their transcription is inversely regulated during parasite development (52–54). Nevertheless, these transcriptional controls do not seem to act, at least primarily, on promoter activity. Several lines of evidence point to this conclusion.

First, in the procyclic form, transcription originating from the VSG promoters can be easily detected despite a strong down regulation of the VSG unit (44, 55, 56; Fig. 5A). This transcription is weak and abortive, pointing to a clear control of RNA elongation (see below), but it is difficult to estimate to what extent a repression of initiation is also involved in this process.

Second, if a VSG promoter is present in an episomal plasmid or targeted to the silent intergenic region of the ribosomal locus, it exhibits full activity in the procyclic form (47, 48, 56; Fig. 5B, C). These observations demonstrate that all the components required for transcription initiation by the VSG promoter are present at this stage, and recently proteins binding specifically to the second box of the VSG promoter have been detected in procyclic parasites (49b). Therefore, the stage-specific regulation observed in the VSG unit is exerted in *cis* and does not involve diffusible transcriptional repressors. Conversely, a ribosomal promoter targeted to the VSG unit appears to be fully active in the procyclic form (56), indicating that the down regulation occuring at this locus is specific to the transcriptional complex recruited by the VSG promoter.

Similar findings have been obtained for the procyclin promoters in the bloodstream form. At this stage, abortive transcription can be detected in the procyclin units and the procyclin promoters appear to be as active as the VSG promoters

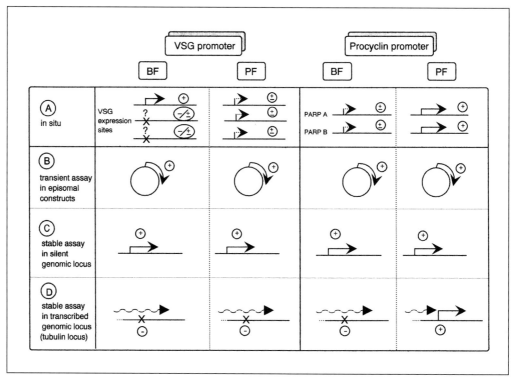

Fig. 5 Activity of the VSG and procyclin promoters. The extent of promoter activity is indicated by ±, and transcription is schematically represented by arrowed lines. The wavy lines in panel D represent transcription by the endogenous pol II. Crosses indicate inactive promoters. BF, bloodstream form; PF, procyclic form.

when contained in plasmid constructs or when targeted in a silent genomic locus (44; L. Vanhamme, M. Berberof, and E. Pays, submitted; Fig. 5).

There is another level of control of VSG gene transcription. In the bloodstream form, only one of the six to 20 VSG expression sites is transcriptionally active at a time (Fig. 5A, left). The mechanism of this selective activation is not understood. The very high sequence conservation at the 5' ends of the different expression sites has made it difficult to determine whether this is achieved through promoter control or premature RNA termination (43, 44). In this respect, it is interesting to note that the promoters from different VSG expression sites of the same trypanosome clone, whether active or not in their original cells, appear to be similarly functional in transient expression assays following transfection into either developmental form (43, 47).

There is one particular case where control of transcription initiation is believed to occur. During the metacyclic stage, all the metacyclic VSG expression sites are activated simultaneously, although ultimately only one VSG gene is transcribed per cell. It has been proposed that this activation is triggered by specific components from the fly (24, 25, 27), although whether by the presence of a stimulatory factor or the release of an inhibitory factor is unknown. In this respect, it is worth noting that metacyclic promoters can be re-activated in the bloodstream form. Indeed, a metacyclic-like VSG unit has been found to be active late during chronic infection (26). This may be a result of an accidental disruption of the normal controls operating on this unit.

2.3 Influence of the chromosomal context

In trypanosomes, the only known examples of transcriptional regulation, i.e. the procyclin and VSG expression sites, appear to depend strongly on the chromosomal context of their promoters.

In addition to the studies where regulated expression of a VSG promoter was abrogated by episomal expression or insertion into the rDNA locus, a VSG promoter has been inserted into a pol II transcription unit. There, it is by-passed by the ongoing pol II polymerase, at both stages of the parasite life cycle (57; Fig. 5D, left). In contrast, the procyclin promoter placed in the same situation is active, but only in the procyclic form (7; M. Berberof, L. Vanhamme, and E. Pays, submitted; Fig. 5D, right). These results indicate that the ability of these promoters to recruit their cognate polymerase is dependent on the developmental stage and genomic context. In the case of the procyclin promoter, it is possible that this ability is linked to a stage-specific termination of upstream pol II activity, since a pol II transcription unit is present immediately upstream from a procyclin unit (29, 30, 44). This does not seem to be the case with the VSG units (16, 17, 43, 44).

The genomic context appears to be particularly important in the determination of the (in)activation of the different VSG expression sites in the bloodstream form (Fig. 5A, left). It is believed that 'position effects', similar to those operating in yeast telomeres (58), reversibly influence the transcriptional status of the VSG unit so that only one is active at any time (59). The inactivation of expression sites may be either triggered or temporarily locked by an unusual DNA modification found only in the silent telomeres of the bloodstream form (60, 61). The modified base is thought be β-D-glucosyl-hydroxymethyluracil (62). While it is conceivable that the presence of this modified base influences the chromatin structure so that transcription is repressed, it remains to be explained why at any one time, a single expression site escapes this modification. It is important to note that no modification is detected in the telomeres of the procyclic form, where a general down regulation of the VSG expression sites occurs (61). Together, the data suggest that the controls acting on the VSG units may be different in the two stages of the life cycle. In support of this hypothesis, the properties of chromatin have been found to differ markedly between the developmental forms of the parasite. Both the level of condensation and the histone composition appear to show stage-specific variations (63). However it is not known to what extent these differences can influence gene expression.

3. Transcription elongation and processing of the primary transcripts

In trypanosomes, processes occuring after transcription initiation appear to represent the main control points for gene expression.

3.1 RNA elongation

As already mentioned, at least part of the transcriptional control of the VSG and procyclin units is exerted at the level of RNA elongation. A progressive blocking of RNA polymerase activity occurs in the VSG unit in the procyclic form (44, 55, 56) and in the procyclin units in the bloodstream form (44; L. Vanhamme, M. Berberof, and E. Pays, in press; Figs 3 and 5A). In the VSG unit, a major attenuation region is found around 700bp from the start site, but no terminator sequence has been identified (56). The fact that transcription is not down regulated when a pol I promoter is inserted into this locus (56) strongly suggests that *cis*-acting chromosomal elements specifically control the transcription driven by the VSG promoter. Interestingly, this control is dependent on the same combination of factors as that triggering the differentiation from the bloodstream into the procyclic form. These factors include temperature and *cis*-aconitate. A cold shock from 37°C to 27°C and the addition of *cis*-aconitate lead to a cumulative reduction of transcription in the VSG unit of the bloodstream form (44, 64, 65). The reverse is true for the procyclin units, where RNA elongation is stimulated by the same combination of factors in the same cells (L. Vanhamme, M. Berberof, and E. Pays, in press). What effects these changes in transcript elongation? Data from several laboratories show that a transient inhibition of protein synthesis inhibits transcription of the VSG unit and stimulates the production of procyclin mRNA (44, 52, 66, 67). Such an inhibition actually occurs during differentiation from the bloodstream to the procyclic form, when the VSG is rapidly replaced by procyclin (68). Thus, it is hypothesized that short-lived, bloodstream-specific proteins are able to stimulate elongation in the VSG unit and to inhibit the synthesis of procyclin mRNA.

3.2 *Trans*-splicing and polyadenylation

A cleavage in the intergenic region of the primary transcript probably occurs as soon as this sequence is synthesized. The released RNA ends are processed by *trans*-splicing and polyadenylation, both events being dependent on the recognition of polypyrimidine tracts (see Chapter 7). *Trans*-splicing and polyadenylation typically occur in a limited region that may contain a few possible processing sites. In addition, alternative processing between quite distant sites has also been observed (33, 69–71). It is clear that determination of the final level of mRNA can be regulated at several steps of these processes.

First, the relative abundance of pyrimidines immediately 5' to the splice site could determine the relative efficiency of splicing and, hence, the final level of

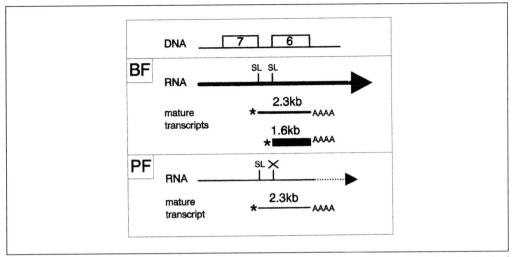

Fig. 6 Stage-specific splicing of the *ESAG 6* transcripts. *ESAG 6* encodes the membrane-anchored subunit of the transferrin receptor of *T. brucei*. In the bloodstream form (BF), alternative splicing generates two transcripts, the shortest of which (1.6kb) is the functional mRNA. In the procyclic form (PF), the 1.6kb species is not produced despite the generation of the largest RNA (2.3kb). The relative thickness of the arrowed lines indicates the abundance of the mature transcripts. SL, site of addition of the spliced leader RNA (asterisk).

mRNA (72–74). This may contribute to differences in mRNA amounts between different genes of the same unit at a given stage (73). However, in some cases of alternative splicing the relative pyrimidine content does not appear to predict the highly selective choice of some splice sites over others (33, 71).

Secondly, alternative processing may be stage-regulated (33, 55, 75). This is the case of the transcripts of *ESAG 6* (55, 70; Fig. 6). This gene is only expressed in the bloodstream form, although it is still transcribed in the procyclic form (44, 55). Splicing of *ESAG 6* transcripts in bloodstream forms occurs with approximately equal probability at two alternative sites, as judged by the respective amounts of mRNA when RNA degradation is prevented by UV irradiation (67, 70). The largest RNA, spliced very close downstream from the polyadenylation site of the preceding gene (*ESAG 7*), contains a long 5'-untranslated region, which is unusual in trypanosomes. Presumably because of the presence of this region, this RNA is highly unstable and quickly degraded in bloodstream forms. The small transcript is spliced very close to the initiation codon, and represents the functional *ESAG 6* mRNA, which is moderately abundant. Remarkably, in the procyclic form only the largest RNA can be detected, even if RNA degradation is blocked by UV irradiation (55). This observation indicates that *trans*-splicing at the second site is stage-regulated. So far, it is difficult to estimate the significance and relative importance of this type of control. As the 3'-UTR appears to be critical in the determination of the relative mRNA abundance (see below), use of alternative polyadenylation sites may also be potentially important in the regulation of gene expression.

Thirdly, there is evidence that polyadenylation itself can be stage-regulated. This has been observed in the case of the primary transcripts for the spliced leader RNA, termed 'mini exon-derived RNA', which are specifically polyadenylated in the stumpy form (76). Interestingly, a significant increase in the length of the procyclin mRNA, probably attributable to an increase of the poly(A) tail, was observed in stumpy forms triggered to differentiate into procyclic forms (54). These processes may increase the stability or translational efficiency of transcripts, although this remains to be demonstrated.

3.3 The untranslated regions and RNA amounts

The location of the processing sites determines the length and sequence of the 5' and 3' UTRs of the mRNAs. There is a growing body of evidence that suggests that these UTRs, in particular the 3' UTR, are involved in the developmental control of gene expression.

In several cases such as those of the hsp83 genes of *Leishmania* (77) and the genes encoding VSG, procyclin, and aldolase of *T. brucei* (47, 78), the 3' UTR sequence of the respective mRNAs has been shown to confer stage-specific expression on a reporter gene in transient activity assays of plasmid constructs. Using the same methods, the 5' UTR is ineffective, except in one case (77). The role of the 3' UTR of the VSG mRNA has been investigated by the analysis of the transcription of reporter constructs stably integrated in the genome (M. Berberof, L. Vanhamme, and E. Pays, submitted). A sequence of approximately 100 nucleotides has been identified as necessary and sufficient to achieve stage-specific regulation. This sequence reduced the production of mRNA in the procyclic form, apparently through a decreased efficiency in the maturation process at this stage. In contrast, it favoured the accumulation of mRNA in the bloodstream form, through an increase in RNA stability. The mechanisms and factors responsible for these effects are not known.

4. RNA translation and protein stability

Gene expression can be regulated downstream from the synthesis of mRNA. In trypanosomes, there is so far no evidence that the export of mRNA from the nucleus to the cytoplasm is subject to developmental control. Conversely, some observations suggest that mRNA translatability and protein stability can be modulated.

4.1 Translational controls

The arguments supporting a role for differential mRNA translatability in the developmental regulation of trypanosomatid gene expression are indirect.

The 3' UTRs of the different procyclin mRNAs contain a 16 nucleotide motif as their only conserved sequence. This sequence forms part of a stem–loop structure the correct folding of which is required for optimal gene expression (79). As the presence or absence of this 16-mer does not affect the amount, polyadenylation, stability, and nucleocytoplasmic distribution of a reporter mRNA (79), it is likely that this effect occurs at the level of mRNA translation. To date, the precise role of the 16-mer in modulating procyclin synthesis is unknown.

Several indirect observations suggest a link between mRNA translatability and stability. Shortly after the triggering of differentiation of stumpy forms into procyclic forms, procyclin mRNA is estimated to be 20-fold more abundant than in later stages (54). This abundance probably correlates with the increased requirement for procyclin synthesis at a time when the whole cell surface must be rapidly covered by this new antigen. This effect is likely to be modulated, at least in part, by increased RNA stability, which can in turn be achieved by increasing the length of the poly(A) tail (54; see above). It is tempting to relate these observations to the need for procyclin mRNA translation during this period.

Differences have also been found to exist between the relative amounts of mRNAs and the proteins they encode. For example, aldolase, a glycolytic enzyme, is 30-fold more abundant in bloodstream than in procyclic forms, while the difference in mRNA levels between stages is only six-fold (80). Similarly, the synthesis of the Nrk (Nek1 related kinase) kinase increases dramatically when long slender bloodstream forms differentiate into stumpy forms, under conditions where no change of the mRNA levels is detected (81). These observations suggest that the mRNA may be differentially translated during development.

4.2 Protein stability

That the turnover rate of a protein may vary depending on the developmental stage has been demonstrated in the case of cytochrome *c* (82). This protein is not detectable in the bloodstream form, while its mRNA is still present at 30% of the level found in the procyclic form. A similar observation applies to cytochrome *c* reductase (83). Such a difference between RNA and protein levels can be explained by a drastic reduction in the protein half-life in the bloodstream form (82). The increased degradation rate of cytochrome *c* at this stage could be linked to the repression of the mitochondrion. Presumably the cytochrome *c* cannot assemble in the inactive organelle.

5. Antigenic variation and novel mechanisms of gene regulation

All parasites have to face the same challenge: growth without killing their host. African trypanosomes meet these requirements through continuous changes of their VSG coat, in a process called antigenic variation. Each bloodstream trypanosome synthesizes a single type of VSG, but this VSG is spontaneously replaced by a

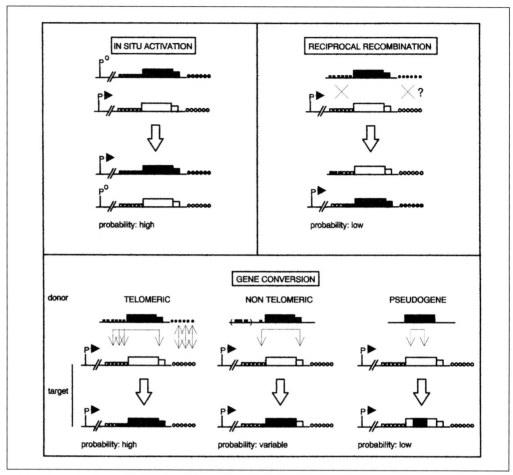

Fig. 7 The mechanisms of antigenic variation. The active telomeric VSG gene expression site is marked with a black arrowhead. Small circles represent telomeric repeats; small squares, 76bp repeats; P, promoter. Note that some VSG genes do not have telomeric sequences; these chromosome internal genes are preceded by variable numbers of 76bp repeats. See text for details of these mechanisms.

different one at a frequency of up to 10^{-2} per generation. The immune response against the VSG is sufficient to kill the majority of parasites after a few days. However, the emergence of new variants allows the propagation of infection through the development of new waves of parasitaemia. Thus, the continuous interplay between antigenic variation of the parasite and antibody response of the host maintains a relatively constant number of trypanosomes for long periods in the bloodstream. Two genetic mechanisms have been found to mediate antigenic variation (Fig. 7): activation of different VSG expression sites and alteration of the VSG gene occupying the active expression site. These processes have been extensively reviewed (9, 59, 84, 85).

5.1 *In situ* (in)activation of VSG expression sites

The mechanisms involved in the successive activation of the different VSG expression sites are presumably coupled to those which dictate the selective activity of a single expression site at one time. These mechanisms do not involve detectable rearrangements within the expression site or its immediate environment. However, the influence of distant DNA rearrangements cannot be excluded. A special type of DNA modification, apparently restricted to transcriptionally silent telomeric DNA (60, 61; see above), may also play a role in this process.

In situ (in)activation of VSG expression sites does not require rearrangements within and around the transcription promoter (44, 48). Although the inactivation may be associated with homologous recombination between two VSG promoter copies, leading to the excision of an intervening 13kb fragment (16, 17, 86), this type of event is not always observed (44, 48). At the other end of the expression site, the change of transcriptional status is often linked to unusual size variations of the telomere (87, 88). These variations, probably due to recombination between telomeric repeats of different chromosomes, may possibly be important in transcriptional switching. Indeed, in higher eukaryotes, the size of the arrays of telomeric repeats influences the structure of chromatin (89). In addition, the 'position effects' which control transcription in the telomeres of yeast are also dependent on the length of these repeats (58). However, in trypanosomes there is so far no experimental data to evaluate these hypotheses.

An inescapable feature of the system controlling the activity of the VSG expression sites in the bloodstream form is the high selectivity of activation: except for a transient period during antigen switching (90, 91), only a single site is active at a time. Therefore, it appears that the nuclear factor or location required to activate the expression site is unique. It has been proposed that this factor or location is in the nucleolus (92), on the basis of the hypothesis that the transcription of the VSG unit is performed by a ribosomal polymerase (36).

It is important to note that the alternative activation of the different VSG expression sites not only achieves a variation of the surface antigen, but also leads to the alternative expression of different sets of ESAGs. Since the ESAGs are not identical between expression sites, slightly different ESAG products may be synthesized following antigenic variation. The ESAGs encode proteins which are crucial for growth, such as adenylate cyclase and a transferrin receptor. Therefore, the expression of variant forms of these proteins may explain why antigenic variation is sometimes linked to changes in the growth properties of the cell. This system also provides the parasite with a means to achieve some antigenic variation of its surface receptors (93).

5.2 VSG gene rearrangements

Most antigenic variation events studied in *T. brucei* are due to DNA rearrangements within the VSG expression site. These rearrangements appear to be a manifestation of the powerful homologous recombination machinery revealed by the development of DNA transfection experiments (6–8). When transfected into trypanosomes, the double-stranded ends of exogenous DNA can invade the genome and recombine efficiently in areas of sequence homology. Based on these observations, it may be proposed that the recombinations occuring in the VSG expression site are triggered by the generation of free DNA ends due to endonucleolytic cleavage (94). Two types of recombination events can induce antigenic variation: gene conversion and reciprocal recombination (Fig. 7).

Gene conversion involves the replacement of the VSG gene, or at least the replacement of the region encoding the surface epitopes, by a copy of another VSG gene (94). Thus, this recombination takes place between a donor (copied) and a target (replaced) gene. The distribution of these respective roles may be linked to the transcriptional status of the recombination partners. The active VSG gene is the only one contained in chromatin with an open structure. It is therefore more sensitive to the endonucleolytic activity which may trigger homologous recombination with one of the several hundred transcriptionally inactive VSG genes (94). The length of the sequence copy–replacement appears to be dictated by the relative location of the regions of homology between the two DNA partners (94, 95). This length can vary between dozens of kilobases and only a few base pairs. However, the preferred size is about 3kb. This represents the distance between the conserved blocks which usually flank the VSG open reading frames: the upstream arrays of 76bp repeats and the 3'-terminal region of the gene. When shorter than the gene, the copy–replacement process generates hybrid genes constructed with segments of both the target and the donor. These segmental conversion events can be quite complex and may even involve several donors for a single target. Mosaic genes can be assembled in such a way. The expression of these new constructs extends the antigenic variation potential of the parasite and thus allows the development of long-lasting, chronic infection (95–99).

Reciprocal recombination differs from gene conversion by its symmetry. In the two cases where it has been observed, exchange between a silent telomere and the one harbouring the active VSG expression site occurred via crossing over in the 76bp repeat region, between the promoter and the VSG gene (Fig. 7). Since there are many telomeric VSG genes, each apparently provided with extensive arrays of 76bp repeats, this type of event could, in principle, occur frequently in antigenic variation. However this is not the case, possibly because of a low probability of DNA cleavage in silent telomeres.

5.3 Programming of VSG expression

The VSGs are not expressed randomly during chronic infection. In the first days, the early, or predominant, antigen types appear, followed by the intermediate and subsequently the late variants. This loose programming prevents the parasite from exhausting its antigen repertoire too rapidly. It can be largely explained by the relative probability of activation of the different genes of the repertoire (97).

The telomeric VSG genes appear to be expressed earlier than the others. Two different reasons probably account for this observation. First, the alternative *in situ* activation of the different telomeric VSG expression sites, which does not depend on DNA recombination, seems to occur with high frequency. It is probable that at least several of the expression sites are equally competent for transcription, and are stochastically selected. Interestingly however, some expression sites seem to be preferred later in infection (97). Second, since the DNA recombinations depend on the recognition of sequence homology between the donor and target sequences, the genes showing the highest level of homology with the expression site will tend to be activated earlier. As the expression site is telomeric, VSG genes present in a telomeric environment are favoured. In particular, the telomeric repeats may increase the probability of recombination through interactions leading to the pairing of chromosome ends (100).

The later and hierarchical activation of the non-telomeric VSG genes is probably also dictated by the relative extent of sequence homology that they share with the expression site. These genes are generally flanked by reduced and variable numbers of 76bp repeats. Therefore, this number may determine the probability of recombination between each gene and the expression site. VSG genes without 76bp repeats have to rely on unusual blocks of homology to recognize and recombine with the expression site, which is relatively improbable and, thus, infrequent. Finally, VSG pseudogenes, either incomplete or interrupted by stop codons, also depend on infrequent sequence homology, actually within the open reading frame of the gene, to reconstitute a functional gene by recombination. It is not surprising that such segmental gene conversion events, which involve improbable sequence recognition and complex DNA rearrangements, typically occur very late in infection (95, 97–99).

The relative probability of gene activation is the major parameter for VSG programming, since a late-expressed VSG can be converted into an early antigen simply by changing its gene environment to one with increased homology with the expression site (97, 100). However, another facet of VSG programming, namely the fact that the immediate succession of variants is sometimes predictable, is less readily explained. A mathematical model has been presented to explain these observations (101). This model postulates that the switching intermediates, the trypanosomes harbouring a mixed VSG coat with the old and the new variant, are not equally viable. It is conceivable that not every pair of VSG molecules is able to form a protective coat, so that some switching intermediates may be favoured over others. This would lead to a preferred succession of variants. While theoretically possible, this hypothesis has not yet been demonstrated experimentally.

5.4 Point mutations

Point mutations in VSG genes may achieve antigenic variation (102). It has been recently demonstrated that such mutations can be generated during the process of gene conversion.

The analysis of independent conversion events templated by the same VSG gene has revealed that different point mutations appear during the synthesis of the gene copy (103). These mutations are not random, but exhibit a strand bias (104). Curiously, they seem to be restricted to the open reading frame of the conversion domain. Two explanations, not necessarily mutually exclusive, may account for these observations: either the DNA polymerase at work during gene conversion is prone to errors, or the DNA template is responsible for these errors.

Some considerations tend to support the second hypothesis. Indeed, point mutations have not been observed in all the cases of gene conversion analysed to date (95, 96, 98), suggesting that these alterations were not due to the intrinsic properties of the DNA polymerase. The mutations were only observed when telomeric VSG genes were used as conversion donors (103, 104). As mentioned above, an unusual DNA modification differentiates the telomeric genes from the others (60–62). Therefore, it is tempting to speculate that this modification is responsible for the misreading by the DNA polymerase. This hypothesis predicts that the telomeric VSG genes should accumulate point mutations not only during gene conversion, but also during DNA replication.

5.5 Evolution of VSGs

The VSG gene repertoires are only minimally conserved between trypanosome stocks (105). Their evolution both reflects the genetic processes that mediate gene switching and contributes to the antigenic variation potential of the parasite (9, 105)

As detailed above, the process of gene conversion can generate new VSG genes, both through the reassortment of segments from different gene donors and through point mutations. There is a way in which the parasite can store these newly created genes in its repertoire. When *in situ* activation is used as a mechanism for antigenic variation, the VSG gene constructed in the formerly active expression site is conserved, thus extending the family of the donor gene (Fig. 8). This new gene can persist and eventually be translocated to a more stable, internal position on the

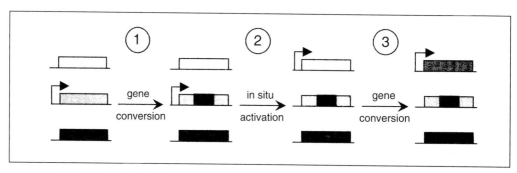

Fig. 8 Hyper-evolution of the VSG gene repertoire. The mechanisms of antigenic variation lead to rapid changes in VSG genes. Gene conversion creates new genes (event 1, shaded gene) that can be stored in the repertoire by *in situ* inactivation of the VSG gene expression site (event 2). Reciprocally, VSG genes present in activated expression sites are prone to be lost by gene conversion (event 3, gene in white). Compare the VSG gene repertoire before event 1 and after event 3.

chromosome (105). Conversely, if the *in situ* activation is followed by gene conversion, the gene present in the formerly active expression site is replaced by the new gene copy, and is thus lost. This continuous re-shaping of the VSG gene repertoire by the successive gains and losses of genes has been observed during antigenic variation (94).

6. Conclusions

In trypanosomes, the transcription of genes generally occurs in polycistronic units, and the adaptative changes of individual gene expression are not primarily regulated at the level of transcription initiation. Transcriptional regulation has only been observed in the units for the genes of the major stage-specific antigens of *T. brucei*, and this regulation is at least partly exerted on RNA elongation. Control of gene expression occurs essentially by the stage-specific modulation of the production of mRNA from the primary polycistronic transcripts. This modulation is mediated by intergenic sequences, in particular the 3' UTRs. These sequences appear to determine the final mRNA abundance by conditioning the rate of RNA processing or the stability of the mature transcripts, or both. Environmental conditions can drastically influence this process, as exemplified by the super-induction of the procyclin mRNA under cold shock or transient inhibition of protein synthesis (66, 67), or the selective accumulation of the glucose transporter THT2 mRNA upon glucose starvation (106). The availability of *in vitro*, synchronously differentiating parasites and DNA transfection protocols will allow the detailed dissection of these events. However, the regulatory factors involved in the modulation of gene expression in trypanosomes remain to be discovered. The same conclusion applies to the factors responsible for the processes of homologous recombination and transcriptional switching of telomeres, which underlie the powerful antigenic variation of these parasites.

Acknowledgements

We thank D. Nolan and the editors for useful comments on the manuscript, and M. Berberof for help in compiling the Figures. The work in our laboratory was supported by the Belgian FRSM and FRC-IM, a research contract with the Communauté Française de Belgique (ARC 89/94–134) and by the Agreement for Collaborative Research between ILRAD (Nairobi) and Belgian Research Centres. L. V. is a senior research assistant of the Belgian National Fund for Scientific Research.

References

1. Matthews, K. R. and Gull, K. (1994) Cycles within cycles: the interplay between differentiation and cell division in *Trypanosoma brucei. Parasitol. Today*, **10**, 473.
2. Borst, P. and Cross, G. A. M. (1982) Molecular basis for trypanosome antigenic variation. *Cell*, **29**, 291.

3. Roditi, I. and Pearson, T. W. (1990) The procyclin coat of African trypanosomes. *Parasitol. Today*, **6**, 79.
4. Ziegelbauer, K., Quinten, M., Schwarz, H., Pearson, T. W., and Overath, P. (1990) Synchronous differentiation of *Trypanosoma brucei* from bloodstream to procyclic forms *in vitro*. *Eur. J. Biochem.*, **192**, 373.
5. Bellofatto, V. and Cross, G. A. M. (1989) Expression of a bacterial gene in a trypanosomatid protozoan. *Science* **244**, 1167.
6. ten Asbroek, A. L. M. A., Ouellette, M., and Borst, P. (1990) Targeted insertion of the neomycin phosphotransferase gene into the tubulin gene cluster of *Trypanosoma brucei*. *Nature*, **348**, 174.
7. Lee, M. G. S. and Van der Ploeg, L. H. T. (1990) Homologous recombination and stable transfection in the parasitic protozoan *Trypanosoma brucei*. *Science*, **250**, 1583.
8. Eid, J. and Sollner-Webb, B. (1991) Stable integrative transformation of *Trypanosoma brucei* that occurs exclusively by homologous recombination. *Proc. Nat. Acad. Sci. USA* **88**, 2118.
9. Pays, E., Vanhamme, L., and Berberof, M. (1994) Genetic controls for the expression of surface antigens in African trypanosomes. *Annu. Rev. Microbiol.*, **48**, 25.
10. Paindavoine, P., Rolin, S., Van Assel, S., Geuskens, M., Jauniaux, J. C., Dinsart, C., Huet, G., and Pays, E. (1992) A gene from the VSG expression site encodes one of several transmembrane adenylate cyclases located on the flagellum of *Trypanosoma brucei*. *Mol. Cell. Biol.*, **12**, 1218.
11. Steverding, D., Stierhof, Y. D., Chaudhri, M., Ligtenberg, M., Schell, D., Beck-Sickinger, A. G., and Overath, P. (1994) ESAG 6 and 7 products of *Trypanosoma brucei* form a transferrin binding protein complex. *Eur. J. Cell Biol.*, **64**, 78.
12. Salmon, D., Geuskens, M., Hanocq, F., Hanocq-Quertier, J., Nolan, D., Ruben, L., and Pays, E. (1994) A novel heterodimeric transferrin receptor encoded by a pair of VSG expression site-associated genes in *Trypanosoma brucei*. *Cell*, **78**, 75.
13. Ligtenberg, M. J., Bitter, W., Kieft, R., Steverding, D., Janssen, H., Calafat, J., and Borst, P. (1994) Reconstitution of a surface transferrin binding complex in insect form *Trypanosoma brucei*. *EMBO J.*, **13**, 2565.
14. Smiley, B. L., Stadnyk, A. W., Myler, P. J., and Stuart, K. (1990) The trypanosome leucine repeat gene in the variant surface glycoprotein expression site encodes a putative metal-binding domain and a region resembling protein-binding domains of yeast, *Drosophila*, and mammalian proteins. *Mol. Cell. Biol.*, **10**, 6436.
15. Revelard, P., Lips, S., and Pays, E. (1990) A gene from the VSG gene expression site of *Trypanosoma brucei* encodes a protein with both leucine-rich repeats and a putative zinc finger. *Nucleic. Acids Res.*, **18**, 7299.
16. Gottesdiener, K., Chung, H. M., Brown, S. D., Lee, M. G. S., and Van der Ploeg, L. H. T. (1991) Characterization of VSG gene expression site promoters and promoter-associated DNA rearrangement events. *Mol. Cell. Biol.*, **11**, 2467.
17. Zomerdijk, J. C. B. M., Kieft, R., Duyndam, M., Shiels, P. G., and Borst, P. (1991) Antigenic variation in *Trypanosoma brucei*: a telomeric expression site for variant-specific surface glycoprotein genes with novel features. *Nucleic. Acids Res.*, **19**, 1359.
18. Gottesdiener, K. M. (1994) A new VSG expression site-associated gene (ESAG) in the promoter region of *Trypanosoma brucei* encodes a protein with 10 potential transmembrane domains. *Mol. Biochem. Parasitol.*, **63**, 143.
19. Lodes, M. J., Smiley, B. L., Stadnyk, A. W., Bennett, J. L., Myler, P. J., and Stuart, K. (1993) Expression of a retroposon-like sequence upstream of the putative *Trypanosoma*

brucei variant surface glycoprotein gene expression site promoter. *Mol. Cell. Biol.*, **13,** 7036.

20. Alexandre, S., Paindavoine, P., Tebabi, P., Pays, A., Halleux, S., Steinert, M., and Pays, E. (1990) Differential expression of a family of putative adenylate/guanylate cyclase genes in *T. brucei. Mol. Biochem. Parasitol.*, **43,** 279.
21. Berberof, M., Pays, A., and Pays, E. (1991) A similar gene is shared by both the VSG and procyclin gene transcription units of *Trypanosoma brucei. Mol. Cell. Biol.*, **11,** 1473.
22. Koenig-Martin, E., Yamage, M., and Roditi, I. (1992) *Trypanosoma brucei*: a procyclin-associated gene encodes a polypeptide related to ESAG 6 and 7 proteins. *Mol. Biochem. Parasitol.*, **55,** 135.
23. Graham, S. V. and Barry, J. D. (1991) Expression site-associated genes transcribed independently of variant surface glycoprotein genes in *Trypanosoma brucei. Mol. Biochem. Parasitol.*, **47,** 31.
24. Lenardo, M. J., Esser, K. M., Moon, A. M., Van der Ploeg, L. H. T., and Donelson, J. E. (1986) Metacyclic variant surface glycoprotein genes of *Trypanosoma brucei* subsp. *rhodesiense* are activated *in situ*, and their expression is transcriptionally regulated. *Mol. Cell. Biol.*, **6,** 1991.
25. Graham, S. V., Matthews, K. R., and Barry, J. D. (1993) *Trypanosoma brucei*: unusual expression site-associated gene homologies in a metacyclic VSG gene expression site. *Exp. Parasitol.*, **76,** 96.
26. Alarcon, C. M., Jin Son, H., Hall, T., and Donelson, J. E. (1994) A monocistronic transcript for a trypanosome variant surface glycoprotein. *Mol. Cell. Biol.*, **14,** 5579.
27. Graham, S. V., Matthews, K. R., Shiels, P. G., and Barry, J. D. (1990) Distinct developmental stage-specific activation mechanisms of trypanosome VSG genes. *Parasitology*, **101,** 361.
28. Clayton, C. E., Fueri, J. P., Itzakhi, J. E., Bellofatto, V., Sherman, D. R., Wisdom, G. S., Vijayasarathy, S., and Mowatt, M. R. (1990) Transcription of the procyclic acidic repetitive protein genes of *Trypanosoma brucei. Mol. Cell. Biol.*, **10,** 3036.
29. Rudenko, G., Le Blancq, S., Smith, J., Lee, M. G. S., Rattray, A., and Van der Ploeg, L. H. T. (1990) Procyclic acidic repetitive (PARP) genes located in an unusually small α-amanitin-resistant transcription unit: PARP promoter activity assayed by transient DNA transfection in *Trypanosoma brucei. Mol. Cell. Biol.*, **10,** 3492.
30. Koenig, E., Delius, H. Carrington, M., Williams, R. O., and Roditi, I. (1989) Duplication and transcription of procyclin genes in *Trypanosoma brucei. Nucleic. Acids Res.*, **17,** 8727.
31. Pays, E. (1993) Genome organization and control of gene expression in trypanosomatids. In *The eukaryotic genome - organisation and regulation*(ed P. M. A.. Broda, S. G. Oliver, and P. F. G Sims,), p. 127. Cambridge University Press.
32. Gibson, W. C., Swinkels, B. S., and Borst, P. (1988) Post-transcriptional control of the differential expression of phosphoglycerate kinase genes in *Trypanosoma brucei. J. Mol. Biol.*, **201,** 315.
33. Revelard, P., Lips, S., and Pays, E. (1993) Alternative splicing within and between alleles of the ATPase gene 1 locus of *Trypanosoma brucei. Mol. Biochem. Parasitol.*, **62,** 93.
34. Cully, D. F., Ip, H. S., and Cross, G. A. M. (1985) Coordinate transcription of variant surface glycoprotein genes and an expression site associated gene family in *Trypanosoma brucei. Cell* **42,** 173.
35. Pays, E., Tebabi, P., Pays, A., Coquelet, H., Revelard, P., Salmon, D., and Steinert, M. (1989) The genes and transcripts of an antigen gene expression site from *T. brucei. Cell*, **57,** 835.

36. Chung, H. M., Lee, M. G. S., and Van der Ploeg, L. H. T. (1992) RNA polymerase I-mediated protein-coding gene expression in *Trypanosoma brucei*. *Parasitol. Today*, **8**, 414.
37. White, T. C., Rudenko, G., and Borst, P. (1986) Three small RNAs within the 10 kb trypanosome rRNA transcription unit are analogous to domain VII of other eukaryotic 28S rRNAs. *Nucleic. Acids Res.*, **14**, 9471.
38. Nakaar, V., Dare, A. O., Hong, D., Ullu, E., and Tschudi, C. (1994) Upstream tRNA genes are essential for expression of small nuclear and cytoplasmic RNA genes in trypanosomes. *Mol. Cell. Biol.*, **14**, 6736.
39. Ben Amar, M. F., Jefferies, D., Pays, A., Bakalara, N., Kendall, G., and Pays, E. (1991) The actin gene promoter of *Trypanosoma brucei*. *Nucleic. Acids Res.*, **19**, 5857.
40. Saito, R. M., Elgort, M. G., and Campbell, D. A. (1994) A conserved upstream element is essential for transcription of the *Leishmania tarentolae* mini-exon gene. *EMBO J.*, **13**, 5460.
41. Grondal, E. J. M., Evers, R., Kosubek, K., and Cornelissen, A. W. C. A. (1989) Characterization of the RNA polymerases of *Trypanosoma brucei*: trypanosomal mRNAs are composed of transcripts derived from both RNA polymerase II and III. *EMBO J.*, **8**, 3383.
42. Wong, A. K. C., Curotto de Lafaille, M. A., and Wirth, D. F. (1994) Identification of a cis-acting gene regulatory element from the lemdr1 locus of *Leishmania enriettii*. *J. Biol. Chem.*, **269**, 26497.
43. Zomerdijk, J. C. B. M., Ouellette, M., ten Asbroek, A. L. M. A., Kieft, R., Bommer, A. M. M., Clayton, C. E., and Borst, P. (1990) The promoter for a variant surface glycoprotein gene expression site in *Trypanosoma brucei*. *EMBO J.*, **9**, 2791.
44. Pays, E., Coquelet, H., Tebabi, P., Pays, A., Jefferies, D., Steinert, M., Koenig, E., Williams, R. O., and Roditi, I. (1990) *Trypanosoma brucei*: constitutive activity of the VSG and procyclin gene promoters. *EMBO J.*, **9**, 3145.
45. Sherman, D. R., Janz, L., Hug, M., and Clayton, C. (1991) Anatomy of the PARP promoter of *Trypanosoma brucei*. *EMBO J.*, **10**, 3379.
46. Brown, S. D., Huang, J., and Van der Ploeg, L. H. T. (1992) The promoter for the procyclic acidic repetitive protein (PARP) genes of *Trypanosoma brucei* shares features with RNA polymerase I promoters. *Mol. Cell. Biol.*, **12**, 2644.
47. Jefferies, D., Tebabi, P., and Pays, E. (1991) Transient activity assays of the *Trypanosoma brucei* VSG gene promoter: control of gene expression at the post-transcriptional level. *Mol. Cell. Biol.*, **11**, 338.
48. Zomerdijk, J. C. B. M., Kieft, R., Shiels, P. G., and Borst, P. (1991) Alpha-amanitin resistant transcription units in trypanosomes: a comparison of promoter sequences for a VSG gene expression site and for the ribosomal RNA genes. *Nucleic. Acids Res.*, **19**, 5153.
49a. Janz, L. and Clayton, C. (1994) The PARP and rRNA promoters of *Trypanosoma brucei* are composed of dissimilar sequence elements that are functionally interchangeable. *Mol. Cell. Biol.*, **14**, 5804.
49b. Vanhamme, L., Pays, A., Tebabi, P., Alexandre, S., and Pays, E. (1995) Specific binding of proteins to the non-coding strand of a crucial element of the variant surface glycoprotein, procyclin and ribosomal promoters of *Trypanosoma brucei*. *Mol. Cell. Biol.*, **15**, in press.
50. Brown, S. D. and Van der Ploeg, L. H. T. (1994) Single-stranded DNA-protein binding in the procyclic acidic repetitive protein (PARP) promoter of *Trypanosoma brucei*. *Mol. Biochem. Parasitol.*, **56**, 109.
51. Clayton, C. (1992) Developmental regulation of gene expression in *Trypanosoma brucei*. *Progr. Nucleic. Acids. Res. Mol. Biol.*, **43**, 37.

52. Ehlers, B., Czichos, J., and Overath, P. (1987) RNA turnover in *Trypanosoma brucei*. *Mol. Cell. Biol.*, **7,** 1242.
53. Roditi, I., Schwarz, H., Pearson, T. W., Beecroft, R. P., Liu, M. K., Richardson, J. P., Bühring, H. J., Pleiss, J., Bülow, R., Williams, R. O., and Overath, P. (1989) Procyclin gene expression and loss of the variant surface glycoprotein during differentiation of *Trypanosoma brucei*. *J. Cell Biol.*, **108,** 737.
54. Pays, E., Hanocq-Quertier, J., Hanocq, F., Van Assel, S., Nolan, D., and Rolin, S. (1993) Abrupt RNA changes precede the first cell division during the differentiation of *Trypanosoma brucei* bloodstream forms into procyclic forms *in vitro*. *Mol. Biochem. Parasitol.*, **61,** 107.
55. Pays, E., Coquelet, H., Pays, A., Tebabi, P., and Steinert, M. (1989) *Trypanosoma brucei*: posttranscriptional control of the variable surface glycoprotein gene expression site. *Mol. Cell. Biol.*, **9,** 4018.
56. Rudenko, G., Blundell, P. A., Taylor, M. C., Kieft, R., and Borst, P. (1994) VSG gene expression site control in insect form *Trypanosoma brucei*. *EMBO J.*, **13,** 5470.
57. Jefferies, D., Tebabi, P., Le Ray, D., and Pays, E. (1993) The *ble* resistance gene as a new selectable marker for *Trypanosoma brucei*: fly transmission of stable procyclic transformants to produce antibiotic resistant bloodstream forms. *Nucleic. Acids Res.*, **21,** 191.
58. Sandell, L. and Zakian, V. (1992) Telomeric effect in yeast. *Trends Cell Biol.*, **2,** 10.
59. Borst, P., Gommers-Ampt, J. H., Ligtenberg, M. J. L., Rudenko, G., Kieft, R., Taylor, M. C., Blundell, P. A., and Van Leeuwen, F. (1993) Control of antigenic variation in African trypanosomes. *Cold Spring Harbor Symp. Quant. Biol.*, **58,** 105.
60. Bernards, A., Van Harten-Loosbroek, N., and Borst, P. (1984) Modification of telomeric DNA in *Trypanosoma brucei*, a role in antigenic variation? *Nucleic. Acids Res.*, **12,** 4153.
61. Pays, E., Delauw, M. F., Laurent, M., and Steinert, M. (1984) Possible DNA modification in GC dinucleotides of *Trypanosoma brucei* telomeric sequences; relationship with antigen gene transcription. *Nucleic. Acids Res.*, **12,** 5235.
62. Gommers-Ampt, J. H., Van Leeuwen, F., de Beer, A. L. J., Vliegenthart, J. F. G., Dizdaroglu, M., Kowalak, J. A., Crain, P.F., and Borst, P. (1993) ß-D-glucosyl-hydroxymethyluracil: a novel modified base present in the DNA of the parasitic protozoan *Trypanosoma brucei*. *Cell*, **75,** 1129.
63. Schlimme, W., Burri, M., Bender, K., Betschart, B., and Hecker, H. (1993) *Trypanosoma brucei brucei*: differences in the nuclear chromatin of bloodstream forms and procyclic culture forms. *Parasitology*, **107,** 237.
64. Kooter, J. M., Van der Spek, H. J., Wagter, R., d'Oliveira, C. E., Van der Hoeven, F., Johnson, P. J., and Borst, P. (1987) The anatomy and transcription of a telomeric expression site for variant-specific surface antigens in *Trypanosoma brucei*. *Cell*, **51,** 261.
65. Alexandre, S., Guyaux, M., Murphy, N. B., Coquelet, H., Pays, A., Steinert, M., and Pays, E. (1988) Putative genes of a variant-specific antigen gene transcription unit in *Trypanosoma brucei*. *Mol. Cell. Biol.*, **8,** 2367.
66. Dorn, P.L., Aman, R. A., and Boothroyd, J. C. (1991) Inhibition of protein synthesis results in super-induction of procyclin (PARP) RNA levels. *Mol. Biochem. Parasitol.*, **44,** 133.
67. Coquelet, H., Steinert, M., and Pays, E. (1991) Ultraviolet irradiation inhibits RNA decay and modifies ribosomal RNA processing in *T. brucei*. *Mol. Biochem. Parasitol.*, **44,** 33.
68. Bass, K. E. and Wang, C. C. (1992) Transient inhibition of protein synthesis accompanies

differentiation of *Trypanosoma brucei* from bloodstream to procyclic forms. *Mol. Biochem. Parasitol.*, **56**, 129.
69. Layden, R. E. and Eisen, H. (1988) Alternate *trans*-splicing in *Trypanosoma equiperdum*: implications for splice site selection. *Mol. Cell. Biol.*, **8**, 1352.
70. Coquelet, H., Tebabi, P., Pays, A., Steinert, M., and Pays, E. (1989) *Trypanosoma brucei*: enrichment by UV of intergenic transcripts from the variable surface glycoprotein gene expression site. *Mol. Cell. Biol.*, **9**, 4022.
71. Vassella, E., Braun, R., and Roditi, I. (1994) Control of polyadenylation and alternative splicing of transcripts from adjacent genes in a procyclin expression site: a dual role for polypyrimidine tracts in trypanosomes? *Nucleic. Acids Res.*, **22**, 1359.
72. Huang, J. and Van der Ploeg, L. H. T. (1991) Requirement of a polypyrimidine tract for trans-splicing in trypanosomes: discriminating the PARP promoter from the immediately adjacent 3' splice acceptor site. *EMBO J.*, **10**, 3877.
73. Kapotas, N. and Bellofatto, V. (1993) Differential response to RNA trans-splicing signals within the phosphoglycerate kinase gene cluster in *Trypanosoma brucei*. *Nucleic. Acids Res.*, **21**, 4067.
74. Hug, M., Hotz, H. R., Hartmann, C., and Clayton, C. (1994) Hierarchies of RNA-processing signals in a trypanosome surface antigen mRNA precursor. *Mol. Cell. Biol.*, **14**, 7428.
75. Erondu, N. E. and Donelson, J. E. (1992) Differential expression of two mRNAs from a single gene encoding an HMG1-like DNA binding protein of African trypanosomes. *Mol. Biochem. Parasitol.*, **51**, 111.
76. Pellé, R. and Murphy, N. B. (1993) Stage-specific differential polyadenylation of miniexon derived RNA in African trypanosomes. *Mol. Biochem. Parasitol.*, **59**, 277.
77. Aly, R., Argaman, M., Halman, S., and Shapira, M. (1994) A regulatory role for the 5' and 3' untranslated regions in differential expression of hsp83 in *Leishmania*. *Nucleic. Acids Res.*, **22**, 2922.
78. Hug, M., Carruthers, V. B., Hartmann, C., Sherman, D. S., Cross, G. A. M., and Clayton, C. (1993) A possible role for the 3' untranslated region in developmental regulation in *Trypanosoma brucei*. *Mol. Biochem. Parasitol.*, **61**, 87.
79. Hehl, A., Vassella, E., Braun, R., and Roditi, I. (1994) A conserved stem-loop structure in the 3' untranslated region of procyclin mRNAs regulates expression in *Trypanosoma brucei*. *Proc. Nat. Acad. Sci. USA*, **91**, 370.
80. Clayton, C. E. (1985) Structure and regulated expression of genes encoding fructose biphosphate aldolase in *Trypanosoma brucei*. *EMBO J.*, **4**, 2997.
81. Gale, M.,Jr, Carter, V., and Parsons, M. (1994) Translational control mediates the developmental regulation of the *Trypanosoma brucei* Nrk protein kinase. *J. Biol. Chem.*, **269**, 31659.
82. Torri, A. F., Bertrand, K. I., and Hajduk, S. L. (1993) Protein stability regulates the expression of cytochrome C during the developmental cycle of *Trypanosoma brucei*. *Mol. Biochem. Parasitol.*, **57**, 316.
83. Priest, J. W. and Hajduk, S. L. (1994) Developmental regulation of *Trypanosoma brucei* cytochrome C reductase during bloodstream to procyclic differentiation. *Mol. Biochem. Parasitol.*, **65**, 291.
84. Cross, G. A. M. (1990) Cellular and genetic aspects of antigenic variation in trypanosomes. *Annu. Rev. Immunol.*, **8**, 83.
85. Van der Ploeg, L. H. T., Gottesdiener, K., and Lee, M. G. S. (1992) Antigenic variation in African trypanosomes. *Trends Genet.*, **8**, 452.

86. Gottesdiener, K., Goriparthi, L., Masucci, J., and Van der Ploeg, L.H.T. (1992) A proposed mechanism for promoter-associated DNA rearrangement events at a variant surface glycoprotein gene expression site. *Mol. Cell. Biol.*, **12**, 4784.
87. Pays, E., Laurent, M., Delinte, K., Van Meirvenne, N., and Steinert, M. (1983) Differential size variations between transcriptionally active and inactive telomeres of *Trypanosoma brucei*. *Nucleic. Acids Res.*, **11**, 8137.
88. Myler, P. J., Aline, R., Scholler, J. K., and Stuart, K. D. (1988) Changes in telomere length associated with antigenic variation in *Trypanosoma brucei*. *Mol. Biochem. Parasitol.*, **29**, 243.
89. Tommerup, H., Dousmanis, A., and de Lange, T. (1994) Unusual chromatin in human telomeres. *Mol. Cell. Biol.*, **14**, 5777.
90. Baltz, T., Giroud, C., Baltz, D., Roth, A., Raibaud, A., and Eisen, H. (1986) Stable expression of two variable surface glycoproteins by cloned *Trypanosoma equiperdum*. *Nature*, **319**, 602.
91. Esser, K. M. and Schoenbechler, M. (1985) Expression of two variant surface glycoproteins on individual African trypanosomes during antigen switching. *Science*, **229**, 190.
92. Shea, C., Lee, M. G. S., and Van der Ploeg, L. H. T. (1987) VSG gene 118 is transcribed from a co-transposed pol-I like promoter. *Cell*, **50**, 603.
93. Borst, P. (1991) Transferrin receptor, antigenic variation and the prospect of a trypanosome vaccine. *Trends Genet.*, **7**, 307.
94. Pays, E. (1985) Gene conversion in trypanosome antigenic variation. *Progr. Nucleic. Acids Res. Mol. Biol.*, **32**, 1.
95. Thon, G., Baltz, T., Giroud, C., and Eisen, H. (1990) Trypanosome variable surface glycoproteins: composite genes and order of expression. *Genes Dev.*, **9**, 1374.
96. Pays, E., Houard, S., Pays, A., Van Assel, S., Dupont, F., Aerts, D., Huet-Duvillier, G., Gomes, V., Richet, C., Degand, P., Van Meirvenne, N., and Steinert, M. (1985) *Trypanosoma brucei*: the extent of conversion in antigen genes may be related to the DNA coding specificity. *Cell*, **42**, 821.
97. Pays, E. (1989) Pseudogenes, chimaeric genes and the timing of antigen variation in African trypanosomes. *Trends Genet.*, **5**, 389.
98. Kamper, S. M. and Barbet, A. F. (1992) Surface epitope variation via mosaic gene formation is potential key to long-term survival of *Trypanosoma brucei*. *Mol. Biochem. Parasitol.*, **53**, 33.
99. Barbet, A. F. and Kamper, S. M. (1993) The importance of mosaïc genes to trypanosome survival. *Parasitol. Today*, **9**, 63.
100. Van der Werf, A., Van Assel, S., Aerts, D., Steinert, M., and Pays, E. (1990) Telomere interactions may condition the programming of antigen expression in *Trypanosoma brucei*. *EMBO J.*, **9**, 1035.
101. Agur, Z., Abiri, D., and Van der Ploeg, L. H. T. (1989) Ordered appearance of antigenic variants of African trypanosomes explained in a mathematical model based on a stochastic switch process and immune-selection against putative switch intermediates. *Proc. Natl. Acad. Sci. USA*, **86**, 9626.
102. Baltz, T., Giroud, C., Bringaud, F., Eisen, H., Jacquemot, C., and Roth, C.W. (1991) Exposed epitopes in a *Trypanosoma equiperdum* variant surface glycoprotein altered by point mutations. *EMBO J.*, **10**, 1653.
103. Lu, Y., Hall, T., Gay, L.S., and Donelson, J. E. (1993) Point mutations are associated with a gene duplication leading to the bloodstream reexpression of a trypanosome metacyclic VSG. *Cell*, **72**, 397.

104. Lu, Y., Alarcon, C. M., Hall, T., Reddy, L. V., and Donelson, J. E. (1994) A strand bias occurs in point mutations associated with variant surface glycoprotein gene conversion in *Trypanosoma rhodesiense*. *Mol. Cell. Biol.*, **14,** 3971.
105. Pays, E. (1986) Variability of antigen genes in African trypanosomes. *Trends Genet.*, **2,** 21.
106. Bringaud, F. and Baltz, T. (1993) Differential regulation of two distinct families of glucose transporter genes in *Trypanosoma brucei*. *Mol. Cell. Biol.*, **13,** 1146.

7 | *Trans*-splicing in trypanosomatid protozoa

ELISABETTA ULLU, CHRISTIAN TSCHUDI and ARTHUR GÜNZL

1. Introduction

Trans-splicing is an intermolecular RNA processing reaction in which exons from two separate RNA molecules are joined together to form a mature species (Fig. 1). In one form of *trans*-splicing, the partners of this reaction are the pre-mRNA and a small RNA, the spliced leader (SL) RNA. The SL sequence, which represents the 5′ end of the SL RNA, is joined to the 3′ splice site located just at the boundary of the mature mRNA within the polycistronic pre-mRNA. The first evidence for this type of *trans*-splicing (or SL addition) was obtained in trypanosomatid protozoa in 1982 (1), when it was noted that an identical 39-nucleotide sequence is present at the very 5′ end of transcripts coding for different variant surface glycoproteins in *Trypanosoma brucei*. This sequence, the SL or mini-exon, was subsequently found at the 5′ terminus of every mRNA of trypanosomatids. Critical evidence in support of the *trans*-splicing model was the detection of various predicted splicing intermediates in *T. brucei* cells (2–4). It is now clear that *trans*-splicing and *cis*-splicing (removal of intervening sequences from pre-mRNA) are fundamentally similar processes. More recently, the list of organisms endowed with *trans*-splicing has grown to include nematodes (5) as well as trematodes (6, 7) and euglenoids (8).

2. Mechanism of *trans*-splicing

Like *cis*-splicing, *trans*-splicing can be described as a two-step reaction (for review, see Ref. 9; Fig. 1). First, the SL RNA is cleaved at the 5′ splice site to give the free 39-nucleotide SL and the newly generated 5′ end of the intron is joined via a 2′-5′ phosphodiester bond to the branch site nucleotide located upstream of the 3′ splice site in the pre-mRNA. Because this is an intermolecular reaction, this branched intermediate is Y shaped, and is analogous to the lariat structure identified in *cis*-splicing. In the second step, the two exons are ligated creating the *trans*-spliced product, and the Y intron is released, and subsequently debranched by a specific enzyme (10). It has clearly been documented that *trans*-splicing is indeed catalysed by the mechanism outlined in Fig. 1 in nematodes (11) and strong circumstantial evidence suggests this mechanism occurs in trypanosomatids as well. This evidence includes the

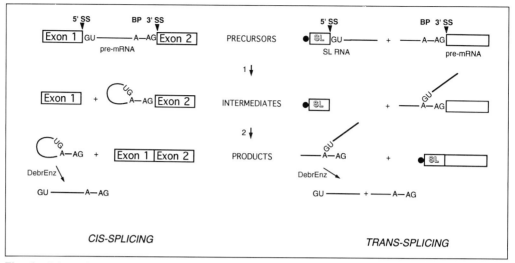

Fig. 1 Schematic representation of pre-mRNA splicing pathways. Exons are shown as boxes. Introns, which begin with the conserved GU dinucleotide at the 5′ splice site (5′ SS) and end at the conserved AG dinucleotide at the 3′ splice site (3′ SS), are indicated by solid lines. The exact 5′ end of the pre-mRNA in the trans-splicing pathway is not known. BP, branch point. DebrEnz, debranching enzyme. The filled circle represents the cap 4 structure of the SL RNA.

presence of consensus sequences for 5′ and 3′ splice sites conforming to the GU/AG rule at the appropriate positions in the SL RNA and in the pre-mRNA, respectively. Even more compelling has been the identification in *T. brucei* cells of intermediates predicted from the *trans*-splicing pathway (2–4), namely Y-branched molecules consisting of the SL intron and pre-mRNA, and the linear form of the SL intron.

2.1. *Trans*-spliceosomal U-snRNPs

The postulation of the *trans*-splicing pathway in *T. brucei* immediately raised two pertinent questions: how do the SL RNA and the pre-mRNA recognize each other and how are they held in place during *trans*-splicing? The first clue came from the demonstration that these organisms have small nuclear ribonucleoprotein particles (snRNPs) containing homologues of the U2, U4, and U6 small nuclear RNAs (U-snRNAs) and that these trypanosome U-snRNPs are essential for *trans*-splicing (12). It is established that five U-snRNPs (U1, U2, U4, U5, and U6) play a key role in *cis*-splicing and that they assemble with the pre-mRNA and a number of protein factors to form the spliceosome (for review, see Ref. 9). Given the high degree of sequence conservation that exists between trypanosome snRNAs and their vertebrate counterparts (13), it is presumed that their basic functions are similar in *cis*- and *trans*-splicing. During *cis*-splicing, these snRNAs engage in a complex series of dynamic RNA–RNA interactions, including snRNA–pre-mRNA and snRNA–snRNA interactions (for a detailed description, see Ref. 14). In *cis*-splicing, both U1 and U5 snRNAs participate in the early identification of the intron 5′ splice site.

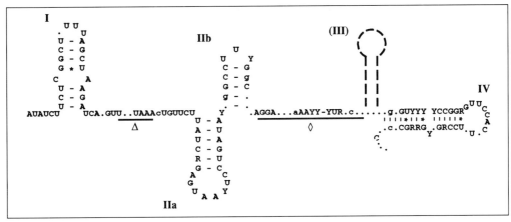

Fig. 2 Trypanosomatid U2 snRNA consensus sequence and secondary structure. The consensus sequence was derived by aligning the U2 snRNA sequences from *T. brucei gambiense* (19), *T. brucei brucei* strains 427 (24) and IsTaT 1.1 (15), *T. cruzi* (27), *Leptomonas collosoma* (27), *T. congolense* (23), *Leishmania mexicana amazonensis* (23), and *L. enrietti* (52). Capital letters represent invariant positions, small-case letters denote positions with one mismatch, dots mark variable positions. Y, pyrimidine residues; and R, purine residues. The stem–loop structures are numbered according to Ares and Igel (26). The dashed stem–loop III is not present in trypanosomes. The branch point recognition-analogous region (Δ) and the Sm binding-analogous site (◊) are underlined.

Since neither U1 nor U5 snRNA homologues have been found in trypanosomes (15), how is the 5′ splice site recognized? There is some circumstantial evidence that a trypanosome U5 snRNA homologue exists, but its function has not yet been tested directly (16). It has also been speculated that the SL RNA itself is endowed with U1- and U5-like functions (17, 18), thus accounting for the apparent absence of these components. Alternatively, it is possible that a different 5′ splice site identification mechanism is operating in trypanosomes, since only a single SL exon exists.

Like their eukaryotic counterparts, U-snRNAs of trypanosomes are complexed with proteins and, with the exception of U6 snRNA, possess a trimethylguanosine cap which is one of the hallmarks of snRNAs (15, 19, 20). In evolutionary diverse organisms such as vertebrates, yeast, and nematodes, U-snRNPs contain a group of core proteins that can be immunoprecipitated by anti-Sm antiserum (21). What sets the *T. brucei* U-snRNPs apart from all the other eukaryotic snRNPs is their lack of cross-reactivity with anti-Sm antibodies (22), as well as their lack of a conserved sequence motif resembling the eukaryotic Sm-binding site $RAU_{n\geq3}GR$ (15, 23). Nevertheless, analysis of the polypeptide composition of *T. brucei* U2 and U4/U6 snRNPs has shown that each snRNP contains a set of five common proteins and a few polypeptides specific for each RNP (22). Interestingly, the common proteins are also part of the SL RNP (22). Binding of the common proteins to trypanosome U2 and U4 snRNA has been shown to be restricted to the 3′ half of the molecule and to be independent of the region analogous to the Sm binding site (Ref. 24 and Fig. 2). Although at present no protein sequence is available, it is likely that the *T. brucei* common proteins represent Sm-homologues. This notion is strengthened by the finding that the U2-specific 40kDa protein, which is the only trypanosomal snRNP protein which has been characterized by gene cloning, shows homology to the human U2A' protein (25).

The sequence of U2 snRNA genes has been determined for eight different trypanosomatids and the consensus sequence is shown in Fig. 2. The trypanosomatid secondary structure conforms to the prototypical eukaryotic U2 snRNA structure (26, 27). The sequences from stem–loop I to stem–loop IIb are highly conserved not only among different trypanosomatids, but also among all U2 snRNAs analysed to date (13). However, stem–loop III is absent in trypanosomatids (26, 27). Notably, the branch point recognition region, which is invariant in metazoan and yeast U2 snRNAs (9), is not conserved in trypanosomatid U2 snRNAs (23, 27). Since this sequence has been shown to interact by hydrogen bonding with complementary sequences of the branch site (9), it is possible that, in trypanosomes, branch-point selection proceeds by a different mechanism.

For U6 snRNA, sequences have been obtained from three trypanosomatid species (15, 20, 28). These sequences are almost identical, indicating that U6 snRNA is the most conserved U-snRNA in these parasites. A comparison of U6 snRNA sequences of *T. brucei* and higher eukaryotes reveals that the central domain is strongly conserved whereas the 5' and 3' ends show limited similarity (29).

Thus far, the only trypanosomatid U4 snRNA sequence is from *T. brucei* (15). Association of the U4 and U6 snRNAs (as predicted from other systems) has been confirmed by various methods including co-purification of U6 snRNA during affinity purification of U4 snRNA (22) and *in vivo* UV cross-linking (30).

Recently, *in vivo* psoralen cross-linking of *T. brucei* SL RNA led to the identification of a new snRNA, called spliced leader-associated RNA (16). It has been suggested that this may represent the U5 snRNA homologue, but a definite identification and a possible role in *trans*-splicing remains to be determined.

2.2 U-snRNA interactions in *trans*-splicing

Little is known about the interaction of U-snRNAs and the splicing substrates during trypanosomatid *trans*-splicing. UV treatment of *T. brucei* cells cross-links the 5' terminal domain of U2 snRNA to terminal sequences of U6 snRNA (30). The same or a similar interaction, termed U2–U6 helix II, has been shown to be required for *cis*-splicing in mammalian cells (14). Interestingly, in *Ascaris lumbricoides*, the 3' terminal U6 sequences involved in U2–U6 helix II formation can be cross-linked to the Sm binding-motif region of the SL RNA (31). Presently, it is not clear whether this interaction occurs in trypanosomes, since sequences at the 3' end of trypanosomatid U6 snRNAs show no obvious complementarity to the Sm-analogous region of their corresponding SL RNAs. Furthermore, Watkins and Agabian were unable to detect any cross-links between the SL RNA and U2, U4, or U6 snRNAs (30). Finally, interaction between U6 snRNA and SL RNA has been proposed on the basis of phylogenetic evidence (28).

3. The substrates of *trans*-splicing

3.1 The spliced leader RNA: structure and function

Spliced leader RNAs are ubiquitously found in kinetoplastid organisms (Fig. 3). The length of the SL sequence is conserved among all trypanosomatids, except for the SL RNA of *Leptomonas collosoma* which has a two nucleotide insertion at position nine. In contrast, the introns of the various SL RNAs are of variable length. Comparison of the sequences of the 15 kinetoplastid SL RNAs shown in Fig. 3 reveals several regions of invariant sequences. In particular, in the SL sequence, 24 nucleotide positions out of 39 are identical, and for approximately 15 nucleotides downstream of the 5′ splice site, there is a high degree of conservation. The 5′ splice site consensus sequence UG^GUAUGN (^ indicates the splice site) is almost invariant among all the SL RNAs analysed to date, but does not completely conform to the mammalian [AG^GU(A/G)AGU] or yeast [AG^GUAUGU] consensus (9).

Overall the SL intronic sequences appear to be much less conserved than the SL sequence itself. There is, however, a sequence motif, $RAY_{3-8}GR$ (Fig. 3), which is present in all SL RNAs towards the 3′ end of the molecule. This motif is reminiscent of the Sm-binding site of higher eukaryotic small nuclear RNAs. Indeed, the *T. cruzi*

```
              <      Spliced leader       > <         SL Intron              >
              1                           39
       .                                 .
TBR    AACUAACGCUAUUAUUAGAACAGUUUCUGUACUAUAUUG GUAUGAGAAGCUCCC----52 nt----AAUCUGG---------    140
TCO    AACUAAAGCUUUAUAUUAGAACAGUUUCUGUACUAUAUUG GUAUGAGAAGCUCCC----64 nt----AACCUGG---------    ?
TCR    AACUAACGCUAUUAUUGAUACAGUUUCUGUACUAUAUUG  GUACGCGAAGCUUCC----25 nt----AAUUUCUUUUGA----   110
TVI    AACUAAAGCUUUAUAUUAGAACAGUUUCUGUACUAUAUUG GUAUGAGAAGCUCCC----14 nt----AAUUUUGG--------   127
TRA    AACUAACGCUAUAUAUUGAUACAGUUUCUGUACUAUAUUG GUAUGCAGCGCUUCC----21 nt----AAUUUUGG---------    98
LEN    AACUAACGCUAUAUAAGUAUCAGUUUCUGUACUUUAUUG  GUAUGCGAAACCUUC----17 nt----GAUUUUGG--------    85
LME    AACUAACGCUAUAUAAGUAUCAGUUUCUGUACUUUAUUG  GUAUGCGAAACUUCC----15 nt----GAUUUUGG--------    96
CFA    AACUAACGCUAUAUAAGUAUCAGUUUCUGUACUUUAUUG  GUAUAAGAAGCUUCC----13 nt----AAUUUUGA--------    95
LCO    AACUAAAACUUUUUGAAGAACAGUUUCUGUACUUCAUUG  GUAUGUAGAGACUUC----13 nt----AAUUUUGG--------    95
              A^A
LSE    AACUAACGCUAUAUAAGUAUCAGUUUCUGUACUUUAUUG  GUAUGAGAAGCUUCC----21 nt----AAUUUUGGG-------    86
TBO    AACUUACGCUAUAAAAGUCACAGUUUCUGUACUAUAUUG  GUAUUAGAAGCUUUC---- 3 nt----GAUCUUCGG-------    95
HSA    AACUAACGCUAUAUUGUUACAGUUUCUGUACUUUAUUG   GUAUGAGAAGCUUAA----19 nt----AAUUUUGG--------    98
ES9    AACUAACGCUAUAUAAGUAUCAGUUUCUGUACUUUAUUG  GUAUGCGAAACUUCC----19 nt----GAUUUUGG--------    95
ES6    AACUAACGCUAUAUAAGUAUCAGUUUCUGUACUUUAUUG  GUAUAAGAAGCUUCC----16 nt----AAUUUUGA--------    91
BCA    AACUUACGCUAUAAAAGAUACAGUUUCUGUACUAUAUUG  GUAUGAGAAGCUUUC---- 7 nt----AAUUUUUUUGA-----   105

CONS   AACUaAcgCUaUaNaagNaaCAGUUUCUGUACUNUAUUG  GUAUgNgaagcUUcC----(N)n-----RA-(Y)n--GR----
                                          5' splice site              Sm-like motif
```

Fig. 3 Comparison of SL RNA sequences from kinetoplastid protozoans. The SL RNA sequences were derived from the corresponding gene sequences from the following organisms: TBR, *T. brucei* (75–77); TCO, *T. congolense* (78); TCR, *T. cruzi* (75); TVI, *T. vivax* (75); TRA, *T. rangeli* (79); LEN, *Leishmania enriettii* (80); LME, *L. mexicana* (51); CFA, *C. fasciculata* (81); LCO, *Leptomonas collosoma* (82); LSE, *L. seymouri* (83); TBO, *Trypanoplasma borreli* (84); HSA, *Herpetomonas samuelpessoai* (85); ES9, *Endotrypanum schaudinni* LV59 (86); ES6, *E. schaudinni* LV86 (86); BCA, *Bodo caudatus* (87). The approximate size of the SL RNA is indicated at the end of each line. Only the conserved sequences of the intron are shown. The consensus sequence (CONS) shows nucleotides that are invariant (capital letters) and highly conserved (lower-case letters). N, any nucleotide; R, purines; Y, pyrimidines.

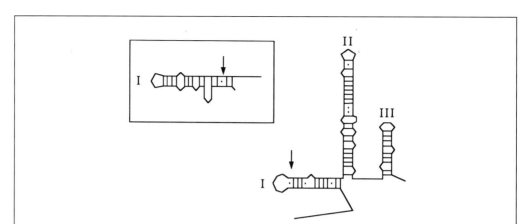

Fig. 4 A putative secondary structure of the *T. brucei* SL RNA. Two different base-pairing schemes have been proposed and these two models differ in the way stem–loop I is folded (33). Form 1 is shown in the insert. The 5' splice site is indicated by an arrow and the three stem–loop regions are numbered I–III.

and *Crithidia fasciculata* SL RNA is recognized by HeLa cell Sm proteins *in vitro* and the binding is dependent on the presence of the Sm-like motif (32). In *T. brucei*, the SL RNA is complexed with a set of core proteins some of which are shared with the U2 and U4–U6 snRNPs and are reminiscent of Sm proteins, whereas others are specific to the SL RNA particle (22). It is intriguing that the Sm-like motif is shared by all SL RNAs but is missing in the kinetoplastid U-snRNAs.

Despite sequence and length variation, SL RNAs from all known kinetoplastids (and also from nematodes) can form similar secondary structures consisting of three stem–loops (32, 33; Fig. 4). Stem–loops II and III are separated by a single-stranded region containing the Sm-like motif, which in the nematode SL RNAs is a true Sm binding site (34). For stem–loop I, which consists of the SL sequence plus a few nucleotides downstream of the 5' splice site, there are two potential folding patterns which are referred to as form 1 and form 2 (33; Fig. 4). *In vitro* studies with a synthetic *Leptomonas collosoma* SL RNA have shown that the two structures are of nearly equal stability (33). However, when the conformation of *L. collosoma* and *T. brucei* SL RNAs was probed *in vivo*, the form 2 secondary structure predominated (88), indicating that specific protein(s) are likely to be important in this interaction. These experiments also showed that the SL RNA has conformational flexibility, but it is not known whether this is a pre-requisite for *trans*-splicing activity.

In vitro, the *Ascaris* SL sequence is dispensable for *trans*-splicing activity of the SL RNA (35). The relevant sequences reside in the SL intron and it is known that sequences around the Sm binding site base-pair with U6 snRNA (31). This interaction is thought to bring the SL RNA 5' splice site to the catalytic centre of the *trans*-spliceosome. In trypanosomatids, however, there is no apparent complementarity between the corresponding regions of SL and U6 snRNAs, and cross-linking studies have so far failed to reveal such an interaction *in vivo* (16, 30). Binding of short 2'-O-methyl RNA oligonucleotides to the SL RNA in *T. brucei* permeable cells has revealed only two accessible regions whose masking inhibits use of the SL RNA in *trans*-splicing. These regions are nucleotides 7–19 in the SL sequence and nucleotides 40–43 just downstream of the 5' splice site (36). Masking the first region inhibits methylation of the SL 5' end.

3.1.1 The cap 4 of the spliced leader RNA

In contrast to nematode and trematode SL RNAs which are capped with 2,2,7,-trimethyl-guanosine (6, 38), trypanosomatid SL RNAs are capped with 7-methyl-guanosine. Interestingly, the four nucleotides adjacent to the cap are also modified with methyl groups, thus by convention this is called a cap 4 structure (39–42). In *T. brucei* and *C. fasciculata*, this structure has been recently determined to be m^7guanosine-ppp-N^6,N^6,2'-O-trimethyladenosine-p-2'-O-methyladenosine-p-2'-O-methylcytosine-p-N^3,2'-O-methyluridine (40). The N^6,N^6,2'-O-trimethyladenosine and the N^3,2'-O-methyluridine modifications are unique to trypanosomes. Remarkably, use of the SL RNA in *trans*-splicing requires proper modification of the cap structure. This conclusion was first suggested by experiments conducted in permeable *T. brucei* cells (37), where it was shown that inhibition of methyltransferases with S-adenosyl-homocysteine led to the synthesis of undermethylated SL RNA which was not used in *trans*-splicing. This result was further supported by preventing methylation of the SL cap by a different route, namely by binding a short antisense oligonucleotide to nucleotides 7–19 of the SL sequence (36), and by *in vivo* studies with the methylation inhibitor sinefungin (43). The SL-specific methyltransferases that catalyse cap modification in trypanosomes are of obvious interest as chemotherapeutic targets. Moreover, their activity might play a pivotal role in regulating the availability of *trans*-splicing 'competent' SL RNA and thus these enzymes might be major players in regulating *trans*-splicing activity as a whole.

The present evidence does not allow us to distinguish whether the modified nucleotides of the SL cap 4 structure are directly involved in the *trans*-splicing reaction or whether their effect is exerted indirectly, perhaps *via* interaction with some other cellular component, or by aiding the folding of an active SL RNA structure. Recent data indicate that the last possibility is unlikely since the secondary structures of undermethylated and fully methylated SL RNAs are the same *in vivo* (88). The SL modifications could be part of a bipartite nuclear location signal, analogous to the trimethylguanosine cap structure–Sm core domain signal for nuclear targeting of vertebrate U-snRNPs (44). However, evidence for a cytoplasmic phase of SL RNP maturation is lacking. The fast kinetics of utilization of the SL RNP in *trans*-splicing (2–4 minutes), both *in vivo* (4) and in permeable trypanosome cells (45), argue against a nuclear–cytoplasmic–nuclear circuit for SL RNP morphogenesis. Another possibility is that the cap 4 structure or a part thereof could be a recognition signal for some component of the splicing apparatus. It has been shown that the presence of a m^7guanosine cap enhances *cis*-splicing of pre-mRNAs *in vitro* (46) and in *Xenopus* oocytes (47) and that a nuclear cap binding complex plays a role in pre-mRNA recognition (48).

3.1.2 Function of the spliced leader sequence

Since the primary structure of the SL has been highly conserved during evolution of kinetoplastid protozoa (Fig. 3), one would expect this sequence to play a major role in some aspect of mRNA biogenesis, metabolism, and function. A high degree of conservation is also observed among nematode (49) and, to a lesser extent, among trematode SL sequences (7). It is clear that in *Ascaris*, the 22-nucleotide SL sequence is essential at the DNA level as part of the SL RNA gene promoter (50). In the case of the SL RNA gene of *Leishmania amazonensis*, *in vivo* studies indicate that part of the promoter is contained within the first 10 nucleotides of the SL gene portion (51), whereas similar studies in *L. seymouri* have found no evidence for promoter elements in the SL sequence (V. Bellofatto, personal communication). Although these studies need to be confirmed *in vitro*, they clearly indicate that transcription factor binding cannot solely account for the high conservation of the 39-nucleotide SL of kinetoplastids. One important possibility to consider is that different portions of the SL might be involved in different or overlapping functions and yet the SL sequence might be conserved as a whole. tRNA molecules and their genes are excellent examples of highly structured short RNA sequences in which different segments of the molecule are essential for different purposes. In the case of the SL, for instance, the nucleotides encompassing the SL cap 4 structure, namely nucleotides 1–4, are absolutely conserved among all SL RNAs analysed to date (Fig. 3), and this is most probably a reflection of the essential role of the cap 4 structure for the use of the SL RNA in *trans*-splicing (36, 37, 43). At the same time, it is conceivable that some of the same nucleotides are crucial for the selection of the transcription start site. A functional analysis of the SL sequence in *T. brucei* with the aid of short antisense RNA oligonucleotides has shown that masking nucleotides 7–19 results in complete inhibition of cap 4 biosynthesis (36). Since the masked nucleotides are located downstream from the sites of modification, it is plausible that recognition of the SL sequence by SL-specific methyltransferases requires either a specific nucleotide sequence or a specific structural motif in the SL RNA.

Finally, it should be considered that during *trans*-splicing the SL sequence might transiently interact with other small nuclear RNAs or the pre-mRNA, or both, *via* complementary sequences. For instance, a few years ago it was noted that part of the SL sequence, and in particular the invariant nucleotides 21–26 (Fig. 3) and sequences in the 5′ half of U2 snRNA (nucleotides 37–42; Fig. 2) have the potential to form base-pairing interactions (15, 19, 52). Although this interaction is phylogenetically conserved among trypanosomatids (23, 27, 52), it has not been detected by *in vivo* cross-linking studies (30).

3.2 Structure of the pre-mRNA

Transcription of protein-coding genes in trypanosomatids generates a primary transcript that is polycistronic in structure and is co-linear with the corresponding chromosomal gene arrangement. Intergenic regions are usually a few hundred

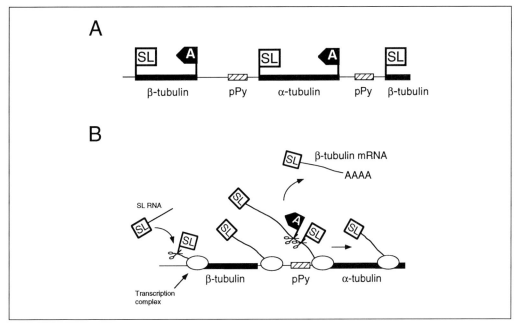

Fig. 5 A: RNA processing sites in the *T. brucei* tubulin gene cluster. Tubulin genes are organized in approximately 15 tandem repeat units of alternating α- and β-tubulin coding regions and the primary transcript from this gene cluster appears to be polycistronic. Coding regions are shown as boxes and intergenic regions as solid lines. The 3′ splice site (SL flag), the poly(A) addition site (A flag) and the polypyrimidine tract (pPy) are indicated. B: Co-transcriptional RNA processing of tubulin pre-mRNA. The model assumes that the SL RNA is not rate limiting. The line emanating from the transcription complex represents the growing pre-mRNA chain. Scissors indicate the RNA processing sites.

nucleotides long but can be as short as 90–100 nucleotides. Within their boundaries are contained the *cis*-acting signals required for correct processing of the pre-mRNA, namely the poly(A) site and the 3′ splice-acceptor site (Fig. 5A). Monocistronic mRNAs are generated by cleaving away the intergenic regions and at the same time modifying both ends of the mRNA coding region. The SL sequence is added to the 5′ end by *trans*-splicing and a poly(A) tail is polymerized at the 3′ end at the poly(A) site(s). Although no consensus sequences like the AAUAAA sequence of higher eukaryotes are present upstream of trypanosomatid poly(A) addition sites, polyadenylation always takes place before or after an adenosine residue and for accuracy, requires a downstream 3′ splice acceptor site (53–57; see Chapter 6).

The structure of many (but not all) trypanosomatid 3′ splice-acceptor sites is closely related to that of the consensus 3′ splice acceptor site of mammalian introns (53, 54, 58, 59). The building blocks are the branch site sequence (60), a polypyrimidine tract of variable length, and the invariant AG dinucleotide at the 3′ splice site. In *cis*-splicing, the length of the polypyrimidine tract, random purine insertions, and the distance of the polypyrimidine tract (pPy) from the branch point sequence have all been shown to affect splicing (9). In addition, it appears that there is a strong preference for U over C, particularly in tracts of less than nine nucleotides. In *T. brucei*, sequence analysis of 3′ splice acceptor sites (58) shows that pPy tracts upstream of the invariant AG are rather variable in length and sequence content (except that there is a bias for U over C, and Us occur most frequently between -20 and -10). A similar situation is also found in *Leishmania* (53). However, functional studies indicate that not all pPy tracts function equally (54, 58). In the case of the α-tubulin 3′ splice site (30 pyrimidines, 40% U), the length of the pPy tract can be substantially decreased without affecting *trans*-splicing (54). On the other hand, the pPy tract upstream of the procyclic acidic repetitive protein (PARP) 3′ splice site (29 pyrimidines, 72% U) is extremely sensitive to substitutions as small as six base pairs (58).

One clear function of the polypyrimidine tract upstream of the *T. brucei* α-tubulin 3′ splice site is the identification of the 3′ splice site itself (54). It has been proposed that the relative richness of pyrimidines in the 40-nucleotide region upstream from the AG dinucleotide determines the strength or frequency of usage of a 3′ splice site (58). However, there has been no systematic investigation of whether or not this is the case. Furthermore, it is not clear to what extent the sequence of the branch site, the position of the branch site with respect to the polypyrimidine tract, and the position of both of these elements with respect to the AG dinucleotide contribute to the efficiency of 3′ splice site selection.

The branch site consensus in yeast and mammals is UACUAA*C and YNYURA*C, respectively (* indicates the site of branch formation; Ref. 9). In the case of trypanosome pre-mRNA *trans*-splicing, no conserved branch site sequence exists. Only two branch sites have been characterized in *T. brucei* (60) and the sequences show imperfect similarity to the mammalian consensus and no obvious complementarity to the branch site recognition sequence of U2 snRNA.

4. Functional role and consequences of *trans*-splicing

The primary role of *trans*-splicing is to create 5′ ends for individual mRNAs derived from polycistronic pre-mRNAs (Chapter 6, Refs. 61, 62). Unlike the situation in other eukaryotes, transcription of trypanosome protein coding genes is not coupled to specific initiation at the mRNA cap site. Recent studies in which protein coding genes have been inserted into the rDNA locus demonstrate that messenger RNA synthesis in *T. brucei* can be performed by RNA polymerase I and is, therefore, not mechanistically restricted to one kind of RNA polymerase and is independent from a specific initiation site (63, 64).

Although the structure of the polycistronic pre-mRNA 5′ end is not known, it is well established that the 'cap' of the mature mRNA is SL-derived and is not further modified after *trans*-splicing (39–42). Thus, *trans*-splicing can also be considered a *trans*-capping reaction. In eukaryotes, the mRNA cap serves a protective role and several lines of evidence show that this holds true for trypanosome mRNA as well. First, inhibition of *trans*-splicing by destruction of one of the U-snRNAs in permeable *T. brucei* cells has uncovered a degradation pathway specific for unspliced tubulin pre-mRNA (12). Second, unspliced mutant pre-mRNA that does not *trans*-splice efficiently also does not accumulate *in vivo* (58, 59). A role in translation is also probable. It seems likely that the m^{7}G moiety of the SL cap is recognized by the ribosome at the time of assembly of the translation initiation complex, as is the case for the m^{7}G cap of other eukaryotic mRNAs. It remains to be determined whether the SL sequence as a whole, including its cap 4 structure, is essential for translation by trypanosome ribosomes. It is possible that in trypanosomatids, the translation apparatus co-evolved with the SL sequence, making translation initiation dependent on a specific mRNA leader sequence. Lastly, exit of mRNA from the nucleus might require the addition of the SL, this being a hallmark of the mature mRNA molecule.

5. The physiology and regulation of *trans*-splicing

Much of what we know about the physiology of *trans*-splicing in trypanosomes (Fig. 5B) has come from analysis of this reaction in detergent-permeabilized cells of *T. brucei* and, more recently, from DNA transfection studies. First, *trans*-splicing kinetics at three different 3' splice sites, namely those of the α- and β-tubulin and the actin pre-mRNAs, are very rapid and consistent with co-transcriptional addition of the SL (65); thus, pre-mRNAs are processed as soon as the 3' splice sites emerge from the transcription complex and, therefore, it is unlikely that polycistronic pre-mRNAs are synthesized *in toto*. This is also supported by the finding that cleavage at the polyadenylation site of hsp70 pre-mRNA occurs very rapidly in crude nuclear preparations (66). Consistent with a fast rate of *trans*-splicing for most pre-mRNAs is recent work that shows that inhibition of pre-mRNA synthesis results in comparable inhibition of *trans*-splicing of the SL RNA, thus indicating a lack of a substantial pool of unspliced pre-mRNA, which can enter the *trans*-splicing pathway (65). In contrast to this observation is the finding that in trypanosomatid steady-state RNAs, there are detectable amounts of dicistronic and polycistronic RNAs. Given the fast rate of *trans*-splicing (65), it is likely that these molecules represent dead-end products of RNA processing pathways.

So far, studies have not revealed any major differences in *trans*-splicing rates among the analysed pre-mRNAs. However, in the permeable cell system only abundant pre-mRNAs can be studied. The question of whether or not the rate of SL addition varies among different 3' splice acceptor sites is therefore still open. Since unspliced transcripts are degraded, regulation of the rate of SL addition could play a pivotal role in modulating the output of mature mRNA from a given gene. This could be accomplished if the affinities for components of the *trans*-splicing machinery varied among different 3' splice-acceptor sites.

Secondly, inhibition of *trans*-splicing by destruction of one of the U-snRNAs does not lead to accumulation of unspliced α-tubulin pre-mRNA because the pre-mRNA is rapidly turned over (12). The ribonuclease activity involved in this degradation pathway is specific for the pre-mRNA, because *trans*-spliced products are stable under identical conditions. In the light of the existence of a discard pathway for unspliced pre-mRNA (which has also been described in *cis*-splicing; 67), we would then postulate that it is the balance between the rate of the biosynthetic reaction (SL addition) and the rate of degradation of pre-mRNA (discard pathway) which ultimately governs the amount of mature mRNA 5′ ends derived from a given gene.

Thirdly, there is a hierarchical order of reactions in trypanosome pre-mRNA processing with *trans*-splicing preceding polyadenylation. This conclusion is based on the observations that: first, inhibition of *trans*-splicing by destruction of U2 snRNA results in inhibition of 3′ end formation and polyadenylation of both the α-tubulin pre-mRNA and the majority of trypanosome pre-mRNA (65); and, second, in permeable cells, *trans*-spliced RNAs exist which have not yet acquired a poly(A) tail, but no polyadenylated RNAs have been detected that are not *trans*-spliced (65). Recent studies using DNA transfection support the model that both in *T. brucei* (54–57) and in *Leishmania* (53), polyadenylation requires active *trans*-splicing (or minimally, recognition of the 3′ splice-acceptor site) and that these two reactions are coupled. A common polypyrimidine tract, which is part of the 3′ splice acceptor site, governs both polyadenylation upstream and *trans*-splicing downstream (54–57). The bifunctional nature of the polypyrimidine tract in trypanosome RNA processing suggests at least two possible mechanisms through which *trans*-splicing and polyadenylation might be coupled (Fig. 6). The polypyrimidine sequences could be sequentially recognized by different components of the *trans*-splicing and polyadenylation machinery. Thus, by virtue of sharing a functional element the reactions appear to be coupled, but they are not. Another possibility is that the primary role of the pyrimidine-rich sequences is to bind a factor (protein or RNP) that is shared by the polyadenylation and *trans*-splicing machineries, thus effectively coupling the two processes physically and functionally. If this were true, it would be tempting to speculate that in trypanosomes there is a unique RNA processing machinery with simultaneous *trans*-splicing and polyadenylation functions. Support for a model coupling polyadenylation and *trans*-splicing comes from the analysis of *cis*-splicing and polyadenylation at terminal exons of mammalian pre-mRNA (68, 69) and of polyadenylation and *trans*-splicing in *C. elegans* polycistronic transcription units (61). However, it is anticipated that in trypanosomes, *trans*-splicing and polyadenylation must also be possible independently of each other. This might be the case for genes that are located at the beginning and at the end of a polycistronic transcription unit. Perhaps, these genes possess some feature that mark them as 'terminal'.

Since in trypanosome cells no transcriptional regulation for housekeeping protein-coding genes has yet been uncovered (see Chapter 6), it is thought that modulation of gene expression is primarily achieved post-transcriptionally by modulating the rates of *trans*-splicing and polyadenylation, the rates of degradation of pre-mRNA and mRNA, and the rate of export of mRNA from the nucleus. It is plausible that coupling 3′ and 5′ end formation might ensure that equal amounts of different mRNAs are produced from polycistronic pre-mRNAs or might increase the rate of production of mature 5′ and 3′ ends. The understanding of how trypanosome *trans*-splicing and polyadenylation contribute to regulation and how

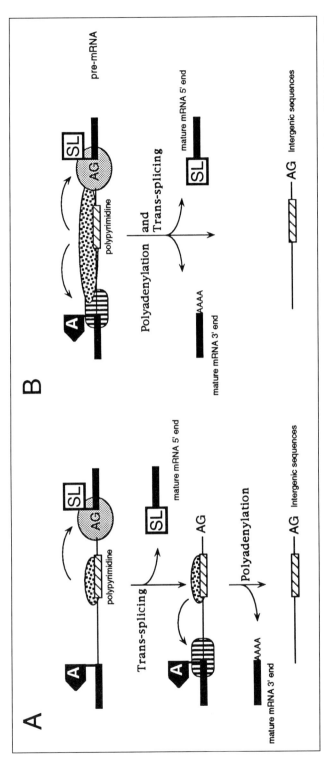

Fig. 6 Hypothetical models for the function of polypyrimidine tracts in polyadenylation and *trans*-splicing. Factors interacting with the poly(A) addition site (A flag), the polypyrimidine tract, and the 3' splice site (SL flag following the AG dinucleotide) are indicated by different shadings and are hypothetical. A: The polypyrimidine tract is recognized twice: once for *trans*-splicing and once for polyadenylation. B: The polypyrimidine tract is recognized by a factor that is shared by the *trans*-splicing and polyadenylation machineries.

these machineries are functionally linked will require a better definition of the functional elements involved in 5' and 3' end formation as well as the identification of factors which specifically interact with the polypyrimidine tract and neighbouring sequences.

6. Evolutionary considerations

The evolutionary origin of *trans*-splicing has not yet been clarified and is still a matter of speculation. Laird (70) proposed an evolutionary pathway in which snRNP-mediated *trans*- and *cis*-splicing may have originated independently from primordial *trans*- and *cis*-self-splicing systems, respectively. Others have expressed the view that the SL RNA represents a molecule in which an exon is fused to an snRNA-like sequence (32, 71) and consequently suggested that *trans*-splicing may have evolved directly from snRNP-mediated *cis*-splicing (71) or, alternatively, that the SL RNA represents an intermediate product in the evolutionary course from group II self splicing to snRNP-mediated *cis*-splicing (32). At present, given the limited number of eukaryotes in which splicing mechanisms have been studied, it is difficult to draw conclusions about the evolutionary relationship between *trans*- and *cis*-splicing. Nevertheless, it is useful to consider the following.

First, *trans*-splicing must have been present very early in eukaryotic evolution. *Trans*-splicing has been found in flagellated protozoa, namely trypanosomatids and euglenoids (8), which diverged very early from each other and together from the eukaryotic lineage (72).

Second, the similarities in mechanisms and function between trypanosome and nematode *trans*-splicing support the notion that *trans*-splicing has a monophyletic origin. Furthermore, *trans*- and *cis*-splicing proceed by similar reaction mechanisms and use a set of common U-snRNA co-factors, with presumed indistinguishable functions. Hence, both mechanisms probably have a common origin and did not evolve independently from self-splicing progenitors.

Third, spliceosomal snRNAs and snRNP-mediated *cis*-splicing has been reported in a variety of protozoa including *Dictyostelium*, *Plasmodium*, *Tetrahymena*, dinoflagellates, and amoebae (13). However, all these organisms diverged significantly later from the main lineage than trypanosomes or *Euglena* (73). Considering the phylogenetic record, it seems unlikely that *trans*-splicing originated directly from *cis*-splicing. Another possibility is that flagellate protozoa lost *cis*-splicing in the course of genome streamlining, as might have been the case for prokaryotes. In this instance, though, one would expect to find indications of snRNP-mediated *cis*-splicing somewhere among early diverged protozoa. A closer look at snRNP-mediated RNA splicing mechanisms in protozoa will be necessary to clarify the issue.

Finally, one should keep in mind that *trans*- and *cis*-splicing serve different functions. *Cis*-splicing and the corresponding intron–exon structure of genes may have strongly contributed to the evolution of new genes (74). In contrast, *trans*-splicing provides elementary functions to a cell. It could well be that *trans*-splicing was the first capping mechanism by which mRNAs were stabilized. Therefore, it is possible that in evolutionary terms, *cis*-splicing originated as a side-track to *trans*-splicing. Once new mechanisms evolved that accomplished the function of *trans*-splicing, for example co-transcriptional capping of mRNA, *trans*-splicing may have been lost in some organisms, but not in others during the course of evolution.

References

1. Boothroyd, J. C. and Cross, G. A. M. (1982) Transcripts coding for variant surface glycoproteins of *Trypanosoma brucei* have a short, identical exon at their 5' end. *Gene*, **20**, 281.
2. Murphy, W. J., Watkins, K. P., and Agabian, N. (1986) Identification of a novel Y branch structure as an intermediate in trypanosome mRNA processing: evidence for trans splicing. *Cell*, **47**, 517.
3. Sutton, R. E. and Boothroyd, J. C. (1986) Evidence for trans splicing in trypanosomes. *Cell*, **47**, 527.
4. Laird, P. W., Zomerdijk, J. C. B. M., de Korte, D., and Borst, P. (1987) *In vivo* labelling of intermediates in the discontinuous synthesis of mRNAs in *Trypanosoma brucei*. *EMBO J.*, **6**, 1055.
5. Krause, M. and Hirsh, D. (1987) A trans-spliced leader sequence on actin mRNA in C. *elegans*. *Cell*, **49**, 753.
6. Rajkovic, A., Davis, R. E., Simonsen, J. N., and Rottman, F. M. (1990) A spliced leader is present on a subset of mRNAs from the human parasite *Schistosoma mansoni*. *Proc.Natl.Acad.Sci.USA*, **87**, 8879.
7. Davis, R. E., Singh, H., Botka, C., Hardwick, C., Meanawy, M. A., and Villanueva, J. (1994) RNA trans-splicing in *Fasciola hepatica*. *J.Biol.Chem.*, **269**, 20026.
8. Tessier, L. H., Keller, M., Chan, R. L, Fournier, R., Weil, J.-H., and Imbault, P. (1991) Short leader sequences may be transferred from small RNAs to pre-mature mRNAs by *trans*-splicing in *Euglena*. *EMBO J.*, **10**, 2621.
9. Moore, M. J., Query, C. C. and Sharp, P. A. (1993) Splicing of precursors to mRNA by the spliceosome. In *The RNA world*. (ed. R. F. Gesteland and J. F. Atkins),p. 303. Cold Spring Harbor Laboratory Press.
10. Sutton, R. E. and Boothroyd, J. C. (1988) Trypanosome trans-splicing utilizes 2'-5' branches and corresponding debranching activity. *EMBO J.*, **7**, 1431.
11. Hannon, G. J., Maroney, P. A., Denker, J. A., and Nilsen, T. W. (1990) Trans splicing of nematode pre-mRNA *in vitro*. *Cell*, **61**, 1247.
12. Tschudi, C. and Ullu, E. (1990) Destruction of U2, U4, or U6 small nuclear RNA blocks *trans* splicing in trypanosome cells. *Cell*, **61**, 459.
13. Reddy, R. and Busch, H. (1988) Small nuclear RNAs: RNA sequences, structure, and modifications. In *Structure and function of major and minor small nuclear ribonucleoprotein particles*. (ed. M. L. Birnstiel),p. 1. Springer-Verlag, Berlin.
14. Nilsen, T.W. (1994) RNA–RNA interactions in the spliceosome: unraveling the ties that bind. *Cell*, **78**, 1.
15. Mottram, J., Perry, K. L., Lizardi, P. M., Luhrmann, R., Agabian, N., and Nelson, R. G. (1989) Isolation and sequence of four small nuclear U RNA genes of *Trypanosoma brucei* subsp. *brucei*: identification of the U2, U4, and U6 RNA analogs. *Mol. Cell. Biol.*, **9**, 1212.
16. Watkins, K. P., Dungan, J. M., and Agabian, N. (1994) Identification of a small RNA that interacts with the 5' splice site of the *Trypanosoma brucei* spliced leader RNA *in vivo*. *Cell*, **76**, 171.

17. Bruzik, J. P. and Steitz, J. A. (1990) Spliced leader RNA sequences can substitute for the essential 5' end of U1 RNA during splicing in a mammalian in vitro system. *Cell*, **62**, 889.
18. Steitz, J. A. (1992) Splicing takes a holiday. *Science*, **257**, 888.
19. Tschudi, C., Richards, F. F., and Ullu, E. (1986) The U2 RNA analogue of *Trypanosoma brucei gambiense*: implications for a splicing mechanism in trypanosomes. *Nucleic Acids Res.*, **14**, 8893.
20. Tschudi, C., Krainer, A. R., and Ullu, E. (1988) The U6 small nuclear RNA from *Trypanosoma brucei*. *Nucleic Acids Res.*, **16**, 11375.
21. Lührmann, R. (1988) snRNP proteins. In *Structure and function of major and minor small nuclear ribonucleoprotein particles*. (ed. M. L. Birnstiel), p. 71. Springer-Verlag, Berlin.
22. Palfi, Z., Günzl, A., Cross, M., and Bindereif, A. (1991) Affinity purification of *Trypanosoma brucei* small nuclear ribonucleoproteins reveals common and specific protein components. *Proc. Natl. Acad. Sci. USA*, **88**, 9097.
23. Tschudi, C., Williams, S. P., and Ullu, E. (1990) Conserved sequences in the U2 snRNA-encoding genes of Kinetoplastida do not include the putative branchpoint recognition region. *Gene*, **91**, 71.
24. Günzl, A., Cross, M., and Bindereif, A. (1991) Domain structure of U2 and U4/U6 small nuclear ribonucleoprotein particles from *Trypanosoma brucei*: identification of trans-spliceosomal specific RNA–protein interactions. *Mol. Cell. Biol.*, **12**, 468.
25. Cross, M., Wieland, B., Palfi, Z., Günzl, A., Röthlisberger, U., Lahm, H.-W., and Bindereif, A. (1993) The trans-spliceosomal U2 snRNP protein 40K of *Trypanosoma brucei*: cloning and analysis of functional domains reveals homology to a mammalian snRNP protein. *EMBO J.*, **12**, 1239.
26. Ares, M., Jr and Igel, A. H. (1990) Lethal and temperature-sensitive mutations and their suppressors identify an essential structural element in U2 small nuclear RNA. *Genes Dev.*, **4**, 2132.
27. Hartshorne, T. and Agabian, N. (1990) A new U2 RNA secondary structure provided by phylogenetic analysis of trypanosomatid U2 RNAs. *Genes Dev.*, **4**, 2121.
28. Xu, G.-L., Wieland, B., and Bindereif, A. (1994) Trans-spliceosomal U6 RNAs of *Crithidia fasciculata* and *Leptomonas seymouri*: deviation from the conserved ACAGAG sequence and potential base pairing with spliced leader RNA. *Mol. Cell. Biol.*, **14**, 4565.
29. Guthrie, C. and Patterson, B. (1988) Spliceosomal snRNAs. *Annu. Rev. Genet.*, **22**, 387.
30. Watkins, K. P. and Agabian, N. (1991) In vivo UV cross-linking of U snRNAs that participate in trypanosome trans-splicing. *Genes Dev.*, **5**, 1859.
31. Hannon, G. J., Maroney, P. A., Yu, Y.-T., Hannon, G. E., and Nilsen, T. W. (1992) Interaction of U6 snRNA with a sequence required for function of the nematode SL RNA in trans-splicing. *Science*, **258**, 1775.
32. Bruzik, J. P., VanDoren, K., Hirsh, D., and Steitz, J. (1988) Trans splicing involves a novel form of small nuclear ribonucleoprotein particles. *Nature*, **335**, 559.
33. LeCuyer, K. A. and Crothers, D. M. (1993) The *Leptomonas collosoma* spliced leader RNA can switch between two alternate structural forms. *Biochemistry*, **32**, 5301.
34. Maroney, P. A., Hannon, G. J., Denker, J. A., and Nilsen, T. W. (1990) The nematode spliced leader RNA participates in *trans*-splicing as an Sm snRNP. *EMBO J.*, **9**, 3667.
35. Maroney, P. A., Hannon, G. J., Shambaugh, J. D., and Nilsen, T. W. (1991) Intramolecular base pairing between the nematode spliced leader and its 5' splice site is not essential for *trans*-splicing *in vitro*. *EMBO J.*, **10**, 3869.
36. Ullu, E. and Tschudi, C. (1993) 2'-O-methyl RNA oligonucleotides identify two func-

tional elements in the trypanosome spliced leader ribonucleoprotein particle. *J. Biol. Chem.*, **268**, 13068.
37. Ullu, E. and Tschudi, C. (1991) Trans splicing in trypanosomes requires methylation of the 5′ end of the spliced leader RNA. *Proc. Natl. Acad. Sci. USA*, **88**, 10074.
38. Van Doren, K. and Hirsh, D. (1988) Trans-spliced leader RNA exists as small nuclear ribonucleoprotein particles in *Caenorhabditis elegans*. *Nature*, **335**, 556.
39. Perry, K., Watkins, K. P., and Agabian, N. (1987) Trypanosome mRNAs have unusual 'cap 4' structure acquired by addition of a spliced leader. *Proc. Natl. Acad. Sci. USA*, **84**, 8190.
40. Bangs, J. D., Crain, P. F., Hashizume, T., McCloskey, J. A., and Boothroyd, J. C. (1992) Mass spectrometry of mRNA cap 4 from trypanosomatids reveals two novel nucleosides. *J. Biol. Chem.*, **267**, 9805.
41. Freistadt, M. S., Cross, G. A. M., Branch, A. D., and Robertson, H. D. (1987) Direct analysis of the mini-exon donor RNA of *Trypanosoma brucei*: detection of a novel cap structure also present in messenger RNA. *Nucleic Acids Res.*, **15**, 9861.
42. Freistadt, M. S., Cross, G. A. M., and Robertson, H. D. (1988) Discontinuously synthesized mRNA from *Trypanosoma brucei* contains the highly methylated 5′ cap structure, m7 GpppA*A*C(2′-O)mU*A. *J. Biol. Chem.*, **263**, 15071.
43. McNally, K. P. and Agabian, N. (1992) *Trypanosoma brucei* spliced-leader RNA methylations are required for *trans* splicing *in vivo*. *Mol. Cell. Biol.*, **12**, 4844.
44. Fischer, U., Sumpter, V., Sekine, M., Satoh, T., and Lührmann, R. (1993) Nucleo-cytoplasmic transport of U snRNPs: definition of a nuclear location signal in the Sm core domain that binds a transport receptor independently of the m_3G cap. *EMBO J.*, **12**, 573.
45. Ullu, E. and Tschudi, C. (1990) Permeable trypanosome cells as a model system for transcription and trans-splicing. *Nucleic Acids Res.*, **18**, 3319.
46. Konarska, M. M., Padgett, R. A., and Sharp, P. A. (1984) Recognition of cap structure in splicing in vitro of mRNA precursors. *Cell*, **38**, 731.
47. Inoue, K., Ohno, M., Sakamoto, H., and Shimura, Y. (1989) Effect of the cap structure on pre-mRNA splicing in *Xenopus* oocyte nuclei. *Genes Dev.*, **3**, 1472.
48. Izaurralde, E., Lewis, J., McGuigan, C., Jankowska, M., Darzynkiewicz, E., and Mattaj, I. W. (1994) A nuclear cap binding protein complex involved in pre-mRNA splicing. *Cell*, **78**, 657.
49. Bektesh, S. L., vanDoren, K. V., and Hirsh, D. (1988) Presence of the *Caenorhabditis elegans* spliced leader on different mRNAs and in different genera of nematodes. *Genes Dev.*, **2**, 1277.
50. Hannon, G. J., Maroney, P. A., Ayers, D. G., Shambaugh, J. D., and Nilsen, T. W. (1990) Transcription of a nematode *trans*-spliced leader RNA requires internal elements for both initiation and 3′ end formation. *EMBO J.*, **9**, 1915.
51. Agami, R., Aly, R., Halman, S., and Shapira, M. (1994) Functional analysis of cis-acting DNA elements required for expression of the SL RNA gene in the parasitic protozoan *Leishmania amazonensis*. *Nucleic Acids Res.*, **22**, 1959.
52. Miller, S. I. and Wirth, D. F. (1988) Trans splicing in *Leishmania enriettii* and identification of ribonucleoprotein complexes containing the spliced leader and U2 equivalent RNAs. *Mol.Cell.Biol.*, **8**, 2597.
53. Lebowitz, J. H., Smith, H. Q., Rusche, L., and Beverley, S. M. (1993) Coupling of poly(A) site selection and *trans*-splicing in *Leishmania*. *Genes Dev.*, **7**, 996.
54. Matthews, K. R., Tschudi, C., and Ullu, E. (1994) A common polypyrimidine-rich motif governs trans-splicing and polyadenylation of tubulin polycystronic pre-mRNA in trypanosomes. *Genes Dev.*, **8**, 491.

55. Hug, M., Hotz, H.-R., Hartmann, C., and Clayton, C. (1994) Hierarchies of RNA-processing signals in a trypanosome surface antigen mRNA precursor. *Mol. Cell. Biol.*, **14**, 7428.
56. Vassella, E., Braun, R., and Roditi, I. (1994) Control of polyadenylation and alternative splicing of transcripts from adjacent genes in a procyclin expression site: dual role for polypyrimidine tracts in trypanosomes? *Nucleic Acids Res.*, **22**, 1359.
57. Schürch, N., Hehl, A., Vasella, E., Braun, R., and Roditi, I. (1994) Accurate polyadenylation of procyclin mRNAs in *Trypanosoma brucei* is determined by pyrimidine-rich elements in the intergenic region. *Mol. Cell. Biol.*, **14**, 3668.
58. Huang, J. and Van der Ploeg, L. H. T. (1991) Requirement of a polypyrimidine tract for *trans*-splicing in trypanosomes: discriminating the PARP promoter from the immediately adjacent 3' splice-acceptor site. *EMBO J.*, **10**, 3877.
59. Curotto de LaFaille, M. A., Laban, A., and Wirth, D. F. (1992) Gene expression in *Leishmania*: analysis of essential 5' DNA sequences. *Proc. Natl. Acad. Sci. USA*, **89**, 2703.
60. Patzelt, E., Perry, K. L., and Agabian, N. (1989) Mapping of branch sites in *trans*-spliced pre-mRNAs of *Trypanosoma brucei*. *Mol. Cell. Biol.*, **9**, 4291.
61. Spieth, J., Brooke, G., Kuersten, S., Lea, K., and Blumenthal, T. (1993) Operons on C. elegans: polycistronic mRNA precursors are processed by *trans*-splicing of SL2 to downstream coding regions. *Cell*, **73**, 521.
62. Zorio, D. A. R., Cheng, N. N., Blumenthal, T., and Spieth, J. (1994) Operons as a common form of chromosomal organization in *C. elegans*. *Nature*, **327**, 270.
63. Zomerdijk, J. C. B. M., Kieft, R., and Borst, P. (1991) Efficient production of functional mRNA mediated by RNA polymerase I in *Trypanosoma brucei*. *Nature*, **353**, 772.
64. Rudenko, G., Chung, H.-M. M., Pham, V. P., and Van der Ploeg, L. H. T. (1991) RNA polymerase I can mediate expression of CAT and *neo* protein-coding genes in *Trypanosoma brucei*. *EMBO J.*, **10**, 3387.
65. Ullu, E., Matthews, K. R., and Tschudi, C. (1993) Temporal order of RNA-processing reactions in Trypanosomes: rapid *trans* splicing precedes polyadenylation of newly synthesized tubulin transcripts. *Mol. Cell. Biol.*, **13**, 720.
66. Huang, J. and Van der Ploeg, L. H. T. (1991) Maturation of polycistronic pre-mRNA in *Trypanosoma brucei*: analysis of trans splicing and poly(A) addition at nascent RNA transcripts from the *hsp70* locus. *Mol. Cell. Biol.*, **11**, 3180.
67. Burgess, S. M. and Guthrie, C. (1993) A mechanism to enhance mRNA splicing fidelity: the RNA-dependent ATPase Prp16 governs usage of a discard pathway for aberrant lariat intermediates. *Cell*, **73**, 1377.
68. Niwa, M. and Berget, S. M. (1991) Mutation of the AAUAAA polyadenylation signal depresses *in vitro* splicing of proximal but not distal introns. *Genes Dev.*, **5**, 2086.
69. Niwa, M., Rose, S. D., and Berget, S. M. (1990) *In vitro* polyadenylation is stimulated by the presence of an upstream intron. *Genes Dev.*, **4**, 1552.
70. Laird, P. W. (1989) Trans splicing in trypanosomes—archaism or adaptation? *Trends Genet.*, **5**, 204.
71. Sharp, P. A. (1987) Trans splicing: variation on a familiar theme? *Cell*, **50**, 147.
72. Sogin, M. L., Gunderson, J. H., Elwood, H. J., Alonso, R. A., and Peattie, D. A. (1989) Phylogenetic meaning of the kingdom concept: an unusual ribosomal RNA from *Giardia lamblia*. *Science*, **243**, 75.
73. Cavalier-Smith, T. (1993) Kingdom protozoa and its 18 phyla. *Microbiol. Rev.*, **57**, 953.
74. Sharp, P. A. (1994) Split genes and RNA splicing. *Cell*, **77**, 805.
75. De Lange, T., Berkvens, T. M., Veeman, H. J. G., Frasch, A. C. C., Barry, J. D., and Borst,

P. (1984) Comparison of the genes coding for the common 5' terminal sequence of messenger RNAs in three trypanosome species. *Nucleic Acids Res.*, **12**, 4431.
76. Campell, D. A., Thornton, D. A., and Boothroyd, J. C. (1984) Apparent discontinuous transcription of *Trypanosoma brucei* variant surface antigen genes. *Nature*, **311**, 350.
77. Dorfman, D. M. and Donelson, J. E. (1984) Characterization of the 1.35 kilobase DNA repeat unit containing the conserved 35 nucleotides at the 5'-termini of variable surface glycoprotein mRNAs in *Trypanosoma brucei*. *Nucleic Acids Res.*, **12**, 4907.
78. Cook, G. A. and Donelson, J. E. (1989) Miniexon gene repeats of *Trypanosoma (Nannomonas) congolense* have internal repeats of 190 base pairs. *Mol. Biochem. Parasitol.*, **25**, 113.
79. Aksoy, S., Shay, G. L., Villanueva, M. S., Beard, C. B., and Richards, F. F. (1992) Spliced leader RNA sequences in *Trypanosoma rangelei* are organized within the 5S rRNA-encoding genes. *Gene*, **113**, 239.
80. Miller, S. I., Landfear, S. M., and Wirth, D. F. (1986) Cloning and characterization of a *Leishmania* gene encoding a RNA spliced leader sequence. *Nucleic Acids Res.*, **14**, 7341.
81. Gabriel, A., Sisodia, S. S., and Cleveland, D. W. (1987) Evidence of discontinuos transcription in the trypanosomatid *Crithidia fasciculata*. *J. Biol. Chem.*, **262**, 16192.
82. Milhausen, M., Nelson, R. G., Sather, S., Selkirk, M., and Agabian, N. (1984) Identification of a small RNA containing the trypanosome spliced leader: a donor of shared 5' sequences of trypanosomatid mRNAs? *Cell*, **38**, 721.
83. Bellofatto, V., Cooper, R., and Cross, G. A. M. (1988) Discontinuous transcription in *Leptomonas seymouri*: presence of intact and interrupted mini-exon gene families. *Nucleic Acids Res.*, **16**, 7437.
84. Maslov, D. A., Elgort, M. G., Wong, S., Peckova, H., Lom, J., Simpson, L., and Campbell, D. A. (1993) Organization of min-exon and 5S rRNA genes in the kinetoplastid *Trypanoplasma borreli*. *Mol. Biochem. Parasitol.*, **61**, 127.
85. Aksoy, S. (1992) Spliced leader RNA and 5S rRNA genes in *Herpetomonas* spp. are genetically linked. *Nucleic Acids Res.*, **20**, 913.
86. Fernandes, O., Degrave, W., and Campbell, D. A. (1993) The mini-exon gene: a molecular marker for *Endotrypanum schaudinni*. *Parasitology*, **107**, 219.
87. Campbell, D. A. (1992) *Bodo caudatas* medRNA and 5S rRNA genes: tandem arrangements and phylogenetic analyses. *Biochem. Biophys. Res. Commun.*, **182**, 1053.
88. Harris, K. A., Crothers, D. M., and Ullu, E. (1995) *In vivo* chemical structure probing of the spliced leader RNA in *Trypanosoma brucei* and *Leptomonas collosoma* reveals a flexible structure that is independent of cap 4 methylations. *RNA*, **1**, 351.

8 | RNA editing: post-transcriptional restructuring of genetic information

STEPHEN L. HAJDUK and ROBERT S. SABATINI

1. Introduction

The formation of translatable mRNAs in the mitochondrion of kinetoplastid protozoa often requires the precise addition or deletion of uridines by RNA editing. The sites of uridine insertion and deletion are specified by small guide RNAs (gRNAs) which appear to function in directing the editing reactions. The extent of RNA editing can vary from the modest four uridines inserted into the cytochrome oxidase II mRNA, to the extensive editing seen in the cytochrome oxidase III mRNA of *Trypanosoma brucei*, in which hundreds of uridines are added and dozens are deleted.

In most cases, the precursor mRNAs and the gRNAs needed for their editing are encoded on separate mitochondrial genomes. The gRNAs are encoded by numerous minicircles while the mRNAs are encoded by the less abundant maxicircles. Thus, genetic information from separate genomes is stitched together by the process of RNA editing. In this chapter, the possible molecular basis of RNA editing will be discussed. The biochemical mechanism and the ribonucleoprotein particles (RNPs) involved in these reactions are only now being resolved. The more complex issues of biological function and evolutionary origin of editing will also be considered. We have not attempted to review the editing literature exhaustively here (that has recently been done; see Refs. 1–7) but rather present the alternative views on mechanism, origins and function of what is a truly remarkable biological phenomenon.

1.1 RNA editing defined

The term RNA editing was coined by Rob Benne to describe the phenomenon he observed in the mitochondrion of trypanosomes (8). This form of RNA editing is best described as insertion/deletion editing and occurs in all kinetoplastid parasites

Table 1 RNA Editing

Organism	Transcripts	Changes	Mechanism
Conversion			
Higher plants	Chloroplast mRNAs	C to U	Deamination
Higher plants	Mitochondrial mRNAs	C to U	Deamination
Higher plants	Mitochondrial tRNAs	C to U	Deamination
Mammals	Apolipoprotein b	C to U	Deamination
Mammals	Wilms' tumour susceptibility mRNA	U to C	Amination
Mammals	Glutamate-gated ion channels mRNA	G to I	Deamination
Marsupial	Mitochondrial tRNAs	C to U	Deamination
Insertion/deletion			
Acanthamoeba	Mitochondrial tRNAs	U to G, A A to G	Unknown
Physarum	Mitochondrial mRNAs	A, U, G, C	Unknown
Physarum	Mitochondrial rRNAs	U, C, A	Unknown
Paramyxovirus	P protein mRNA	G	RNA polymerase stuttering
Kinetoplastids	Mitochondrial mRNAs	U	gRNA directed unknown

studied to date. In all cases, editing involves uridine deletion or insertion and presumably proceeds by the same mechanism. Insertion/deletion editing has also been observed in the mitochondrion of the slime mold *Physarum polycephalum* (9–11). However, unlike kinetoplastid editing, addition of G, A, U, and C has been observed and ribosomal RNAs as well as mRNAs are edited in this organism (Table 1).

RNA editing has also been observed in transcripts from other protozoa, mammals, plants and viruses (Table 1). In these examples, editing generally involves nucleotide conversions, albeit by highly divergent mechanisms. In several cases, the mechanism of editing requires the site-specific deamination of cytidine to form uridine within the edited mRNA (12, 13). While most RNA editing thus far described is post-transcriptional, paramyxoviruses produce different transcripts for the same gene by a transcriptional stuttering mechanism that leads to site-specific nucleotide insertion or deletion (14, 15).

Though mechanistically very diverse, all forms of RNA editing have a single common trait that has allowed an encompassing definition for RNA editing to be suggested (1). Benne proposes that RNA editing is any RNA processing reaction in which the internal sequence of an RNA is changed to a normal ribonucleotide (A, U, G or C) that is different from that specified by the gene. These changes may involve deletion, insertion or nucleotide conversion reactions. The phylogenetic and mechanistic separation of insertion/deletion editing and nucleotide conversion editing supports considering each form of editing independently (16). In this chapter we will only consider mitochondrial mRNA editing in the kinetoplastid parasites.

1.2 Mechanisms of RNA processing

What is the chemical mechanism involved in kinetoplastid mRNA editing? This question can be divided into two separate but related parts. First, does editing proceed by multiple rounds of endonuclease hydrolysis of the target mRNA followed by ligation, or by single-step bond-exchange, transesterification reactions? Second, is editing catalysed by protein or RNA? Before discussing the evidence in favour of particular mechanisms for kinetoplastid editing, we will first consider the mechanisms of other, well-defined, RNA processing reactions.

Most biological catalysts (enzymes) are proteins. Enzymes function by folding to form active sites where a substrate binds and the reaction is catalysed. Protein catalysts often require specifically positioned metal ions not only to coordinate folding of the protein but within the active site for catalysis. The discovery by Cech and co-workers that RNAs could also serve as biological catalysts dramatically changed our thinking about biological reactions (17, 18). Catalytic RNAs, called 'ribozymes', have been found to function in a number of RNA processing reactions (18, 19). In addition, there is now direct evidence to support the possibility that protein synthesis is also RNA-catalysed (20).

Reactions such as group I and group II RNA self-splicing, spliceosome mediated mRNA splicing, and RNAse P processing of precursor tRNAs are all RNA-catalysed (18). Conversely, the polyadenylation of mRNAs and the splicing of yeast and mammalian tRNAs are protein-catalysed RNA processing reactions (21, 22). There are a number of surprising parallels between protein and RNA catalysed reactions. For example, both protein and RNA-catalysed splicing reactions require metal ions to co-ordinate enzyme folding and often to facilitate actual catalysis. In addition, both protein and RNA-catalysed reactions show remarkable specificity in splice site selection.

In group I self-splicing reactions, the joining of two exons and removal of the intron proceeds by a two-step transesterification mechanism (23) (Fig. 1A). In the first step, an exogenous guanosine is bound to the active site of the RNA and the 3'-OH serves as an attacking group at the 5' splice site. The intron is removed by a second transesterification reaction in which the 3'-OH of the 5' exon is the attacking nucleophile. In this reaction, the active site of the enzyme and the guanosine binding site are formed by the specific folding of the RNA. Divalent metal ions are required for group I intron self-splicing, and are likely to play a role in formation of the active site of the enzyme. Although superficially distinct, it is likely that both group II and spliceosome-mediated reactions are fundamentally similar to group I intron splicing. Both are RNA catalysed and intron removal is the result of a two-step transesterification mechanism. In mRNA splicing, the ribozyme catalyses the concerted bond exchange reactions leading to the spliced product.

It is important to keep in mind that ribozymes can also catalyse the site-specific hydrolysis of RNAs. The best studied example of this type of reaction is the cleavage of precursor tRNAs by RNAse P (24). At high ionic strength (1M salt), and in the presence of Mg^{2+} and precursor tRNA, the ribozyme positions a water molecule,

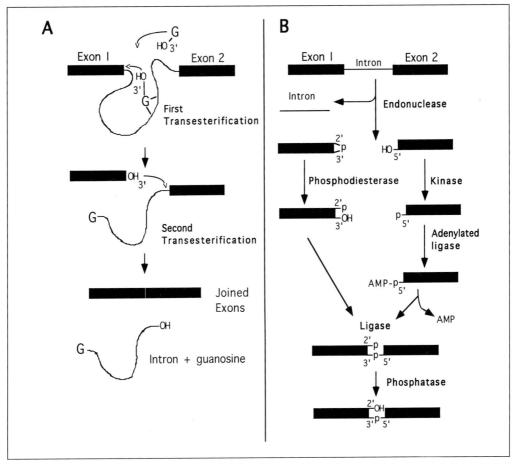

Fig. 1 Examples of RNA- and protein catalysed RNA splicing reactions. A: Group I intron splicing. The incoming guanosine binds to a site in the intron and is positioned to allow a transesterification reaction between the 3'-OH of the guanosine and the phosphodiester bond at the 5' splice site. This nucleophilic attack results in the addition of the guanosine to the intron and the release of exon I with a 3'-OH. A second transesterification joins the two exons and releases the linear intron. B: tRNA splicing. This is a protein-mediated enzymatic process which is initiated by the release of the intron by a pair of endonuclease cleavages, and ends in the joining of the two exons by RNA ligase.

whose OH will serve as the nucleophile, adjacent to the phosphodiester bond to be hydrolysed (24). Thus, like peptide enzymes, ribozymes are capable of catalysing mechanistically different reactions including transesterification and hydrolysis of RNA substrates.

In contrast, introns are removed from precursor tRNAs by a very different mechanism. tRNA splicing is protein mediated and ATP dependent. There is no evidence for the involvement of RNA as a catalyst in any step of tRNA splicing. Excision of a tRNA intron first involves the cleavage of the pre-tRNA at the intron–exon junctions by a specific endoribonuclease. The exons are joined by a multifunctional polypeptide containing polynucleotide kinase, 3′-phosphodiesterase, and RNA ligase activities (25) (Fig. 1B). It is interesting that intron removal involves two protein catalysed reactions: first, specific hydrolysis of the phosphodiester bonds at the intron–exon junctions; and, second, a transesterification reaction resulting in the joining of the exons. As will be discussed later, both protein-and RNA-catalysed hydrolysis and transesterification reactions may play a role in kinetoplastid RNA editing.

1.3 Kinetoplastid RNA editing

When first described by Benne and co-workers (8), RNA editing seemed to violate one of the fundamental elements of the 'central dogma' of molecular biology. The editing of the cytochrome oxidase subunit II mRNA resulted in changes to the coding sequence of the mRNA not directed by a DNA template. Though it is still somewhat tarnished by editing, the discovery of gRNAs and their genes saved the central dogma from complete collapse.

The discovery of RNA editing actually began with a series of straightforward studies on the mitochondrial genome of *T. brucei* (26, 27). These studies led to the identification of an apparent frameshift in the cytochrome oxidase subunit II gene (28). Further elegant and careful studies led to the startling discovery that the cytochrome oxidase subunit II mRNA was changed by the addition of four uridines resulting in the formation of a complete open reading frame for cytochrome oxidase II (8). This initial report was rapidly followed by a bewildering number of examples of mitochondrial mRNAs that differed from the corresponding genes (29, 30). In trypanosomes, the sequence differences always involved the addition or deletion of uridines and, as shown by Feagin and co-workers for cytochrome oxidase III, could result in the addition or deletion of hundreds of uridines at many sites in the mRNA (31).

The post-transcriptional nature of kinetoplastid mRNA editing has been directly established in labelling experiments with isolated mitochondria (32). The analysis of partially edited mRNAs suggests that the overall directionality of editing is 3′ to 5′ along the unedited mRNA precursor molecule (Fig. 2; 33–35). A detailed understanding of the chemistry of uridine insertion and deletion has been more elusive. A major hindrance to these studies has been the difficulty in developing *in vitro* editing systems. Recently, an *in vitro* system for the analysis of uridine deletion been described (36). This important breakthrough should allow the mechanism of uridine deletion to be studied in detail. Unfortunately, *in vitro* systems that efficiently insert uridines remain to be developed. The lack of definitive data has not hindered the generation of models to explain the mechanism of RNA editing in these organisms. Several of these models will be critically evaluated below (Section 3).

By analogy with other RNA processing reactions, the assembly of RNPs is likely to be a necessary step in RNA editing in kinetoplastids. There is now good evidence

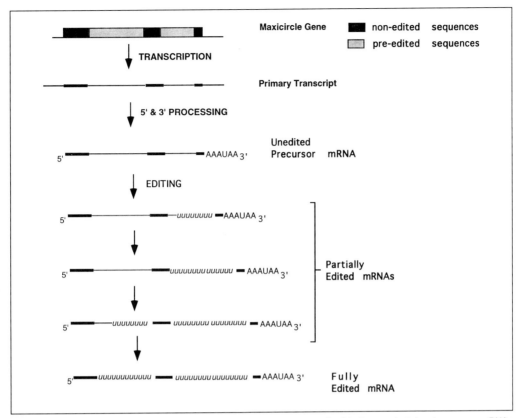

Fig. 2 A simplified view of RNA editing. Maxicircle genes are transcribed to form polycistronic precursor RNAs which are processed by polyadenylation, endonuclease cleavage to form the 5' ends, and editing. Editing is a semi-ordered process moving in an overall 3' to the 5' direction along the mRNA. (Diagram shows only addition editing; deletion occurs concurrently.)

that RNPs involved in editing are present within mitochondrial extracts (37, 38) and that these complexes can be assembled *in vitro* (39–41). In addition, several laboratories are just beginning to characterize the components of these complexes and the pathway of RNP assembly. Despite obvious differences in complexes from different kinetoplastid species, some generalizations about their components can already be made (Section 4).

The important biological questions of why trypanosomes edit mRNAs and how editing has evolved are also being investigated by several groups. While these questions are difficult to address without a detailed understanding of the biochemical mechanisms involved, some progress has been made. We know that editing can be developmentally regulated in some species of kinetoplastids (42–44). From sequence analysis, we conclude that extensive editing of transcripts is associated with the most ancient of the kinetoplastid lineages, suggesting that editing may be a very primitive process (45–47).

2. General aspects of kinetoplastid RNA editing

2.1 The mitochondrial genome of trypanosomes

The production of a functional mitochondrion requires the transcription and processing of approximately 15 mRNAs from mitochondrial genes. These transcripts are translated on mitochondrial ribosomes and produce functional proteins. In addition, several hundred proteins encoded on nuclear genes are synthesized on cytoplasmic ribosomes and post-translationally targeted to the mitochondrion (48).

The arrangement of the mitochondrial genome in trypanosomes is unique (as described in Chapter 5). All the protein-coding genes and the mitochondrial ribosomal RNA genes are found on large circular DNA molecules called maxicircles. Although there are approximately 50 maxicircles per mitochondrion, their sequences are extremely homogeneous. Surprisingly, no tRNA genes have been identified on the maxicircle and it is now apparent that most, if not all, of the mitochondrial tRNAs of trypanosomes are encoded in the nucleus and must be imported (49–51). The complete sequences of the *T. brucei* and the *Leishmania tarentolae* maxicircles are known, and a total of 11 genes for known mitochondrial proteins and the 9S and 12S rRNA genes have been identified (Fig. 3). In addition, there are several unassigned open reading frames and other transcribed regions that lack open reading frames but encode G-rich transcripts, a characteristic of precursors to extensively edited mRNAs. A comparison of the sequence of the maxicircle and corresponding cDNAs has revealed that many of the maxicircle genes lack the complete coding information needed to produce functional mRNAs. In *T. brucei*, all but five of the protein-coding genes are incomplete and encode mRNAs which are modified to varying extents by uridine insertion and deletion (Table 2). In addition to mRNAs and rRNAs, the maxicircles of *L. tarentolae* and *C. fasciculata* also encode gRNAs. A total of 13 gRNA genes have been identified on the maxicircle of one *Leishmania* strain (52). In contrast, the maxicircle of *T. brucei* does not contain any

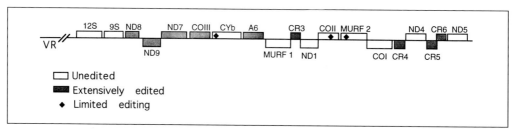

Fig. 3 Transcription maps of the maxicircle and minicircle genomes of *T. brucei*. Linear map of the 22kb maxicircle of *T. brucei*. The genes on the top of the line are transcribed from left to right, while the genes beneath the line are transcribed right to left. The ribosomal RNAs (12S, 9S) have added uridines at their 3' ends. Transcripts from cytochrome *b* (CYb), cytochrome oxidase II (COII), and MURF2 have limited amounts of internal editing (black diamonds). The transcripts from the NADH dehydrogenase subunits 7, 8, and 9 (ND7, ND8, ND9), cytochrome oxidase III (COIII), ATPase subunit 6 (A6), and G-C rich regions 3, 4, 5, 6 (CR3, CR4, CR5, CR6) genes are all extensively edited (shaded boxes). The variable region of the maxicircle is indicated (VR).

Table 2 *T. brucei* maxicircle transcripts

Gene	Number of uridines added	Number of uridines deleted	Edited size (b)	Life-stage edited	Reference
CYb	34	0	1151	Pro	2
A6	447	28	811	Pro/BS	86
COI	0	0		unedited	30
COII	4	0	663	Pro	8
COIII	547	41	969	Pro/BS	31
ND1	0	0		unedited	30
ND4	0	0		unedited	30
ND5	0	0		unedited	30
ND7	553	89	1238	5' Pro/BS; 3' BS	79
ND8	259	46	574	Pro/BS	43
ND9	345	20	649	Pro/BS	81
CR6*	132	28	325	Pro/BS	76
MURF 1	0	0		unedited	30
MURF 2	26	4	1111	Pro/BS	30
CR3	148	13	299	unknown	87
CR4	325	40	567	unknown	87
CR5	210	13	452	unknown	87
9S rRNA	7 uridines added to 3' terminus				88
12S rRNA	8–15 uridines added to 3' terminus				88

* CR6 may encode the small ribosomal protein S12
Pro, procyclic form; BS, bloodstream form

gRNA genes, although the 3' untranslated region of the cytochrome oxidase II mRNA may function as an intramolecular guide sequence (see Section 2.2).

Maxicircles are catenated together with approximately 5–10 000 minicircles to form a huge network structure called the kinetoplast DNA (kDNA). Minicircles range in size from approximately 0.8kb in *Leishmania* to 2.5kb in *Crithidia*. In *T. brucei*, minicircles are about 1kb (Fig. 4). Depending on the species, there may be up to 250 different minicircle sequence classes within a single kDNA network. The genetic function of minicircles was elucidated with the discovery of gRNA genes (53–55). This finding eliminated decades of speculation that denied minicircles a genetic role and instead proposed that they were merely structural elements, perhaps providing a scaffolding for maxicircle replication or segregation. Now it is clear that, while minicircles do not encode proteins directly, they play an important role in RNA maturation by encoding most of the gRNAs necessary for the editing of maxicircle mRNAs.

Examination of the minicircle gRNA genes of *T. brucei* revealed that each gRNA gene is flanked by an imperfect 18bp inverted repeat sequence and that transcription begins 32bp from the upstream repeat (Fig. 4; 53). This precise spacing suggested that these sequences might function as promoters for gRNAs but no experimental evidence is available to support this theory. Moreover, the lack of comparable inverted repeats in the minicircles of *L. tarentolae*, *C. fasciculata*, and

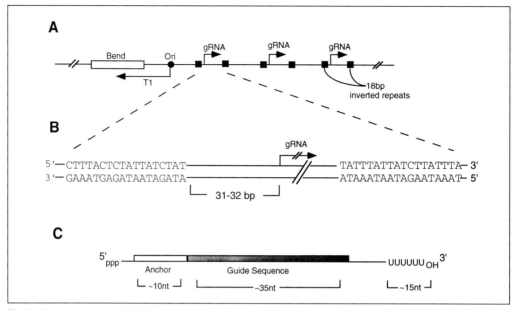

Fig. 4 Genomic organization of minicircle gRNA genes. A: Linear map of a 1kb minicircle of *T. brucei*. The bent helical region of this minicircle (open box) and the origin of replication (Ori) are within the conserved region of the minicircle. The transcript T1 is probably a primer for DNA replication. The gRNA genes (arrows) are flanked by 18bp imperfect inverted repeat sequences (dark boxes). B: A single gRNA transcription unit showing the 18bp repeats, start site for transcription and spacing between upstream repeat and transcription start site. C: General features of gRNAs. 5' triphosphate indicates that these are primary transcripts. The anchor region basepairs with the pre-edited mRNA, the guide sequence directs uridine addition or deletion, and the 3' oligo-(U) tail is added post-transcriptionally.

some *T. brucei* gRNA genes suggests that other sequences might serve as promoters in these minicircles (56, 57).

Since *T. brucei* contains at least 250 different minicircle sequence classes and each minicircle contains three gRNA genes, the minicircles could encode at least 750 different gRNAs. In the old laboratory strains of *L. tarentolae* (UC) used in most editing studies, minicircle sequence heterogeneity is apparently much lower. There are only 15 different minicircle-encoded gRNA genes, consistent with the reduced amount of editing seen in these organisms (58). However, more recently established isolates of *L. tarentolae* (LEM 125) show more extensive heterogeneity of minicircle sequences and a corresponding increase in the extent of RNA editing. For example, editing of maxicircle-encoded transcripts for components of the NADH dehydrogenase complex I is reduced in the UC strain of *L. tarentolae*. These transcripts are extensively edited in the LEM 125 strain correlating with additional minicircle- and maxicircle-encoded gRNAs (58).

The intriguing structure of the kDNA network has been described in Chapter 5. With the knowledge that minicircles do indeed have a coding function, less attention has been paid to the function of the minicircle as an element of the kDNA network. It is worth considering that this unique DNA structure might be involved with the assembly or segregation of the editing machinery within the mitochondria, reminiscent of the sequestration of spliceosomes within the nucleus of mammals.

2.2 Guiding of RNA editing

The discovery of gRNAs established an information base for the editing of mRNAs (52). With the possible exception of the internal guide sequence within the cytochrome oxidase subunit II 3' UTR, all gRNAs appear to be transcribed from their own genes and associate with pre-edited mRNAs within mitochondrial RNPs.

Each gRNA is complementary to the sequence of a portion of an edited mRNA (Fig. 5). Typically, the region of complementary is short (about 45nt) and is defined both by conventional Watson and Crick base pairing (A-U and G-C) and also by G-U base pairing. Thus, while gRNAs are complementary to edited sequences they cannot serve as conventional templates for polymerization of edited sequences.

Near the 5' terminus of all gRNAs is a sequence of approximately 10–15nt which is complementary to unedited mRNA sequences immediately 3' to the editing site (Fig. 5A). The base pairing of this 'anchor' sequence with the mRNA at the editing site is likely to be an important step in editing. This basepairing interaction could be necessary for the assembly of the editing complex or directly involved in the formation of the catalytic site for editing.

Another universal feature of gRNAs is the presence of a 5–15nt oligo-uridine tail at the 3' terminus (Fig. 5A). The oligo(U) tail is added post-transcriptionally by the terminal uridylyl transferase (TUTase) present in kinetoplastid mitochondria. The presence of an oligo(U) tail led to the proposal that gRNAs might serve as the donor of added uridines or the acceptor of deleted uridines during editing (60, 61). In the case of the intramolecular guide sequence of cytochrome oxidase II, uridines

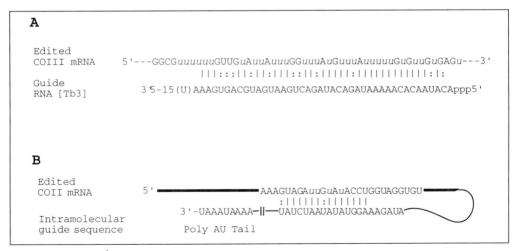

Fig. 5 Guide RNAs direct the addition and deletion of uridines. A: Sequence complementarity of a minicircle encoded gRNA and a portion of edited cytochrome oxidase III (COIII) mRNA. B: Putative intramolecular guide sequence in the 3' UTR of the cytochrome oxidase II (COII) mRNA.

might be donated by the poly A(U) tail of the mRNA (Fig. 5B) (59, 62). Direct evidence that the oligo(U) is involved in donating or accepting uridines is lacking and, as will be discussed later, there is some evidence that free UTP is the source of added uridines in *Leishmania* (Section 3).

3. Biochemical mechanisms of RNA editing

Three different pathways for RNA editing can be proposed which are consistent with recent experimental evidence. All three propose a role for the gRNA in directing editing and allow for an overall 3′ to 5′ directionality of the process. However, the models differ significantly in the proposed chemical reactions which result in internal uridine addition or deletion. These may be analogous to the transesterification events which occur in group I and II splicing or, alternatively, may involve enyzmatic cleavage and ligation in a mechanism that is more analogous to tRNA splicing.

Two of the proposed models are based on the direct involvement of the oligo(U) tail of the gRNA in addition or deletion reactions. Cech proposed the existence of a possible intermediate in editing which would be formed by the covalent joining of the 3′ end of the gRNA with the 3′ half of the mRNA at an editing site (61). Such chimeric molecules have been revealed by polymerase chain reaction analysis of mitochondrial RNA (60, 63) and are formed *in vitro* when complementary gRNAs and mRNAs are incubated with mitochondrial extract (64–66). Two potential pathways for RNA editing involving the chimera intermediate are presented in Fig. 6.

3.1 Cleavage–ligation mechanism

The mitochondrion of kinetoplastids contains a number of enzymatic activities that may be responsible for editing. These activities are consistent with the formation of chimeric gRNA–mRNA intermediates and the resolution of these molecules to form edited mRNAs. According to this model, the sites in the mRNA to be edited contain structural information which is recognized by an editing site-specific endoribonuclease (Fig. 6A; 67–69). Cleavage of pre-edited RNA, but not edited RNA, occurs at or near the 3′-most editing site in the absence of exogenously added gRNA. This suggests that the pre-edited mRNA forms a cleavage recognition site in the absence of gRNA interaction. Consistent with this model are recent observations that pre-edited mRNAs form stable stem–loop structures at editing sites. Editing of the 3′ portion of the site leads to a potential shift in the position of the stem–loop (Fig. 6B). The cleavage of the pre-edited mRNA at the editing site may be important in facilitating correct basepairing of the cognate gRNA with the editing site of the mRNA. This basepairing is likely to be necessary for subsequent steps in this editing scheme.

Following endonuclease cleavage, the gRNA is joined to the mRNA 3′ fragment at the editing site. Formation of the chimeric gRNA–mRNA is mediated by the action of a mitochondrial RNA ligase activity (70, 71). Chimera formation is protein dependent and requires the hydrolysis of ATP at the β-γ position, as predicted if an RNA ligase were to catalyse the reaction (71). In addition, two proteins of 50 and 57kDa have been identified that are adenylated in the presence of ATP and have been shown to be the adenylated form of RNA ligase (71). Similar covalent AMP–enzyme intermediates have been identified in tRNA splicing reactions. The trypanosome RNA ligase intermediates are deadenylated during both ligase reactions and chimera formation with the corresponding release of AMP.

In uridine insertion reactions, once the chimeric molecule is formed, the oligo(U) tail of the gRNA is now basepaired with either an A or G within the gRNA immediately adjacent to the anchor (Fig. 6A, third step). The next unpaired base of the oligo(U) tail is susceptible to cleavage by an endonuclease and the 5′ portion of the mRNA rejoined by the action of RNA ligase. The product of this ligation could have one or more uridines added at the point of chimera formation. Deletion of uridines at the editing sites would follow a similar pathway. After chimera formation, unpaired uridines within the mRNA would be transferred to the oligo(U) tail of the gRNA following endonuclease cleavage.

There is now strong support for the cleavage–ligation mechanism. The presence of an editing site-specific endonuclease activity, RNA ligase, and chimeric gRNA–mRNAs is consistent with this pathway. While the involvement of endonuclease and RNA ligase in complete editing reactions has not been established, recent results indicate that the formation of the cytochrome *b* mRNA–gRNA chimera requires RNA ligase (71). Finally, the presence of the editing site specific endonuclease, ligase activities, and chimera formation all within the same mitochondrial RNP indicates that these might participate in the same reactions (see Section 4). It may be important to keep in mind that, while protein candidates for the RNA ligase have been identified (71), the endonuclease cleavage could be catalysed by either a protein or RNA enzyme.

3.2 Transesterification mechanism

Transesterification is an appealing mechanism for RNA editing (60, 61; Fig. 6B). According to this model, the first step in editing is formation of a base-paired hybrid between the cognate gRNA and the pre-edited mRNA. A gRNA–mRNA chimera is formed by the nucleophilic attack of the 3′-OH of the gRNA oligo(U) tail on the phosphodiester bond at the editing site of the mRNA. This results in a chimera identical to the one described as an intermediate in the cleavage–ligation model. In this proposed pathway, the chimera is resolved by a second transesterification reaction mediated by the 3′-OH of the 5′ portion of the mRNA. The phosphodiester bond attacked in this case would again be defined by a mismatch in the base pairing of the oligo(U) tail to the gRNA at the editing site.

If editing is the result of successive rounds of transesterification, then it is possible that editing is entirely RNA catalysed. While it is clear that proteins and ATP hydrolysis are required for chimera formation *in vitro*, it could be argued that this reflects the involvement of proteins in the correct folding of a ribozyme consisting

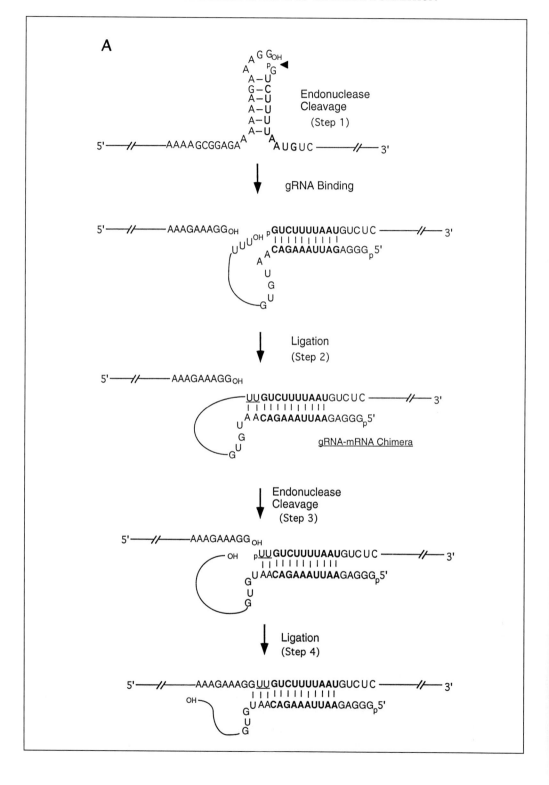

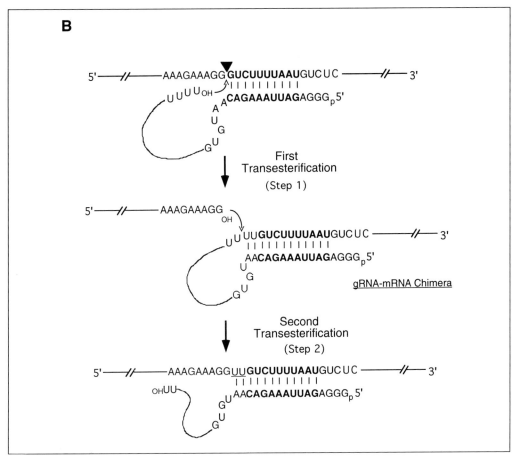

Fig. 6 Schematic drawing two possible mechanisms for kinetoplastid mRNA editing. A: Cleavage ligation model. A gRNA–mRNA chimeric molecule is formed by the sequential cleavage of the unedited mRNA at the editing site (step 1), followed by ligation of the oligo(U) tail of the gRNA to the 3' cleavage product (step 2). The chimera is resolved by a second round of cleavage (step 3), within the oligo(U) tail, and ligation of the 5' fragment of the mRNA (step 4). B: The transesterification model. The 3'-hydroxyl of the oligo(U) tail of the gRNA attacks the phosphodiester bond at the editing site resulting in the formation of a gRNA–mRNA chimeric molecule (step 1). The chimera is resolved by a second round of transesterification with the 3'-hydroxyl of the 5' cleavage product attacking within the oligo(U) tail of the gRNA (step 2). Both models propose the formation of the same intermediate chimeric gRNA–mRNA molecule and the transfer of uridines from the gRNA to the editing site of the mRNA.

of the cognate gRNA–mRNA hybrid. Attempts to reconstitute either editing or chimera formation with purified RNAs, in the absence of protein, have thus far been unsuccessful.

3.3 TUTase addition mechanism

A third but currently less favoured model involves the addition of uridine residues directly to the mRNA by the action of the mitochondrial TUTase (70). In this reaction, pre-edited mRNA is cleaved by the editing site-specific endoribonuclease and uridines are added to the 5' mRNA cleavage product by the mitochondrial TUTase. Following uridine addition, the mRNA fragments are rejoined by RNA ligase. The discovery of gRNA–mRNA chimeras has made this model less appealing. However, Simpson and co-workers have recently described an editing-like activity in mitochondrial RNPs that lack chimera-forming activity (38, 41). This suggests that the uridines added during editing could come directly from a pool of UTP rather then being donated by the oligo(U) tail of the gRNAs.

So, at least for the moment, we are left with three models for editing. Two involve a putative chimeric gRNA–mRNA intermediate while the other relies on the introduction of uridines by the combined action of endonuclease, TUTase, and ligase. The major questions rasied by these models are: first, are editing reactions catalysed by RNA or protein or both; second, does the mechanism involve cleavage–ligation or transesterification; and, finally, are the uridines added during editing derived directly from a free mitochondrial pool of UTP or from the oligo(U) tail of the gRNAs? The recent development of an *in vitro* uridine deletion system (36) suggests that the answers to these questions are not too far away.

An understanding of the mechanism of RNA editing would not only provide a basis for the pathways involved but might also provide clues as to the evolutionary origins of this unusual process. An RNA-catalysed transesterification mechanism would be viewed as a clear indication that the mechanisms of RNA editing and splicing represent related and perhaps very ancient processes (see section 5).

4. Involvement of mitochondrial RNPs in RNA editing

RNA editing in kinetoplastids is a complex affair involving at least two RNAs (the cognate gRNA and pre-edited mRNAs) and a number of associated proteins. The role of these proteins and RNAs in catalysis is unclear at this time and the nature of the interactions leading to the formation of the mitochondrial RNPs (dubbed 'editosomes') is only beginning to be elucidated.

Putative mitochondrial editing complexes were first identified in mitochondrial extracts of *T. brucei* (37). Glycerol gradient centrifugation revealed that two mitochondrial RNPs contain gRNAs: complex I at 19S; and complex II at 35–40S. Pre-edited and edited mRNAs were found associated exclusively with complex II. Several enzymatic activities which might be associated with editing (including RNA ligase, TUTase, and chimera-forming activities) co-migrated with both complex I and complex II on glycerol gradients. Whereas ligase and chimera forming activities were shared rather evenly by complexes I and II, TUTase activity was mainly found with complex I. Initial studies failed to reveal the association of the editing site-specific endonuclease in either complex I or complex II (37). However, more recently, an editing site-specific endonuclease has been found associated with complex I (B. K. Adler and S. L. Hajduk, unpublished results). The failure to initially detect this activity may have been due to the presence of competing RNA ligase activity in the same assays. Studies by Göringer and co-workers (72) are largely consistent with the presence of two classes of gRNA-associated mitochondrial RNPs in trypanosomes.

Mitochondrial extracts from *L. tarentolae* also contain two gRNA-containing RNPs which appear to be more heterogeneous than the complex I and II RNPs of *T. brucei*. The 'T' class of mitochondrial RNPs is composed of multiple RNP complexes ranging in size from 80 to 140nm (38). The 'T' complexes contain TUTase activity and superficially resemble the complex I RNPs of *T. brucei*. However, no ligase, endonuclease, or chimera-forming activity was found associated with these complexes (38). The 'G' class of mitochondrial RNPs are large particles of 170–300nm containing gRNAs and an editing-like uridine incorporation activity. The exact nature of this uridine addition activity is not known, although addition is internal and localized to editing domains of mRNAs (38).

It is tempting to speculate that the complex I of *T. brucei* and the T complexes of *L. tarentolae* might be functionally related. Despite several unresolved differences in the size and enzymatic activities associated with these complexes, it seems likely that they function in the maturation of the gRNAs by the addition of the 3' oligo(U)tail. In addition, these 'pre-assembled' complexes might play a role in the selection of the cognate pre-edited mRNAs (37). Again, despite obvious differences, it is also possible that complex II of *T. brucei* and the G complexes of *L. tarentolae* also share a common function. While the *Leishmania* complexes lack the chimera-forming activity displayed by the trypanosome complex II, they do appear capable of internal uridine addition to pre-edited mRNAs. However, further studies are needed before these RNPs can be crowned the editosome.

Specific protein–RNA interactions have also been studied by the *in vitro* assembly of gRNA–RNPs (39–41). A number of different approaches have been taken to identify mitochondrial proteins which stably associate with gRNAs. In the most detailed study, UV cross-linking was used to identify proteins that bind gRNAs *in vitro*. Proteins of 90, 21, and 9kDa bind gRNA in the presence of 100mM KCl, suggesting high affinities for gRNAs (72). An additional five proteins ranging in size from 9 to 124kDa bind to gRNAs at a lower affinity. When different gRNAs, specific for different mRNA editing sites, are incubated with mitochondrial extracts, the same eight proteins are able to bind. This suggests that there may be a very limited number of proteins that recognize and bind gRNAs. While the function of the gRNA-binding proteins is unknown, the binding of the 90kDa protein to gRNAs appears specific for the 3' oligo(U) tail (72).

A model for the involvement of mitochondrial RNPs in the editing of *T. brucei* mRNAs is schematically shown in Fig. 7. In complex I, gRNAs are associated with a set of specific proteins including mitochondrial RNA ligase, TUTase, and 90, 21, and 9kDa gRNA-binding proteins (37,72). The major function of this complex is the maturation of gRNA by the addition of a 3' oligo(U) tail. Complex I can associate with pre-edited mRNAs to form the 35–40S complex II. A pathway of RNP assembly and RNA editing can be suggested based on the composition of the *T. brucei*

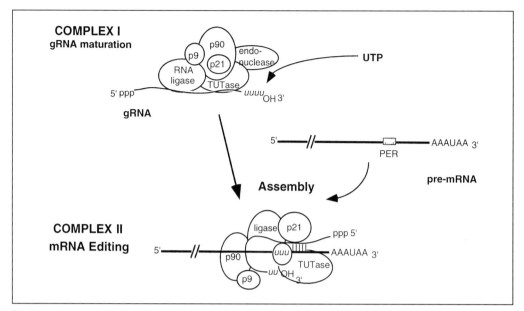

Fig. 7 Proposed model for assembly and maturation of editing complexes. gRNAs are associated with specific proteins in a 19S RNP (complex I) which is the site of uridine tailing of the gRNAs. Complex I can bind pre-mRNAs and associated proteins to form a 35–40S RNP (complex II). The assembly of complex II probably requires base pairing of the anchor region of the gRNA and mRNA immediately 3' to the pre-edited region (PER) of the mRNA. The complex may function to hold the 5' mRNA fragment in position for the second step in editing. Specific proteins of 9, 21, and 90 kDa have been identified by UV cross-linking (39), while RNA ligase, TUTase, and endonuclease activities have been assayed directly (37, 71).

complexes. Initially, UTP is incorporated into the 3' oligo(U) tail of the gRNAs in complex I. It is rapidly chased to a gRNA population in complex II and then to internal sites in the mRNA. Finally, edited mRNAs leave complex II and associate with mitochondrial ribosomes. While highly speculative, this model is supported by the presence of several shared proteins in complexes I and II, and preliminary pulse chase experiments with isolated mitochondria (V. W. Pollard, S. M. Antinucci, and S. L. Hajduk, unpublished).

5. Function and origin of RNA editing

5.1 Role of RNA editing in regulation of gene expression

A small but indispensable fraction of the mitochondrial proteins required for electron transport and ATP production are encoded by the maxicircle. Based entirely on analysis of genomic and cDNA sequences, we can predict with some certainty that RNA editing is obligatory for the formation of mitochondrial mRNAs with correct initiation and termination codons and contiguous open reading frames to encode these mitochondrial proteins. However, the direct demonstration that edited mRNAs are translated into mitochondrial proteins has not been accomplished. A universal feature of proteins encoded in the mitochondria is their extreme hydrophobicity. This has frustrated all attempts at purification and subsequent sequencing of proteins formed by the translation of edited mRNAs. An indirect approach to demonstrate proteins encoded by edited mRNAs is the use of antibodies specific for peptide epitopes specified by edited mRNA sequences. There is a single account of the successful use of this approach. Antibodies prepared against a synthetic peptide corresponding to the carboxyl terminal domain of cytochrome oxidase subunit II were used to identify a protein by western blot analysis (73). Only the edited form of the cytochrome oxidase II mRNA could be translated to form this portion of the protein since the unedited mRNA contains a termination codon which would result in a truncated protein lacking the carboxyl terminal domain. The unequivocal demonstration that edited mRNAs actually specify proteins remains an important future experiment.

Recent results indicate that RNA editing may also influence the relative abundance of mitochondrial mRNAs. Much of the maxicircle is composed of tightly packed genes in which the 5' terminus of a downstream gene and the 3' terminus of an upstream gene are separated by only a few nucleotides, or are immediately juxtaposed to one another, or actually overlap (74). Several studies suggest that the maxicircle is initially transcribed as one or several polycistronic transcripts (75). Editing of a polycistronic, precursor RNA spanning the sequence that overlaps the 3' end of S12 and the 5' end of ND5 produces a transcript with edited S12 mRNA while effectively destroying the 5' portion of the ND5 mRNA (76). Conversely, if processing of the polycistronic transcript to form the 5' terminus of the ND5 precedes editing then the S12 sequence may be truncated. Thus, the rates of editing and 5'-end formation may be important in modulating the relative amounts of individual mRNAs derived from the same transcription unit.

In a number of eukaryotic organisms, selective or alternative mRNA processing (either by splice site-selection or differential polyadenylation of mRNAs) leads to the production of different mRNAs encoded by the same gene. It is appealing to speculate that editing of trypanosome mRNAs might also allow alternative RNAs to be produced. While the production of two functionally distinct molecules has not been demonstrated at the protein level, due to the problems discussed above, the possibility clearly exists. There are a number of examples of either selective or differential editing which could produce multiple mRNAs. The edited and unedited forms of the cytochrome *b* mRNA contain in-frame methionine codons and long opening reading frames (29, 77). Editing of the cytochrome *b* mRNA forms an AUG codon in-frame with a protein coding sequence which shows significant homology to cytochrome *b* from other organisms. Translation of the unedited cytochrome *b* mRNA would produce a protein which differs at the N-terminus from the protein produced from the edited mRNA. In the case of cytochrome oxidase II, editing adds four uridines to the mRNA within the 3' half of the mRNA and alters the reading frame of mRNA 3' to the editing site. The added uridines shift a UUA termination codon out of frame and lengthen the putative cytochrome oxidase II protein product at the C terminus (8).

5.2 RNA editing is developmentally regulated

Many trypanosomatids have complicated lifecycles alternating between a vertebrate host and an insect vector (78; see Chapter 6). The African trypanosomes undergo both morphological and biochemical changes during the transition from mammalian to insect environment. The most striking of these changes involve mitochondrial ATP production (48). In the bloodstream of the mammal, mitochondrial activities are suppressed and ATP production is restricted to substrate level phosphorylation by glycolysis. Following ingestion by the insect vector, the bloodstream trypanosomes differentiate to the procyclic developmental stage in the midgut of the tsetse fly. Procyclics have a complete cytochrome-mediated electron transport system and ATP is produced by oxidative phosphorylation.

Developmental changes in mitochondrial activities correlate with changes in the editing of specific mRNA. For example, in both *T. brucei* and *Trypanosoma congolense*, the abundance of edited cytochrome *b* and cytochrome oxidase II mRNA is low in bloodstream trypanosomes and increases in the procyclic trypanosomes (44, 77). On the other hand, the amounts of edited mRNA from NADH dehydrogenase subunits 8 and 9 are low in the procyclic trypanosomes but increase in the bloodstream stages (43, 79). In the case of NADH dehydrogenase subunit 7, the developmental regulation of editing is not only transcript specific but also domain specific. In bloodstream trypanosomes, a fully edited NADH dehydrogenase subunit 7 transcript is produced, while in procyclics editing is restricted to the 5' domain (79). Differential editing of the 3' domain is predicted to yield two forms of the mRNA for NADH dehydrogenase subunit 7.

It is now clear that the editing of specific mRNAs and even domains within an mRNA can be developmentally modulated. Since the transcription of mitochondrial genes in trypanosomes is constitutive during the lifecycle (75), it is reasonable to speculate that editing may play a central role in controlling mitochondrial biogenesis during the developmental cycle of the African trypanosome. The obvious question is how do trypanosomes accomplish this feat? Perhaps the most appealing possibility is that the extent of editing of a given mRNA is modulated by the abundance of its cognate gRNAs. This possibility has now been examined for a number of gRNAs. Currently, no correlation has been established between the abundance of a gRNA and its edited mRNA (80, 81). If availability of gRNAs does not regulate editing, then what is the mechanism? Without a detailed knowledge of the mechanism of editing it may be premature to speculate. However, one is tempted to predict that developmentally regulated protein factors, specific for particular mRNAs, might either suppress or activate the editing machinery. It will be interesting to explore the factors involved in this process as we gain a better understanding of the mechanism of editing and the components of the editosome.

5.3. Evolutionary origin of kinetoplastid RNA editing

The evolutionary origin of editing is important to consider in relation to both the parasitic life style and the mechanism of editing. Phylogenetic lineages based on nuclear small subunit ribosomal RNA sequences have established that the free Living *Bodo caudatus* and the fish parasite *Trypanoplasma borreli*, both of the family Bodonidae, are the earliest diverging kinetoplastids thus far studied (82). In the family *Trypanosomatidae*, the African and South American trypanosomes appear to have diverged first while *Crithidia* and *Leishmania* diverged more recently (83). The establishment of these phylogenetic lineages allows us to consider whether editing is an ancient trait maintained by the kinetoplastids or a trait acquired recently, perhaps as an adaptation to parasitism. A comparison of the editing patterns of MURF4, NADH dehydrogenase subunit 7, and cytochrome oxidase subunit III has revealed that extensive editing of these transcripts occurs in the more divergent *Trypanosoma* species while limited 5' terminal editing takes place in *Crithidia* and *Leishmania* (83). These observation led Simpson and co-workers to conclude that editing was a primitive process. Support for this view has been obtained from studies on the editing of mitochondrial transcripts from the more ancient *T. borreli* (84, 85). These studies revealed that cytochrome *b*, rRNA protein S12, and cytochome oxidase subunit I mRNAs are extensively edited. For example, the cytochrome *b* mRNA, which is only modestly edited in *T. brucei* with 35 uridines inserted at the 5' terminus, is modified by the addition of 186 uridines and the deletion of 42 uridines throughout the mRNA in *T. borreli*. Similarly dramatic is the case of cytochrome oxidase I. Its mRNA is unedited in all kinetoplastids studied except *T. borreli* where 250 uridines are added and 25 are deleted. Surprisingly, the mRNA for cytochrome oxidase II mRNA is not edited at all in this species (85).

Taken together these data clearly suggest that editing is a primitive process as ancient as the order Kinetoplastida itself or possibly older. This would suggest that editing might be 500 million years old and lends support to the notion that the basic biochemical mechanisms of editing and RNA-catalysed splicing might have a common origin. Though appealing, direct evidence for such a relationship is lacking, while support for a protein-catalysed mechanism for editing continues to mount.

References

1. Benne, R. (1994) RNA editing in trypanosomes. *Eur. J. Biochem.*, **221**, 9.
2. Hajduk, S. L., Harris, M. E., and Pollard, V. W. (1992) RNA editing in kinetoplastid mitochondria. *FASEB J.*, **7**, 54.
3. Simpson, L., Maslov, D. A., and Blum, B. (1993) RNA editing in Leishmania mitochondria. In *RNA editing: the alteration of protein coding sequences of RNA* (ed, R. Benne), p. 53. Ellis Horwood, New York.
4. Stuart, K. (1993) RNA editing in mitochondria of African trypanosomes. In *RNA editing: the alteration of protein coding sequences of RNA* (ed. R. Benne), p.25. Ellis Horwood, New York.
5. Bass, B. L. (1993) RNA editing. New uses for old players in the RNA world. In *The RNA world* (ed. R. F.,Gesteland and J. F. Akins), p. 383. Cold Spring Harbor Laboratory Press.
6. Simpson, L. and Maslov, D. A. (1994) Ancient origin of RNA editing in kinetoplastid protozoa. *Curr. Opin. Genet. Dev.*, **4**, 887.

7. Adler, B. K. and Hajduk, S. L. (1994) Mechanisms and origins of RNA editing. *Curr. Opin. Genet. and Dev.*, **4**, 316.
8. Benne, R., Van Den Burg, J., Brakenhoff, J. P. H., Sloof, P., Van Boom, J. H., and Tromp, M. C. (1986) Major transcript of the frameshifted *coxII* gene from trypanosome mitochondria contains four nucleotides that are not encoded in the DNA. *Cell*, **46**, 819.
9. Mahendran, R., Spottsword, M. R., and Miller, D. L. (1991) RNA editing by cytidine insertion in mitochondria of *Physarum polycephalum*. *Nature*, **349**, 434.
10. Maherdren, R., Spottswood, M. S., Ghate, A., Ling, M., Jeng, K., and Miller, D. L. (1994) Editing of the mitochondrial small subunit rRNA in *Physarum polycephalum*. *EMBO J.*, **13**, 232.
11. Gott, J. M., Visomirski, L. M., and Hunter, J. L. (1993) Substitutional and insertional RNA editing of cytochrome c oxidase subunit 1 mRNA of *Physarum polycephalum*. *J. Biol. Chem.*, **268**, 25483.
12. Powell, L. M., Willis, S. C., Pease, R. J., Edwards, Y. H., Knott, T. J., and Scott, J. (1987) A novel form of tissue specific RNA processing produces apolipoprotein B-48 in intestine. *Cell*, **50**, 355.
13. Gualberto, J. M., Lamattina, L., Bonnard, G., Weil, J. and Grienenberget, J. (1989) RNA editing in wheat mitochondria results in the conservation of protein sequences. *Nature*, **341**, 660.
14. Vidal, S., Curran, J. and Kolakofsky, D. (1990) A stuttering model for paramyxovirus P mRNA editing. *EMBO J.*, **9**, 2017.
15. Jacques, J., Hausmann, S., and Kolakofsky, D. (1994) Paramyxovirus mRNA editing leads to G deletions as well as insertions. *EMBO J.*, **13**, 5494.
16. Gray, M. W. and Covello, P. S. (1993) RNA editing in plant mitochondria and chloroplasts. *FASEB J.*, **7**, 64.
17. Cech, T. R., Zaug, A. J., and Grabowski, P. J. (1981) In vitro splicing of the ribosomal RNA precursor of Tetrahymena: Involvement of a guanosine nucleotide in the excision of the intervening sequence. *Cell*, **27**, 487.
18. Cech, T. R. (1993) Structure and mechanism of the large catalytic RNAs: group I and group II introns and ribonuclease P. In *The RNA World* (ed. R. F. Gesteland and J. F. Akins), p. 239. Cold Spring Harbor Laboratory Press.
19. Pan, D. M. and Uhlenbeck, O. C. (1993) Self-cleaving catalytic RNA. *FASEB J.*, **7**, 25.
20. Holler, H. F. (1993) tRNA–rRNA interactions and peptidyl transferase. *FASEB J.*, **7**, 87.
21. Phizichy, E. M. and Greer, C. L. (1993) Pre-tRNA splicing: variation on a theme or exception to the rule? *Trends Biochem. Sci.* **18**, 31.
22. Wahle, E. and Keller, W. (1992) The biochemistry of 3′-end cleavage and polyadenylation of messenger RNA precursors. *Annu. Rev. Biochem.*, **61**, 419.
23. Cech, T. R. (1990) Self-splicing of group I introns. *Annu. Rev. Biochem.* **59**, 543.
24. Pace, N. and Smith, D. (1990) Ribonuclease P: function and variation. *J. Biol. Chem.*, **265**, 3587.
25. Gegenheimer, P., Gabius, H., Peebles, C., and Abelson, J. (1983) An RNA ligase from wheat germ which participates in transfer RNA splicing in vitro. *J. Biol. Chem.*, **258**, 8365.
26. Benne, R., van den Burg, J., Brakenhoff, J., de Vries, B., and Nederlof, P. (1985) Mitochondrial genes in trypanosomes: abnormal initiator triplets, a conserved frameshift in the gene for cytochrome oxidase subunit II and evidence for a novel mechanism of gene expression. In *Achievements and perspectives of mitochondrial research* (ed. E. Quagliariello, E. C. Slater, F. Palmierei, C. Saccone, and A. M. Kroon), p. 325. Elsevier, Amsterdam.
27. Benne, R., De Vries, B. F., van den Burg, J., and Klaver, B. (1983) The nucleotide

sequence of a segment of *Trypanosoma brucei* mitochondrial maxicircle DNA that contains the gene for apocytochrome *b* and some unusual unassigned reading frames. *Nucleic. Acids Res.*. **11**, 6925.

28. Hensgens, L. A., Brakenhoff, J., de Vries, B. F., Sloof, P., Tromp, M., Van Boom, J. H., and Benne, R. (1984) The sequence of the gene for cytochrome *c* oxidase subunit I, a frameshift containing gene for cytochrome c oxidase subunit II and seven unassigned reading frames in *Trypanosoma brucei* mitochondrial maxicircle DNA. *Nucleic. Acids Res.*, **12**, 7327.
29. Feagin, J. E., Jasmer, D. P., and Stuart, K. (1987) Developmentally regulated addition of nucleotides within apocytochrome *b* transcripts in *Trypanosoma brucei*. *Cell*, **49**, 337.
30. Shaw, J. M., Feagin, J. E., Stuart, K., and Simpson, L. (1988) Editing of kinetoplastid mitochondrial mRNAs by uridine addition and deletion generates conserved amino acid sequences and AUG initiation codons. *Cell*, **53**, 401.
31. Feagin, J. E., Abraham, J. M., and Stuart, K. (1988) Extensive editing of the cytochrome *c* oxidase III transcript in *Trypanosoma brucei*. *Cell*, **53**, 413.
32. Harris, M. E., Moore, D. R., and Hajduk, S. L. (1990) Addition of uridines to edited RNAs in trypanosome mitochondria occurs independently of transcription. *J. Biol. Chem.*, **265**, 11368.
33. Abraham, J. M., Feagin, J. E., and Stuart, K. (1988) Characterization of cytochrome *c* oxidase III transcripts that are edited only in the 3′ region. *Cell*, **55**, 267.
34. Sturm, N. R. and Simpson, L. (1990) Partially edited mRNAs for cytochrome *b* and subunit III of cytochrome oxidase from *Leishmania tarentolae* mitochondria: RNA editing intermediates. *Cell*, **61**, 871.
35. Decker, C. J. and Sollner-Webb, B. (1990) RNA editing involves indiscriminate U changes throughout precisely defined editing domains. *Cell*, **61**, 1001.
36. Seiwert S. D. and Stuart, K. (1994) RNA editing; transfer of genetic information from gRNA to precursor mRNA in vitro. *Science*, **266**, 114.
37. Pollard, V. W., Harris, M. E., and Hajduk, S.L. (1992) Editing complexes from *Trypanosoma brucei* mitochondria. *EMBO J.*, **12**, 4429.
38. Peris, M., Frech, G. C., Simpson, A. M., Bringaud, F., Byrne, E., Bakker, A. and Simpson, L. (1994) Characterization of two classes of ribonucleoprotein complexes possibly involved in RNA editing from *Leishmania tarentolae* mitochondria. *EMBO J.*, **13**, 1664.
39. Goringer, H. U., Koslowsky, D. J., Morales, T. H., and Stuart, K. (1994) The formation of mitochondrial ribonucleoprotein complexes involving guide RNA molecules in *Trypanosoma brucei*. *Proc. Natl. Acad. Sci. USA.*, **91**, 1776.
40. Read, L. K., Goringer, H. U., and Stuart, K. (1994) *Mol. Cell. Biol.*, **14**, 2629.
41. Frech, G. C., Bakalara, N., Simpson, L., and Simpson, A. M. (1995) *In vitro* RNA editing-like activity in a mitochondrial extract from *Leishmania tarentolae*. *EMBO J.*, **14**, 178.
42. Feagin, J. E. and Stuart, K. (1988) Developmental aspects of uridine addition within mitochondrial transcripts of *Trypanosoma brucei*. *Mol. Cell. Biol.*, **8**, 1259.
43. Souza, Z. E., Myler, P. J., and Stuart, K. (1992) Maxicircle CR1 transcripts of *Trypanosoma brucei* are edited and developmentally regulated and encode a putative iron-sulfur protein homologous to an NADH dehydrogenase subunit. *Mol. Cell. Biol.*, **12**, 2100.
44. Read, L. K., Stankey, K. A., Fish, W. R., Muthiani, A. M., and Stuart, K. (1994) Developmental regulation of RNA editing and polyadenylation in four life cycle stages of *Trypanosoma congolense*. *Mol. Biochem. Parasitol.*, **68**, 297.
45. Landweber, L. F. and Gilbert, W. (1993) RNA editing as a source of genetic variation. *Nature*, **363**, 179.

46. Maslov, D. A., Avila, H. A., Lake, J. A., and Simpson, L. (1994) Evolution of RNA editing in kinetoplastid protozoa. *Nature*, **368**, 345.
47. Simpson, L. and Maslov, D. A. (1994) RNA editing and the evolution of parasites. *Science*, **264**, 1870.
48. Priest, J. W. and Hajduk, S. L. (1994) Developmental regulation of mitochondrial biogenesis in *Trypanosoma brucei*. *J. Bioenerg. Biomemb.*, **26**, 179.
49. Simpson, A. M., Suyama, Y., Dewes, H., Campbell, D. A., and Simpson, S. (1989) Kinetoplastid mitochondria contain functional tRNAs which are encoded in nuclear DNA and also contain small minicircle and maxicircle transcripts of unknown function. *Nucleic. Acids Res.*, **17**, 5427.
50. Hancock, K. and Hajduk, S. L. (1990) Mitochondrial tRNAs are nuclear encoded in *Trypanosoma brucei*. *J. Biol. Chem.*, **265**, 19208.
51. Mottram, J. C., Bell, S. D., Nelson, R. G. and Barry, J. D. (1991) tRNAs of *Trypanosoma brucei*. *J. Biol. Chem.*, **266**, 18313.
52. Blum, B., Bakalara, N., and Simpson, L. (1990) A model for RNA editing in kinetoplastid mitochondria: 'Guide' RNA molecules transcribed from maxicircle DNA provide the edited information. *Cell*, **60**, 189.
53. Pollard, V. W., Rohrer, S. P., Michelotti, E. F., Hancock, K., and Hajduk, S. L. (1990) Organization of minicircle genes for guide RNAs in *Trypanosoma brucei*. *Cell*, **63**, 783.
54. Sturm, N. R. and Simpson, L. (1990) Kinetoplast DNA minicircles encode guide RNAs for editing of cytochrome oxidase subunit III mRNA. *Cell*, **61**, 879.
55. Pollard, V. W. and Hajduk, S. L. (1991) *Trypanosoma equiperdum* minicircles encode three distinct primary transcripts which exhibit guide RNA characteristics. *Mol. Cell. Biol.*, **11**, 1668.
56. Van der Spek, H., Arts, G. J., Zwaal, R. R., Van den Burg, J., Sloof, P., and Benne, R. (1991) Conserved genes encode guide RNAs in mitochondria of *Crithidia fasciculata*. *EMBO J.*, **10**, 1217.
57. Riley, G. R., Corell, R. A., and Stuart, K. (1994) Multiple guide RNAs for identical editing of *Trypanosoma brucei* apocytochrome *b* mRNA have an unusual minicircle location and are developmentally regulated. *J. Biol. Chem.*, **269**, 6101.
58. Theimann, O. H., Maslov, D. A., and Simpson, L. (1994) Disruption of RNA editing in *Leishmania tarentolae* by the loss of minicircle-encoded guide RNA genes. *EMBO J.*, **13**, 5689.
59. Kim, K. S., Teixeira, S. M. R., Kirchhoff, L. V., and Donelson, J. E. (1994) Transcription and editing of cytochrome oxidase II RNAs in *Trypanosoma cruzi*. *J. Biol. Chem.*, **269**, 1206.
60. Blum, B., Sturm, N. R., Simpson, A. M., and Simpson L. (1991) Chimeric gRNA-mRNA molecules with oligo(U) tails covalently linked at sites of RNA editing suggest that U addition occurs by transesterification. *Cell*, **65**, 543.
61. Cech, T. (1991) RNA editing: World's smallest introns? *Cell*, **64**, 667.
62. Arts, G. J., Van der Spek, H., Speijer, D., Van den Berg, J., Van Steeg, H., Sloof, P., and Benne, R. (1993) Implications of novel gRNA features for the mechanism of RNA editing in *Crithidia fasciculata*. *EMBO J.*, **12**, 1523.
63. Read, L. K., Corell, R. A. and Stuart, K. (1992) Chimeric and truncated RNAs in *Trypanosoma brucei* suggest transesterifications at non-consecutive sites during RNA editing. *Nucleic. Acids Res.*, **20**, 2341.
64. Harris, M. E. and Hajduk, S. L. (1992) Kinetoplastid RNA editing: *In vitro* formation of cytochrome *b* gRNA–mRNA chimeras from synthetic substrate RNAs. *Cell*, **68**, 1091.
65. Koslowsky, D. J, Goringer, H. U., Morales, T. H., and Stuart, K. (1992) *In vitro* guide

RNA/mRNA chimaera formation in *Trypanosoma brucei* RNA editing. *Nature*, **356**, 807.

66. Blum, B. and Simpson, L. (1992) Formation of gRNA/mRNA chimeric molecules *in vitro*, the initial step of RNA editing, is dependent on an anchor sequence. *Proc. Natl. Acad. Sci. USA*, **89**, 1194.
67. Harris, M. E., Decker, C., Sollner-Webb, B., and Hajduk, S.L. (1992) Specific cleavage of pre-edited mRNAs in trypanosome mitochondrial extracts. *Mol. Cell. Biol.*, **12**, 2591.
68. Simpson, A. M., Bakalara, N. and Simpson, L. (1992) A ribonuclease activity is activated by heparin or by digestion with proteinase K in mitochondrial extracts of *Leishmania tarentolae*. *J. Biol. Chem.*, **267**, 6782.
69. Piller, K. J., Decker, C. J., Harris, M. E., Hajduk, S. L., and Sollner-Webb, B. (1995) Editing domains of *Trypanosoma brucei* mitochondrial RNAs identified by secondary structure. *Mol. Cell. Biol.* **15**, 2916.
70. Bakalara, N., Simpson, A. M., and Simpson, L. (1989) The *Leishmania* kinetoplast mitochondrion contains terminal uridylyltransferase and RNA ligase activities. *J. Biol. Chem.*, **264**, 18679.
71. Sabatini, R. and Hajduk S. L. (1995) RNA ligase and its involvement in guide RNA/mRNA chimera formation: Evidence for cleavage–ligation type mechanism of *Trypanosoma brucei* mRNA editing. *J. Biol. Chem.* **270**, 7233.
72. Koller, J., Norskau, G., Paul, A. S., Stuart, S., and Göringer, H. U. (1994) Different *Trypanosoma brucei* guide RNA molecules associate with an identical complement of mitochondrial proteins *in vitro*. *Nucleic. Acids Res.*, **22**, 1988.
73. Shaw, J. M., Campbell, D., and Simpson, L. (1989) Internal frameshifts within the mitochondrial genes for cytochrome oxidase subunitII and maxicircle unidentified reading frame 3 of *Leishmania tarentolae* are corrected by RNA editing: evidence for translation of the edited cytochrome oxidase II mRNA. *Proc. Natl. Acad. Sci. USA*, **86**, 6220.
74. Simpson, L. (1986) Kinetoplast DNA in trypanosomatid flagellates. *Int. Rev. Cytol.*, **99**, 119.
75. Michelotti, E. F., Harris, M. E., Adler, B. K., Torri, A. F., and Hajduk, S. L. (1992) Mitochondrial ribosomal RNA synthesis, processing and developmentally regulated expression in *Trypanosoma brucei*. *Mol. Biochem. Parasitol.*, **54**, 31.
76. Read, L. K., Myler, P. J., and Stuart, K. (1992) Extensive editing of both processed and preprocessed maxicircle CR 6 transcripts in *Trypanosoma brucei*. *J. Biol. Chem.*, **267**, 1123.
77. Feagin, J. E. and Stuart, K. (1988) Developmental aspects of uridine addition within mitochondrial transcripts of *Trypanosoma brucei*. *Mol. Cell. Biol.*, **8**, 1259.
78. Vickerman, K. (1985) Developmental cycle and biology of pathogenic trypanosomes. *Br. Med. Bull.*, **41**, 105.
79. Koslowsky, D. J., Bhat, G. H., Perrollaz, A. L., Feagin, J. E., and Stuart, K. (1990) The MURF3 gene of *T. brucei* contains multiple domains of extensive editing and is homologous to a subunit of NADH dehydrogenase. *Cell*, **62**, 901.
80. Koslowsky, D., Riley, G. R., Feagin, J. E., and Stuart, K. (1992) Guide RNAs for transcripts with developmentally regulated RNA editing are present in both life-cycle stages of *Trypanosoma brucei*. *Mol. Cell. Biol.*, **12**, 2043.
81. Corell, R. A., Feagin, J. E., Riley, G. R., Strickland, T., Guderian, J. A., Myler, P.J., and Stuart, K. (1993) *Trypanosoma brucei* minicircles encode multiple guide RNAs which can direct editing of extensively overlapping sequences. *Nucleic Acids Res.*, **21**, 4313.
82. Ferandes, A. P., Nelson, K., and Beverley, S. M. (1993) Evolution of nuclear ribosomal RNAs in kinetplastid protozoa: Perspectives on the age and origins of parasitism. *Proc.*

Natl. Acad. Sci. USA, **90**, 11608.
83. Maslov, D. A., Avila, H. A., Lake, J. A., and Simpson, L. (1994) Evolution of RNA editing in kinetoplastid protozoa. *Nature*, **368**, 345.
84. Moslov, D. A. and Simpson, L. (1994) RNA editing and mitochondrial genomic organization in the cryptobiid kinetoplastid protozoan *Trypanoplasma borreli*. *Mol. Cell. Biol.*, **14**, 8174.
85. Lukes, J., Arts, G. J., van den Burg, J., de Haan, A., Opperdoes, F., Sloof, P. and Benne, R. (1994) Novel pattern of editing regions in mitochondrial transcripts of the cryptobiid *Trypanoplasma borreli*. *EMBO J.*, **13**, 5086.
86. Bhat, G. G., Koslowsky, D. J., Feagin, J. E., Smiley, B. L., and Stuart, K. (1990) An extensively edited mitochondrial transcript in kinetoplastids encodes a protein homologous to ATPase subunit-6. *Cell*, **61**, 885.
87. Stuart, K. (1993) The RNA editing process. *Sem. Cell Biol.*, **4**, 251.
88. Adler, B. K., Harris, M. E., Bertrand, K. I., and Hajduk, S. L., (1991) Modification of *Trypanosoma brucei* mitochondrial rRNA by posttranscriptional 3' polyuridine tail formation. *Mol. Cell. Biol.*, **11**, 5878.

9 | Biogenesis of specialized organelles: glycosomes and hydrogenosomes

JÜRG M. SOMMER, PETER J. BRADLEY, C. C. WANG and
PATRICIA J. JOHNSON

1. Introduction

Compartmentalization of cellular functions into organelles is a characteristic of all eukaryotic cells. The development of highly specialized intracellular organelles in response to unusual host environments is most strikingly demonstrated by two structures found in certain parasitic protozoa: the glycosomes of trypanosomatids and the hydrogenosomes of trichomonads. These unique organelles represent a major cellular difference between host and parasite and have thus been studied extensively as possible targets for chemotherapeutic agents. Both of these organelles participate in the production of ATP, although in very different ways. The glycosome increases the rate of glycolysis by at least an order of magnitude, while the hydrogenosome participates in substrate-level phosphorylation, converting pyruvate to acetate, CO_2, and molecular hydrogen in the process. The evolutionary origin of these organelles has been a matter of much speculation in the past, but recent molecular studies of their biogenesis have confirmed that the glycosomes are derived from an ancestral microbody related to the peroxisome, while the hydrogenosomes of trichmonads may share a common progenitor organelle with mitochondria.

As discussed below, the *in vivo* import of peroxisomal proteins into the glycosomes of *Trypanosoma brucei* demonstrates a functional conservation of the import mechanism for proteins bearing a C-terminal targeting signal. Detailed mutational analysis of this targeting signal has identified largely overlapping sets of recognition sequences that specify peroxisomal and glycosomal protein localization, but has also revealed some unique sequences specific for glycosomal targeting. The *in vitro* import of proteins into the hydrogenosomes of *Trichomonas vaginalis*, on the other hand, involves short, cleavable N-terminal sequences similar to those involved in mitochondrial protein import in higher eukaryotes. The structural features of a class of hydrogenosomes found in certain fungi, however, are more characteristic of peroxisomes and glycosomes, raising the possibility that hydrogen-producing organelles may have arisen independently at different times during evolution.

2. The glycosome

2.1 Morphology

Glycosomes are globular organelles with a diameter of 0.2–0.3μm, surrounded by a single phospholipid bilayer membrane and containing no detectable nucleic acids (1). Between 10 and 100 glycosomes appear to be present in all organisms belonging to the order Kinetoplastida, which includes the hemoflagellates *T..brucei* (African trypanosomes), *Trypanosoma cruzi* (American trypanosomes), and *Leishmania*, as well as the invertebrate parasites *Crithidia, Trypanoplasma, Leptomonas,* and *Herpetomonas*, and the plant parasite *Phytomonas*. The blood-borne parasites are able to rapidly adjust to new environments upon infecting their hosts, usually by differentiating into morphologically and functionally distinct cell types. The life cycles of these parasites differ considerably, but generally include transmission by an insect vector to the mammalian host during a blood feeding. In the mammalian host, the parasites may remain and multiply predominantly in the bloodstream, as is the case for *T. brucei*, or they may reside intracellularly, as observed for *T. cruzi* and *Leishmania* (2). Other kinetoplastid protozoa, such as the insect parasite *Crithidia*, may appear as a single cell type, either free-living or within the host intestine.

2.2 Function

Surrounded by an abundance of glucose, the bloodstream form of *T. brucei* has adapted to rely entirely on glycolysis and substrate-level phosphorylation for its energy production. Most mitochondrial functions, including the tricarboxylic acid cycle and the cytochrome-dependent electron transport chain, are completely suppressed in this developmental stage (3). By compartmentalizing the first seven glycolytic enzymes and two glycerol-metabolizing enzymes into the membrane-bound glycosome, the parasite is able to greatly increase the rate of glycolysis and thus overcome the relatively poor energy yield from conversion of glucose to pyruvate as an end product. In the glucose-deficient environment of the tsetse fly midgut, ATP generation by oxidative phosphorylation in the now well developed mitochondrion becomes the main source of energy for the parasite. Accordingly, profiles of the glycosomal and mitochondrial enzymes change dramatically during differentiation from the bloodstream form into the insect (procyclic) form. While the levels of several major glycosomal proteins, including hexokinase, phosphofructokinase, glycosomal phosphoglycerate kinase (gPGK), and aldolase, are reduced during differentiation to the insect form, two other glycosomal proteins, phosphoenolpyruvate carboxykinase (PEPCK) and malate dehydrogenase (MDH), are greatly induced (4). Higher levels of PEPCK and MDH thus allow conversion of the glycolytic products into malate which is used in the tricarboxylic acid cycle. In addition, PEPCK may also function in the decarboxylation of oxaloacetate produced from oxidation of amino acids which are abundant in the fly gut (5). In the intracellular forms of *T. cruzi* and the related *Leishmania*, glycolysis proceeds at a much slower rate than in the bloodstream form of *T. brucei*, and only the first six steps of glycolysis take place inside the glycosome. Glycosomal PEPCK and MDH are present at all stages of the life cycle and their expression is not affected by glucose levels (5, 6). The energy metabolism of these organisms more closely resembles that of the procyclic form of *T. brucei*, with succinate as an end product under aerobic growth conditions.

In addition to the functions related to energy production, several other biochemical pathways, including β-oxidation of fatty acids, pyrimidine biosynthesis, and purine salvage, also appear to be at least partly associated with the glycosomes in some kinetoplastid organisms (7–9).

2.3 Biogenesis

Many of the trypanosomal enzymes involved in glycolysis exist as both cytoplasmic and glycosomal isozymes that are encoded by separate genes. Like other trypanosomatid genes, they are transcribed polycistronically and are processed into mature RNAs encoding individual proteins by *trans*-splicing and polyadenylation. Stage-specific control of gene expression is usually accomplished at the levels of RNA processing, translation initiation, and protein stability, rather than at the level of transcription (see Chapter 6). The glycosomal proteins are synthesized on free polysomes and enter the organelle post-translationally, within 3 to 5 minutes after synthesis (10, 11). No proteolytic cleavage of a signal sequence or other modification has been detected following import of the proteins. Thus, the morphology and kinetics of protein import closely resemble that of other microbodies, including the peroxisomes found in most higher eukaryotes, and plant glyoxysomes.

2.4 Glycosomal targeting signals

The kinetoplastid organisms are believed to have diverged soon after the separation of prokaryotes and eukaryotes (12, 13). The observation that the glycosomes in most trypanosomatids lack many typical peroxisomal proteins, including catalase and other peroxidases, and instead contain an entire set of glycolytic enzymes, raises the possibility that the targeting signals directing proteins to the glycosomes have diverged substantially from those required for import into peroxisomes of higher eukaryotes. Like peroxisomes, the purified glycosomes are fragile organelles, and the lack of protein processing upon translocation has made it difficult to establish *in vitro* assays that measure uptake of glycosomal proteins (14). In order to identify glycosomal targeting signals, fusion proteins consisting of a foreign reporter enzyme and glycosomal protein sequences have been expressed *in vivo* in *T. brucei* and assayed for their subcellular localization by cell fractionation and immunoelectron microscopy.

2.4.1 Phosphoglycerate kinase

In *T. brucei*, a 20 amino acid extension at the C-terminus of gPGK is the most obvious difference between the cytosolic and glycosomal isozymes. The C-terminal tripeptide serine-serine-leucine (SSL) of gPGK, although similar to the C-terminal peroxisomal targeting signal serine-lysine-leucine (SKL) (15), is unable to target the reporter enzymes firefly luciferase or chloramphenicol acetyltransferase (CAT) to mammalian peroxisomes (15, 16). Moreover, when expressed in *Saccharomyces cerevisiae*, gPGK remains in the cytoplasm, suggesting that it lacks a functional peroxisomal targeting signal (13). To test for the presence of a glycosomal targeting signal in gPGK, the entire gPGK coding region was appended to that of luciferase, which had its own peroxisomal targeting signal (SKL) deleted. The *in vivo* expression of this fusion protein in procyclic *T. brucei* was accomplished by providing the gene in a tandem array with the hygromycin phosphotransferase gene on a plasmid construct driven by the *T. brucei* procyclin promoter. Stable, hygromycin-resistant transformants were obtained by targeted insertion of this construct into the tubulin locus of *T. brucei* (17, 18). Cell fractionation and immunoelectron microscopy showed that the luciferaseΔSKL–gPGK fusion protein was efficiently imported into the glycosomes (Fig.1). The C-terminal SSL sequence is essential for import of this fusion protein, since a similar protein lacking this tripeptide is found exclusively in the cytoplasm (18). Furthermore, a C-terminal SSL sequence is sufficient to direct two other cytosolic proteins, CAT and β-glucuronidase (GUS), to the glycosomes, with an import efficiency of more than 50% (16, 18, 19; Fig. 1).Complete import of these fusion proteins is observed when the last 20 amino acids of gPGK are used in lieu of the tripeptide alone, but the increased import is not dependent on the specific sequence immediately upstream of the SSL tripeptide (16, 18). Possibly, the amino acid extension rich in small hydrophobic and hydroxylated amino acids can act as a spacer which results in a more efficient presentation of the C-terminal tripeptide to the import machinery. Thus, SSL is essential, as well as sufficient, for glycosomal import of gPGK in *T. brucei*.

2.4.2 Phosphoenolpyruvate carboxykinase

T. brucei PEPCK has a predicted length of 525 amino acids ending in the tripeptide serine-arginine-leucine (SRL). A fusion protein of luciferaseΔSKL and the full-length *T. brucei* PEPCK is imported into the glycosomes when expressed in procyclic trypanosomes. However, partial import still occurs when the C-terminal SRL tripeptide is removed, suggesting that PEPCK may contain a second, internal glycosomal targeting signal. Further analysis has shown that deletion of the C-terminal 29 amino acids of PEPCK reduces the import only by half, while a deletion of the last 47 amino acids completely abolishes the import (20). The presence of two independent organelle targeting signals in a single protein is not unprecedented; it has been reported also for the *Hansenula polymorpha* PER1 gene product (21), *Candida tropicalis* acyl-CoA oxidase (22) and catalase A of *S. cerevisiae* (23).

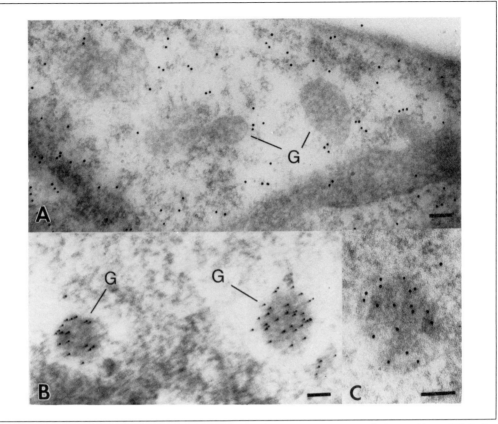

Fig. 1 Cryosections of *T. brucei* procyclic cells expressing β-glucuronidase (A) or a β-glucuronidase–SSL fusion protein (B and C). Sections were probed with an anti-β-glucuronidase antiserum and a colloidal gold-conjugated secondary antibody to reveal the glycosomal (G) localization of this fusion protein. The bar represents 0.1μm. Reprinted with permission from Ref. 18.

2.4.3 Specificity of the C-terminal glycosomal targeting signal

When expressed in mammalian cells and trypanosomes, firefly luciferase is found to be imported into peroxisomes and glycosomes, respectively (17, 24). The peroxisomal targeting was shown to require the last three amino acids of the protein (SKL), and at a C-terminal location this tripeptide is sufficient to target a cytosolic carrier protein to the peroxisomes (15). Mutagenesis of the C-terminal SKL sequence of luciferase indicated that a limited number of conservative substitutions are compatible with targeting to mammalian peroxisomes (15). In order to determine the exact sequence requirements for glycosomal import, SKL variants of the luciferase gene were expressed in *T. brucei* and tested for *in vivo* import into the glycosome. Any small amino acid, except for the negatively charged aspartic and glutamic acids, could substitute for the serine to form a functional targeting signal for import of luciferase into the glycosomes (Fig. 2). The lysine residue in SKL could be substituted with an amino acid capable of hydrogen bonding, while the C-terminal

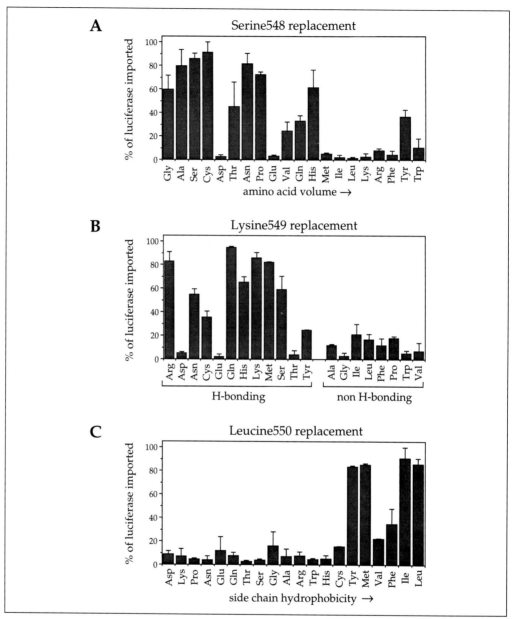

Fig. 2 Expression of mutant firefly luciferase genes encoding replacements of the serine, lysine, or leucine residues in the C-terminal targeting signal in *T. brucei* procyclic cells. The efficiency of import of the mutant proteins is shown as the percent of pelleted (glycosomal) luciferase activity following digitonin-solubilization of the plasma membrane and centrifugation (25). A: The amino acids replacing the serine residue in SKL are ordered as a function of increasing amino acid volume. B: The amino acids replacing the lysine residue in SKL are grouped according to the hydrogen-bonding capability of their amino acid side chains. C: The amino acids replacing the C-terminal leucine are ordered according to their relative hydrophobicity (TOTFT scale (26)). A stop codon in any of these three positions resulted in less than 2% import. Adapted from Ref. 4 with permission from Annual Reviews.

leucine of luciferase could be replaced with a subset of hydrophobic amino acids. Thus, the tripeptides that were found to function as glycosomal targeting signal at the C-terminus of luciferase consist of {A/C/G/H/N/P/S/T}KL, S{H/K/M/N/Q/R/S}L and SK{I/L/M/Y}. An extensive analysis of double or triple amino acid substitutions in SKL has not been done, but it is likely that some other combinations of these three amino acid families would also form functional targeting signals.

2.4.4 C-terminal targeting signals in other trypanosomatid proteins

Based on the mutational analysis of the glycosomal targeting requirements for luciferase, the SKL-like tripeptides found at the ends of four major glycolytic enzymes of *T. brucei* probably act as glycosomal targeting signals for these proteins. Conserved tripeptide signals are also present at the C-termini of the homologous proteins in other trypanosomatids (Table 1).

The genes encoding hypoxanthine–guanine phosphoribosyltransferase (HGPRTase) from four kinetoplastid species have been cloned recently. While the *T. brucei* enzyme ends in AKR, which is not predicted to function as a glycosomal targeting signal, the HGPRTases from *T. cruzi*, *Leishmania donovani* and *Crithidia fasciculata* end in SKY, SKV, and SKL, respectively (27, 28; T. Allen and B. Ullman, unpublished), which are potential glycosomal targeting signals (17). The subcellular local-

Table 1 C-terminal glycosomal protein targeting signals

Protein	Species	C-terminal tripeptide	Evidence[a]	Reference
Glyceraldehyde-phosphate dehydrogenase	T. brucei	-AKL	Luciferase-AKL is glycosomal	29
	T. cruzi	-ARL		30
	L. mexicana	-SKM	Luciferase-SKM is glycosomal	31
	T. borelli	-AKL	Luciferase-AKL is glycosomal	32
Glucosephosphate isomerase	T. brucei	-SHL	Luciferase-SHL is glycosomal	33
	L. mexicana	-AHL		34
Phosphoenolpyruvate carboxykinase	T. brucei	-SRL	Glycosomal import of luc-PEPCK is SRL-dependent	20
	T. cruzi	-ARL		20
Phosphoglycerate kinase	T. brucei	-SSL	Glycosomal import of luc-gPGK is SSL-dependent	18
Hypoxanthine-guanine phosphoribosyltransferase	T. cruzi	-SKY	Luciferase-SKY is partially glycosomal	28
	L. donovani	-SKV	Luciferase-SKV is partially glycosomal	(T.Allen and B.Ullman, unpublished)
	C. fasciculata	-SKL	Luciferase-SKL is glycosomal	(T.Allen and B.Ullman, unpublished)
	T. brucei	-AKR	(AKR is not likely to be a targeting signal)	27

[a] All evidence is from analysis of luciferase (luc) mutants or fusion proteins expressed in *T. brucei*

ization of HGPRTase activity has been tentatively assigned to the glycosomes in *T. brucei* (D. Hammond and C. C. Wang, unpublished) and *Leishmania mexicana* (9), but is still unknown in the other species.

The presence of an SKL-like sequence at the C-terminus of a protein does not necessarily indicate a glycosomal localization. For example, trypanothione reductases (TR) from five kinetoplast species end in potential glycosomal targeting signals. TR is the trypanosomatid equivalent of glutathione reductase, a flavoprotein disulphide reductase that plays a crucial role in protecting cells from oxidative stress in most organisms (35). When compared with the erythrocyte and *E. coli* glutathione reductases, the TRs each contain a 19 to 20 amino acid C-terminal extension ending in SSL, SNL, or ASL (36, 37). Although the amino acid sequences of these C-terminal extensions are at least 75% homologous, they are not required for enzymatic activity (38). All the TR activity was reported in the cytosolic fraction of cell homogenates prepared from either bloodstream forms or procyclic *T. brucei* (39). In *T. cruzi*, the enzyme appears to reside primarily in the mitochondrion, as determined by immunoelectron microscopy (40). The structure of *C. fasciculata* TR has been determined by X-ray crystallography, and in this protein, the C-terminal peptide may be part of the dimer interface and perhaps not be accessible to glycosomal protein receptors in the fully assembled enzyme (41).

In another example, glucosephosphate isomerase of *L. mexicana* promastigotes appears to be 90% cytoplasmic, yet the protein ends in the tripeptide AHL, a likely glycosomal targeting signal (34). The homologous *T. brucei* enzyme, which is entirely glycosomal in the bloodstream form, shares 69% amino acid identity and ends in SHL. Thus, it appears that the glycosomal import of this enzyme is somehow prevented in *Leishmania*, perhaps by minor differences in the sequence upstream of the C-terminal tripeptide that may be incompatible with import, or by another, as yet undetermined mechanism.

2.4.5 N-terminal glycosomal targeting signals

The peroxisomal 3-ketoacyl-CoA thiolases from rat and human contain cleavable N-terminal presequences ranging from 26 to 36 amino acids in length (42, 43). Similarly, watermelon MDH has a 37 amino acid N-terminal presequence which is cleaved upon import into the glyoxysome (44). The thiolase N-terminal sequences have been shown to target a reporter protein (CAT) to the peroxisome (42, 43) and the presequence is essential for peroxisomal targeting when watermelon MDH is expressed in the yeast *H. polymorpha* (44). A consensus motif of R{I/L}xx{I/L/V}xx{H/Q}L has emerged from these studies, where x may be almost any amino acid. This motif is also part of the N-terminal peroxisomal targeting signal of thiolases from several yeasts (45) and of *H. polymorpha* amine oxidase (46). In these species, however, the N-termini are not cleaved upon import into the peroxisome.

Among the glycosomal enzymes, *T. brucei* aldolase lacks a C-terminal tripeptide targeting signal but contains an RVxxLxxQL motif at amino acid positions 4–12 (47). The prediction that this N-terminal sequence of aldolase constitutes a glycosomal targeting signal was recently confirmed by expressing aldolase–CAT fusion constructs in *T. brucei*. When present at the N-terminus of CAT, the first 18 amino acids of aldolase resulted in the import of a substantial amount of fusion protein. Slightly longer or shorter N-terminal fragments of aldolase reduced the import efficiency, indicating that perhaps conformational effects or additional requirements for secondary targeting signals may limit the import of such fusion proteins (47).

2.5 Glycosomal versus peroxisomal import

Proteins destined for the glycosome of trypanosomatids or the peroxisomes of higher eukaryotes are synthesized on free polysomes and imported post-translationally. Both of these organelles share a common mechanism based on the C-terminal SKL tripeptide motif for import of the majority of their proteins. Some peroxisomal proteins are imported by a genetically distinct mechanism that recognizes an N-terminal signal which also may function in trypanosomatids (47).

The results from the C-terminal amino acid replacements in luciferase, and a related study on SKL variants at the C-terminus of CAT (16), indicate that many more substitutions are tolerated in the glycosomal targeting signal than in sequence that directs peroxisomal import in mammalian cells. In particular, three tripeptide sequences (SSL, SNL, and SKI) that allow protein import into the glycosomes have been shown not to act as peroxisomal targeting signals in mammalian cells when present at the end of luciferase or CAT (15–17). The reason for the greater degeneracy in the glycosomal tripeptide targeting signal is not clear, but it may be a general phenomenon that lower eukaryotes are less stringent in their sequence requirements for C-terminal peroxisomal targeting signals (4).

2.6 Mutational analysis of glycosome assembly

The study of microbody assembly has made rapid progress through the identification and cloning of genes required for peroxisome biogenesis in yeast and human cell lines. Receptors that interact with the C-terminal or the N-terminal peroxisomal targeting signal have been identified (48–50). An integral membrane protein that is essential for import of proteins containing either of these signals may constitute part of the translocation machinery (51). Zinc-binding proteins involved in peroxisomal protein import (52, 53) and ubiquitin-conjugating enzymes required for peroxisome proliferation (54, 55) have recently been identified. The finding that native proteins and fully assembled oligomers may be imported into the peroxisome without substantial unfolding or disruption of subunits (56, 57) will probably lead to further insights into a novel mechanism of protein translocation and organelle biogenesis.

Unlike the studies of yeast peroxisome biogenesis, the mutational analysis of glycosomal protein import is limited by the lack of advanced genetic methodologies. Based on the observation that some enzymes involved in β-oxidation of fatty acids and ether lipid biosynthesis are also localized to the glycosomes (7, 58), glycosome-deficient *T. brucei* mutants have been isolated by selection for UV-resistance among mutagenized procyclic forms grown in the presence of pyrene-derivatized fatty acid or fatty alcohol (T. F. Chou and C. C. Wang, unpublished). Electron-microscopic analysis revealed a complete absence of recognizable glycosomes in these mutants. Upon further cultivation, however, these cell lines appeared to be unstable and the glycosome-deficient phenotype was easily reversed, while resistance to the pyrene-derivatized substrates persisted. The nature of this apparent long-term adaptation to the selective pyrene–UV pressure remains to be determined, but at this time, it precludes the cloning of genes involved in glycosome biogenesis by functional complementation of these mutants.

2.7 Glycosomal protein import as a potential target for chemotherapy

Current treatment of trypanosomiasis includes the use of drugs that are carcinogenic or toxic to the host, such as arsenicals (melarsoprol), pentavalent antimonials, and suramin. The unique biochemical features of the glycosomes are expected to provide potential targets for new anti-trypanosomal treatment. Inhibition of the glycolytic pathway is clearly deleterious to the bloodstream form of *T. brucei* (59). Therefore, new chemotherapeutic agents could include specific inhibitors of glycosomal enzymes (60, 61). Alternatively, inhibition of glycosomal protein import could provide an even more specific strategy for anti-trypanosomal therapy. For example, preventing the import of those glycolytic enzymes which depend on their C-terminal tripeptides as targeting signals, while allowing others to be imported normally, would segregate the glycolytic enzymes and probably disrupt the glycolytic pathway. The comparison of glycosomal and peroxisomal protein import has so far revealed some clear differences between the two. These differences could be exploited in the design of specific peptide analogues that interfere with the C-terminal targeting signal for glycosomal, but not for mammalian peroxisomal, import.

3. The hydrogenosome

3.1 Trichomonad hydrogenosomes

Hydrogenosomes were first described in several species of trichomonads, which, like trypanosomatids, are flagellated protozoa believed to have diverged very early from the main branch of eukaryotic evolution (62–64). Trichomonads differ from typical eukaryotes not only by the presence of hydrogenosomes, but also by the absence of two organelles that are the hallmarks of eukaryotes: mitochondria and peroxisomes (Fig. 3). Other eukaryotic organelles, such as the endoplasmic reticulum, Golgi complex, and lysosomes, are, however, present. Numerous granules are seen throughout the cytoplasm; these are the repositories of vast glycogen stores. The nucleus, located at the anterior end of the cell, appears to have normal nuclear pores but lacks nuclear envelope breakdown during mitosis, a feature common to many protists (65). The two most frequently studied trichomonads, *Trichomonas*

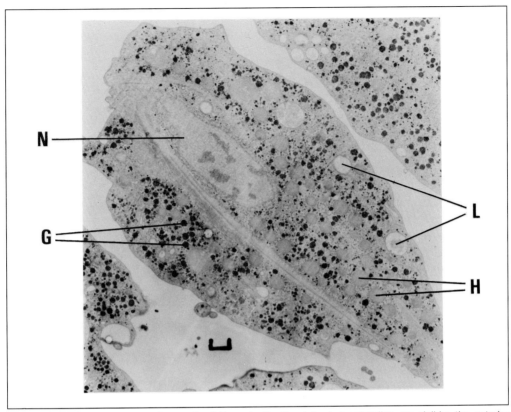

Fig. 3 Transmission electron micrograph of *T. vaginalis*. The following organelles are visible: the anterior nucleus (N), numerous spherical, electron-opaque hydrogenosomes (H), a few variably sized lysosomes (L), and many darkly stained glycogen granules (G). Endoplasmic reticulum and Golgi are not visible in this section, but are also present in trichomonads. The bar represents 1μm.

vaginalis and *Tritrichomonas foetus*, are sexually transmitted parasites that reside in the urogenital tract of humans and cattle respectively. Over 150 million cases of human trichomoniasis are reported each year and the large financial losses due to trichomoniasis in cattle in the developing countries have made these parasites important in both the medical and agricultural communities.

3.2 Morphology

Trichomonad hydrogenosomes are spherical organelles of 0.5–1μm diameter that are bound by two membranes (66). At first, the close proximity of the surrounding membranes and lack of intraorganellar membrane structures resulted in these organelles being defined as single membrane-bound vesicles. Later, freeze-fracture electron microscopy clearly showed the presence of a double membrane (67), thus changing the popular view that hydrogenosomes were microbodies similar to peroxisomes and glycosomes. Hydrogenosomes do not contain detectable DNA or

3.3 Function

The hydrogenosome aids in the production of ATP by compartmentalizing the fermentative metabolism of pyruvate, leading to the production of molecular hydrogen (69). This property of the organelle is responsible for its name: the hydrogenosome. A partial metabolic map of the hydrogenosome is shown in Fig. 4. Pyruvate generated by glycolysis or by conversion from malate is decarboxylated by pyruvate:ferredoxin oxidoreductase to form acetyl-CoA. The electrons are transferred from pyruvate:ferredoxin oxidoreductase to the iron-sulphur protein ferredoxin and subsequently to protons forming molecular hydrogen, a process catalysed by hydrogenase. Acetyl-CoA is converted to acetate with the concomitant conversion of succinate to succinyl-CoA by succinyl-CoA synthetase. The generation of succinyl-CoA is coupled to ATP production via substrate-level phosphoryla-

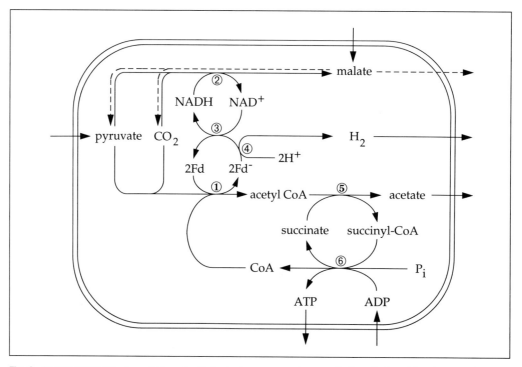

Fig. 4 Partial metabolic map of *T. vaginalis* hydrogenosomes. 1, pyruvate:ferredoxin oxidoreductase; 2, malic enzyme; 3, NAD:ferredoxin oxidoreductase; 4, hydrogenase; 5, acetate: succinate CoA transferase; and 6, succinyl CoA synthetase. Physiological direction of reactions are indicated by solid arrows and the dashed arrows indicate the detection of malate formation *in vitro*. The double line denotes the double-membrane surrounding the hydrogenosome. Modified from Ref. 69, with permission from Elsevier Science Publishers.

tion. While this pathway describes the key biochemical process of energy production in trichomonad hydrogenosomes, other minor pathways are likely to exist. The hydrogenosome is also the site of activation of metronidazole, the only drug currently used for treatment of trichomoniasis in the United States. The efficacy of this drug relies on a conversion of the drug to a cytotoxic form, via a direct interaction with hydrogenosomal proteins (70). Thus, in addition to its metabolic importance, the hydrogenosome plays an essential role in the susceptibility of trichomonads to therapy.

3.4 Biogenesis

The RNAs encoding two hydrogenosomal proteins, ferredoxin and the β-subunit of succinyl CoA synthase, have been shown to partition exclusively with free polysomes, demonstrating that hydrogenosomal proteins are probably imported into the organelle post-translationally (71). Organelles that import proteins post-translationally are thought to reproduce by fission from pre-existing organelles, as opposed to budding from the endoplasmic reticulum. Post-translational translocation of proteins into the hydrogenosome thus supports morphological evidence that hydrogenosomes divide by fission (72).

3.4.1 N-terminal leader sequences

To date, 11 genes encoding six different hydrogenosomal proteins of *T. vaginalis* have been identified (73–78). All eleven genes encode short N-terminal sequences that are not present on the mature protein isolated from hydrogenosomes (Fig. 5). The length of these leader sequences ranges from five to 12 amino acids and they

Protein	Leader sequence	Mature N-terminus
Ferredoxin	M L S Q V C R F	G T I T
β-Succinyl CoA Synthetase 1	M L S A S S N F A R N	F N I L
β-Succinyl CoA Synthetase 2	M L S A S S N F A R N	F N I L
α-Succinyl CoA Synthetase 1	M L A G D F S R N	L K Q P
α-Succinyl CoA Synthetase 2	M L S S S F E R N	L H Q P
α-Succinyl CoA Synthetase 3	M L S S S F E R N	L H Q P
Adenylate kinase	M L S T L A K R F	A S G K
Pyruvate: ferredoxin oxidoreductase 1	M L R N F	S K R V
Pyruvate: ferredoxin oxidoreductase 2	M L R S F	G K R I
Malic enzyme 1	M L T S S V S V P V R N	I C R A
Malic enzyme 2	M L S S S F E R N	I C R S

Fig. 5 Leader sequences on *T. vaginalis* hydrogenosomal proteins. Sequences to the left of the arrow (leader sequences) are encoded by genes of the indicated hydrogenosomal proteins, but are absent from mature proteins that have been isolated from the organelle. Mature proteins begin with the sequences to the right of the arrow.

are strikingly similar in amino acid composition. The most conserved residue is leucine, the amino acid immediately following the starting methionine in all eleven sequences. Arginine is present at the -2 position relative to the cleavage site in all except pyruvate:ferredoxin oxidoreductase, where it is found at position -3. Either phenylalanine or asparagine is found at position -1 from the cleavage site. Also noteworthy is the prevalence of serine in the sequences, particularly at the third amino acid of the sequences. The N-terminal location, the cleavage, and the amino acid composition of leader sequences on hydrogenosomal proteins resemble presequences which target proteins to the matrix of the mitochondria (79, 80). The similarity of the hydrogenosomal leader sequences to each other and to mitochondrial protein targeting signals makes these leaders strong candidates for the import signal on hydrogenosomal proteins.

3.4.2 *In vitro* protein translocation

In contrast to *T. brucei*, no transfection system is available for the *in vivo* study of protein translocation in *T. vaginalis*. Therefore, we have recently established an *in vitro* hydrogenosomal import assay to test whether the N-terminal leader sequences identified on hydrogenosomal proteins can target these proteins to the organelle (P. J. Bradley, C. J. Lahti, and P. J. Johnson, unpublished). The main components of the assay are purified intact hydrogenosomes, radiolabeled full-length hydrogenosomal proteins, a *T. vaginalis* cytosolic extract, and an ATP regenerating system. Following incubation of these components, the hydrogenosomes are re-isolated and the proportion of imported, labeled protein is determined. That the substrate is actually within the organelle is verified by assessing its conversion to the mature size and by protection from externally added, membrane-impermeable protease. Ferredoxin has been used as the prototype transport substrate because its relatively small size (12 kDa) allows clear separation of the precursor and mature proteins by gel electrophoresis. Recombinant proteins produced from constructs that have the leader sequence deleted (except for the starting methionine) are used to test whether protein translocation is leader dependent.

In preliminary experiments, *in vitro* import and cleavage of radiolabeled ferredoxin has been shown to be dependent on the presence of the leader sequence. Furthermore, protein translocation in our *in vitro* assay is also dependent on temperature, ATP, and the presence of intact organelles (P. J. Bradley, C. J. Lahti, and P. J. Johnson, unpublished). Experiments are underway to determine whether the leader sequence alone is sufficient to confer compartmentalization to a non-hydrogenosomal protein. Mutating conserved residues in the leader sequences and testing the effect on binding and translocation may determine the exact features of the signal that are essential for hydrogenosomal protein import.

3.4.3 Mitochondrial versus hydrogenosomal targeting

Hydrogenosomes have been considered the anaerobic equivalent of mitochondria as both are involved in carbohydrate metabolism and their presence in a cell is usually mutually exclusive (81). At first glance, general similarities between the two organelles support the hypothesis that hydrogenosomes are mitochondria that have lost the oxidative phosphorylation capacity in an anaerobic environment. Hydrogenosomes and mitochondria are the only known double membrane-bound organelles involved in pyruvate metabolism. Although not well characterized, hydrogenosomes are presumed to have membrane translocators for uptake of pyruvate and malate and for dispensing of ATP, similar to the ATP/ADP translocator found in mitochondria. One could imagine that having lost the oxidative phosphorylation machinery, the extensive folding observed for the mitochondrial inner membrane would become unnecessary in hydrogenosomes. Similarly, mitochondria appear to have lost over 99% of the DNA of the endosymbiont that gave rise to this organelle, so hydrogenosomes could have just as readily lost the entire genome.

A comparison of hydrogenosomal and mitochondrial protein leader sequences reveals striking similarities:

(1) both sequences have an amino-terminal location on the protein;

(2) both are cleaved from the mature protein found in the organelle;

(3) both have similar amino acid composition.

For example, 40% of known mitochondrial leader sequences begin with methionine-leucine and 16% begin with methionine-leucine-serine (all examined hydrogenosomal leader sequences begin with methionine-leucine and 55% begin with methionine-leucine-serine). Also, a majority of mitochondrial leader sequences contain arginine at either -2 or -10 from the cleavage site (79), and all but two of the eleven characterized hydrogenosomal leader sequences contain arginine at -2 from the cleavage site. Arginine, leucine, and serine typically are over represented in mitochondrial leader sequences and all three amino acids are present in each of these short hydrogenosomal leader sequences. Finally, both hydrogenosomal and mitochondrial leader sequences have the potential to form amphiphilic α-helices (73, 80).

Despite their striking similarity, some marked differences between hydrogenosomal and mitochondrial protein leader sequences also exist. One conspicuous difference is in their length. Leader sequences of mitochondrial proteins are typically 20–80 amino acids long, whereas the leaders identified thus far for hydrogenosomal proteins contain only five to 12 amino acids. Also, mitochondrial leader sequences are invariably devoid of acidic residues, whereas three of the 11 leader sequences described for hydrogenosomal proteins contain an acidic residue (75). Finally, the conservation of primary sequence appears to be greater for hydrogenosomal leaders than for mitochondrial leader sequences.

Notwithstanding these specific differences between leader sequences of hydrogenosomal proteins and mitochondrial proteins, a number of parallels can be drawn between apparent mechanisms used to target proteins to hydrogenosomes and mitochondria:

(1) both types of proteins are synthesized on free polysomes and are post-translationally translocated into the organelle;

(2) an amino-terminal, cleavable leader is required for translocation into both organelles;

(3) translocation is an active process that is ATP and temperature dependent.

These data indicate that the general mechanisms underlying protein targeting and biogenesis of hydrogenosomes and mitochondria must be similar. Given the similarities between these organelles and the fact that trichomonads appear to have evolved from the main line of eukaryotic evolution prior to the appearance of true mitochondria (62), it is tempting to speculate that trichomonad hydrogenosomes and mitochondria found in other eukaryotes evolved from a common progenitor organelle, as opposed to the popular theory that hydrogenosomes are degenerate mitochondria (82).

3.5 Hydrogenosome-like organelles in other organisms

A membrane-bound hydrogenase activity has also been observed in several other anaerobic or aerotolerant protists, including certain fungi (83–85) and ciliates (86–90), all of which have diverged much later in evolution, from ancestors that had well established mitochondria. These organisms all lack mitochondria, with the exception of the free-living amoebo-flagellate *Psalteriomonas lanterna*, which is reported to contain both hydrogenosomes and mitochondria (91). The morphology of hydrogenosomes from different groups is highly variable, and biochemical studies of hydrogenosomes from some of these organisms revealed substantial differences in the pathways utilized for ATP production (reviewed in Ref. 66). For example, in the free-living ciliate *P. lanterna*, the hydrogenosome has membrane structures throughout the matrix, resembling those of mitochondria or chloroplasts. On the other hand, hydrogenosomes of rumen fungi look more similar to peroxisomes or glycosomes and it has been suggested that these hydrogenosomes are in fact derived from microbody-like organelles, since they appear to be bound only by a single membrane. Preliminary experiments indicating that the hydrogenase of the rumen fungus *Neocallimastix* is recognized by an antiserum raised against the C-terminal peroxisomal targeting signal (SKL) seems to support this hypothesis (92).

The morphological differences between hydrogenase-containing organelles in these diverse protists and their sporadic appearance throughout phylogeny thus questions a single lineage hypothesis for this organelle. It is likely that the hydrogenosomes in these organisms have evolved independently and at different times through convergent evolution. Further biochemical characterization of these organelles and comparative analysis of genes encoding hydrogenosomal proteins may provide future insight into the origin of these organelles.

4. Conclusions

Glycosomes and hydrogenosomes are intracellular organelles of unusual biochemical functions involved in energy production in a limited number of protist species.

As such they are both of general biological interest and may ultimately yield a unique target for novel antiparasitic treatments. The biochemical and molecular study of these organelles has begun to answer many questions about their function and biogenesis, as well as their possible evolutionary origins. The best studied prototypes of these organelles include the glycosomes of trypanosomes and the hydrogenosomes of trichomonads. While precursor proteins destined for each of these organelles are synthesized on free polysomes and imported post-translationally, the similarities between these two organelles end there. Glycosomes are part of a group of microbodies bound by a single membrane with a highly conserved mode of protein import using primarily the C-terminal three amino acids of the mature protein as a targeting signal. Hydrogenosomes of trichomonads are bound by two membranes and the import of proteins depends on an N-terminal, cleavable signal peptide.

In vivo expression of recombinant proteins in *T. brucei* has allowed a detailed analysis of the sequence requirement for glycosomal protein import. The dependence of protein import on an SKL-like motif found at the C termini of several glycosomal proteins has unmistakably shown a close evolutionary relationship between the peroxisomes of higher eukaryotes and glycosomes. These studies have also revealed some novel targeting signals that may function exclusively for import into the glycosomes, including the C-terminal SSL sequence of glycosomal PGK. The lack of a suitable transfection system of trichomonads has thus far not allowed direct *in vivo* testing of targeting signals functioning in hydrogenosomal protein import. However, a recently developed *in vitro* import assay has confirmed the observation that hydrogenosomal proteins are made as larger precursor proteins which are processed upon import into the organelle. The cleavable signal sequences found at the N-termini of eleven *T. vaginalis* hydrogenosomal proteins are generally shorter and somewhat more conserved than those found on mitochondrial proteins, but it remains to be seen if a primary sequence requirement for import exists in the hydrogenosomal targeting signal. The major criterion for a leader sequence to function in targeting proteins to the hydrogenosomes thus remains to be determined.

Acknowledgements

We thank our colleagues Drs G. Keller, M. Müller, and B. Ullman for continued support and encouragement and for sharing unpublished results. This work was supported by Public Health Service Grant AI-21786 (to C. C. W.), grant AI 27857 and by a Burroughs–Wellcome New Investigator in Molecular Parasitology Award (to P. J. J.).

References

1. Opperdoes, F. R., Baudhuin, P., Coppens, I., De Roe, C., Edwards, S. W., Weijers, P. J., and Misset, O. (1984) Purification, morphometric analysis, and characterization of the

glycosomes (microbodies) of the protozoan hemoflagellate *Trypanosoma brucei. J. Cell Biol.*, **98**, 1178.
2. de Raadt, P. (1989) Trypanosomiases. In *Tropical medicine and parasitology* (ed. R. Goldsmith and D. Heyneman) p. 256. Appleton and Lange, Norwalk, Conn.
3. Vickerman, K., Tetley, L., Hendry, K. A., and Turner, C. M. (1988) Biology of African trypanosomes in the tsetse fly. *Biol. Cell*, **64**, 109.
4. Sommer, J. M. and Wang, C. C. (1994) Targeting proteins to the glycosomes in African trypanosomes. *Annu. Rev. Microbiol.*, **48**, 105.
5. Urbina, J. A. (1994) Intermediary metabolism of *Trypanosoma cruzi*. *Parasitol. Today*, **10**, 107.
6. Cazzulo, J. J., Franke de Cazzulo, B. M., Engel, J. C., and Cannata, J. J. (1985) End products and enzyme levels of aerobic glucose fermentation in trypanosomatids. *Mol. Biochem. Parasitol.*, **16**, 329.
7. Hart, D. T. and Opperdoes, F. R. (1984) The occurrence of glycosomes (microbodies) in the promastigote stage of four major *Leishmania* species. *Mol. Biochem. Parasitol.*, **13**, 159.
8. Hammond, D. J. and Gutteridge, W. E. (1984) Purine and pyrimidine metabolism in the Trypanosomatidae. *Mol. Biochem. Parasitol.*, **13**, 243.
9. Hassan, H. F. and Coombs, G. H. (1985) Purine phosphoribosyltransferases of *Leishmania mexicana mexicana* and other flagellate protozoa. *Comp. Biochem. Physiol. [b]*, **82**, 773.
10. Clayton, C. E. (1987) Import of fructose bisphosphate aldolase into the glycosomes of *Trypanosoma brucei. J. Cell. Biol.*, **105**, 2649.
11. Hart, D. T., Baudhuin, P., Opperdoes, F. R., and de Duve, C. (1987) Biogenesis of the glycosome in *Trypanosoma brucei*: the synthesis, translocation and turnover of glycosomal polypeptides. *EMBO J.*, **6**, 1403.
12. Michels, P. A. (1986) Evolutionary aspects of trypanosomes: analysis of genes. *J. Mol. Evol.*, **24**, 45.
13. Michels, P. A. M. and Opperdoes, F. R. (1991) The evolutionary origin of glycosomes. *Parasitol. Today*, **7**, 105.
14. Sommer, J. M., Thissen, J. A., Parsons, M., and Wang, C. C. (1990) Characterization of an in vitro assay for import of 3-phosphoglycerate kinase into the glycosomes of *Trypanosoma brucei*. *Mol. Cell. Biol.*, **10**, 4545.
15. Gould, S. J., Keller, G. A., Hosken, N., Wilkinson, J., and Subramani, S. (1989) A conserved tripeptide sorts proteins to peroxisomes. *J. Cell Biol.*, **108**, 1657.
16. Blattner, J., Swinkels, B., Dorsam, H., Prospero, T., Subramani, S., and Clayton, C. (1992) Glycosome assembly in trypanosomes: variations in the acceptable degeneracy of a COOH-terminal microbody targeting signal. *J. Cell Biol.*, **119**, 1129.
17. Sommer, J. M., Cheng, Q. L., Keller, G. A., and Wang, C. C. (1992) In vivo import of firefly luciferase into the glycosomes of *Trypanosoma brucei* and mutational analysis of the C-terminal targeting signal. *Mol. Biol. Cell*, **3**, 749.
18. Sommer, J. M., Peterson, G., Keller, G. A., Parsons, M., and Wang, C. C. (1993) The C-terminal tripeptide of glycosomal phosphoglycerate kinase is both necessary and sufficient for import into the glycosomes of *Trypanosoma brucei*. *FEBS Lett.*, **316**, 53.
19. Fung, K. and Clayton, C. (1991) Recognition of a peroxisomal tripeptide entry signal by the glycosomes of *Trypanosoma brucei*. *Mol. Biochem. Parasitol.*, **45**, 261.
20. Sommer, J. M., Nguyen, T. T., and Wang, C. C. (1994) Phosphoenolpyruvate carboxykinase of *Trypanosoma brucei* is targeted to the glycosomes by a C-terminal sequence. *FEBS Lett.*, **350**, 125.
21. Waterham, H. R., Titorenko, V. I., Haima, P., Cregg, J. M., Harder, W., and Veenhuis, M.

(1994) The *Hansenula polymorpha* PER1 gene is essential for peroxisome biogenesis and encodes a peroxisomal matrix protein with both carboxy- and amino-terminal targeting signals. *J. Cell Biol.*, **127**, 737.
22. Small, G. M., Szabo, L. J., and Lazarow, P. B. (1988) Acyl-CoA oxidase contains two targeting sequences each of which can mediate protein import into peroxisomes. *EMBO J.*, **7**, 1167.
23. Kragler, F., Langeder, A., Raupachova, J., Binder, M., and Hartig, A. (1993) Two independent peroxisomal targeting signals in catalase A of *Saccharomyces cerevisiae*. *J. Cell Biol.*, **120**, 665.
24. Keller, G. A., Gould, S., Deluca, M., and Subramani, S. (1987) Firefly luciferase is targeted to peroxisomes in mammalian cells. *Proc. Natl. Acad. Sci. U S A*, **84**, 3264.
25. Visser, N. and Opperdoes, F. R. (1980) Glycolysis in *Trypanosoma brucei*. *Eur. J. Biochem.*, **103**, 623.
26. Cornette, J. L., Cease, K. B., Margalit, H., Spouge, J. L., Berzofsky, J. A., and DeLisi, C. (1987) Hydrophobicity scales and computational techniques for detecting amphipathic structures in proteins. *J. Mol. Biol.*, **195**, 659.
27. Allen, T. E. and Ullman, B. (1993) Cloning and expression of the hypoxanthine-guanine phosphoribosyltransferase gene from *Trypanosoma brucei*. *Nucleic Acids Res.*, **21**, 5431.
28. Allen, T. E. and Ullman, B. (1994) Molecular characterization and overexpression of the hypoxanthine-guanine phosphoribosyltransferase gene from *Trypanosoma cruzi*. *Mol. Biochem. Parasitol.*, **65**, 233.
29. Michels, P. A., Poliszczak, A., Osinga, K. A., Misset, O., Van, B. J., Wierenga, R. K., Borst, P., and Opperdoes, F. R. (1986) Two tandemly linked identical genes code for the glycosomal glyceraldehyde-phosphate dehydrogenase in *Trypanosoma brucei*. *EMBO J.*, **5**, 1049.
30. Kendall, G., Wilderspin, A. F., Ashall, F., Miles, M. A., and Kelly, J. M. (1990) *Trypanosoma cruzi* glycosomal glyceraldehyde-3-phosphate dehydrogenase does not conform to the 'hotspot' topogenic signal model. *EMBO J.*, **9**, 2751.
31. Hannaert, V., Blaauw, M., Kohl, L., Allert, S., Opperdoes, F. R., and Michels, P. A. (1992) Molecular analysis of the cytosolic and glycosomal glyceraldehyde-3-phosphate dehydrogenase in *Leishmania mexicana*. *Mol. Biochem. Parasitol.*, **55**, 115.
32. Opperdoes, F. R. and Michels, P. A. (1993) The glycosomes of the Kinetoplastida. *Biochimie*, **75**, 231.
33. Marchand, M., Kooystra, U., Wierenga, R. K., Lambeir, A. M., Van Beeumen, J., Opperdoes, F. R., and Michels, P. A. (1989) Glucosephosphate isomerase from *Trypanosoma brucei*. Cloning and characterization of the gene and analysis of the enzyme. *Eur. J. Biochem.*, **184**, 455.
34. Nyame, K., Dothi, C. D., Opperdoes, F. R., and Michels, P. A. M. (1994) Subcellular distribution and characterization of glucosephosphate isomerase in *Leishmania mexicana mexicana*. *Mol. Biochem. Parasitol.*, **67**, 269.
35. Shames, S. L., Fairlamb, A. H., Cerami, A., and Walsh, C. T. (1986) Purification and characterization of trypanothione reductase from *Crithidia fasciculata*, a newly discovered member of the family of disulfide-containing flavoprotein reductases. *Biochemistry*, **25**, 3519.
36. Aboagye-Kwarteng, T., Smith, K., and Fairlamb, A. H. (1992) Molecular characterization of the trypanothione reductase gene from *Crithidia fasciculata* and *Trypanosoma brucei*: comparison with other flavoprotein disulphide oxidoreductases with respect to substrate specificity and catalytic mechanism. *Mol. Microbiol.*, **6**, 3089.

37. Taylor, M. C., Kelly, J. M., Chapman, C. J., Fairlamb, A. H., and Miles, M. A. (1994) The structure, organization, and expression of the *Leishmania donovani* gene encoding trypanothione reductase. *Mol. Biochem. Parasitol.*, **64**, 293.
38. Sullivan, F. X., Sobolov, S. B., Bradley, M., and Walsh, C. T. (1991) Mutational analysis of parasite trypanothione reductase: acquisition of glutathione reductase activity in a triple mutant. *Biochemistry*, **30**, 2761.
39. Smith, K., Opperdoes, F. R., and Fairlamb, A. H. (1991) Subcellular distribution of trypanothione reductase in bloodstream and procyclic forms of *Trypanosoma brucei*. *Mol. Biochem. Parasitol.*, **48**, 109.
40. Meziane-Cherif, D., Aumercier, M., Kora, I., Sergheraert, C., Tartar, A., Dubremetz, J. F., and Ouaissi, M. A. (1994) *Trypanosoma cruzi*: immunolocalization of trypanothione reductase. *Exp. Parasitol.*, **79**, 536.
41. Kuriyan, J., Kong, X. P., Krishna, T. S., Sweet, R. M., Murgolo, N. J., Field, H., Cerami, A., and Henderson, G. B. (1991) X-ray structure of trypanothione reductase from *Crithidia fasciculata* at 2.4-Å resolution. *Proc. Natl. Acad. Sci. USA*, **88**, 8764.
42. Swinkels, B. W., Gould, S. J., Bodnar, A. G., Rachubinski, R. A., and Subramani, S. (1991) A novel, cleavable peroxisomal targeting signal at the amino-terminus of the rat 3-ketoacyl-CoA thiolase. *EMBO J.*, **10**, 3255.
43. Osumi, T., Tsukamoto, T., Hata, S., Yokota, S., Miura, S., Fujiki, Y., Hijikata, M., Miyazawa, S., and Hashimoto, T. (1991) Amino-terminal presequence of the precursor of peroxisomal 3-ketoacyl-CoA thiolase is a cleavable signal peptide for peroxisomal targeting. *Biochem. Biophys. Res. Commun.*, **181**, 947.
44. Gietl, C., Faber, K. N., van der Klei, I. J., and Veenhuis, M. (1994) Mutational analysis of the N-terminal topogenic signal of watermelon glyoxysomal malate dehydrogenase using the heterologous host *Hansenula polymorpha*. *Proc. Natl. Acad. Sci. USA*, **91**, 3151.
45. de Hoop, M. J. and Ab, G. (1992) Import of proteins into peroxisomes and other microbodies. *Biochem. J.*, **286**, 657.
46. Faber, K. N., Keizer-Gunnink, I., Pluim, D., Harder, W., Ab, G., and Veenhuis, M. (1995) The N-terminus of amine oxidase of *Hansenula polymorpha* contains a peroxisomal targeting signal. *FEBS Lett.*, **357**, 115.
47. Blattner, J., Dörsam, H., and Clayton, C. E. (1995) Function of N-terminal import signals in trypanosome microbodies. *FEBS Letters*, **360**, 310.
48. McCollum, D., Monosov, E., and Subramani, S. (1993) The *pas8* mutant of *Pichia pastoris* exhibits the peroxisomal protein import deficiencies of Zellweger syndrome cells—the PAS8 protein binds to the COOH-terminal tripeptide peroxisomal targeting signal, and is a member of the TPR protein family. *J. Cell Biol.*, **121**, 761.
49. Brocard, C., Kragler, F., Simon, M. M., Schuster, T., and Hartig, A. (1994) The tetratricopeptide repeat-domain of the PAS10 protein of *Saccharomyces cerevisiae* is essential for binding the peroxisomal targeting signal SKL. *Biochem. Biophys. Res. Commun.*, **204**, 1016.
50. Dodt, G., Braverman, N., Wong, C., Valle, D., and Gould, S. J. (1994) The cytosolic human peroxisomal targeting signal 1 receptor (PTS1R) rescues the peroxisomal assembly defect in a patient with neonatal adrenoleukodystrophy. *Mol. Biol. Cell*, **5 (S)**, 478.
51. Cregg, J. M., Liu, H., Veenhuis, M., and Tan, X. (1994) The *per3* gene product of *Pichia pastoris* is a peroxisomal integral membrane protein required for peroxisome biogenesis. *Mol. Biol. Cell*, **5 (S)**, 479.
52. Kalish, J. E., Morrell, J., and Gould, S. J. (1994) The *Pichia pastoris pas10* gene encodes a zinc finger and is required for peroxisome assembly. *Mol. Biol. Cell.*, **5 (S)**, 478.
53. Gould, S. J., Theda, C., Morrell, J. C., Berg, J. M., and Kalish, J. E. (1994) A zinc-binding

protein in the peroxisome membrane is required for an early step in peroxisome assembly. *Mol. Biol. Cell*, **5 (S)**, 478.
54. Wiebel, F. F. and Kunau, W. H. (1992) The Pas2 protein essential for peroxisome biogenesis is related to ubiquitin-conjugating enzymes. *Nature*, **359**, 73.
55. Crane, D. I., Kalish, J. E., and Gould, S. J. (1994) The *Pichia pastoris* PAS4 gene encodes a ubiquitin-conjugating enzyme required for peroxisome assembly. *J. Biol. Chem.*, **269**, 21835.
56. Glover, J. R., Andrews, D. W., and Rachubinski, R. A. (1994) *Saccharomyces cerevisiae* peroxisomal thiolase is imported as a dimer. *Proc. Natl. Acad. Sci. USA*, **91**, 10541.
57. McNew, J. A. and Goodman, J. M. (1994) An oligomeric protein is imported into peroxisomes in vivo. *J. Cell Biol.*, **127**, 1245.
58. Opperdoes, F. R. (1984) Localization of the initial steps in alkoxyphospholipid biosynthesis in glycosomes (microbodies) of *Trypanosoma brucei*. *FEBS Lett.*, **169**, 35.
59. Clarkson, A. B. J. and Brohn, F. H. (1976) Trypanosomiasis: an approach to chemotherapy by the inhibition of carbohydrate catabolism. *Science*, **194**, 204.
60. Willson, M., Callens, M., Kuntz, D. A., Perié, J., and Opperdoes, F. R. (1993) Synthesis and activity of inhibitors highly specific for the glycolytic enzymes from *Trypanosoma brucei*. *Mol. Biochem. Parasitol.*, **59**, 201.
61. Willson, M., Lauth, N., Perie, J., Callens, M., and Opperdoes, F. R. (1994) Inhibition of glyceraldehyde-3-phosphate dehydrogenase by phosphorylated epoxides and alpha-enones. *Biochemistry*, **33**, 214.
62. Leipe, D. D., Gunderson, J. H., Nerad, T. A., and Sogin, M. L. (1993) Small subunit ribosomal RNA+ of *Hexamita inflata* and the quest for the first branch in the eukaryotic tree. *Mol. Biochem. Parasitol.*, **59**, 41.
63. Lindmark, D. G. and Müller, M. (1973) Hydrogenosome, a cytoplasmic organelle of the anaerobic flagellate *Tritrichomonas foetus*, and its role in pyruvate metabolism. *J. Biol. Chem.*, **248**, 7724.
64. Lindmark, D. G. and Müller, M. (1975) Hydrogenosomes in *Trichomonas vaginalis*. *J. Parasitol.*, **61**, 552.
65. Honigberg, B. M. and Brugerolle, G. (1990) Structure. In *Trichomonads parasitic in humans*. (ed. B. M. Honigberg), p. 5. Springer-Verlag, New York.
66. Müller, M. (1993) The hydrogenosome. *J. Gen. Microbiol.*, **139**, 2879.
67. Honigberg, B. M., Volkmann, D., Entzeroth, R., and Scholtyseck, E. (1984) A freeze-fracture electron microscope study of *Trichomonas vaginalis* Donné and *Tritrichomonas foetus* (Riedmüller). *J. Protozool.*, **31**, 116.
68. Müller, M. (1988) Energy metabolism of protozoa without mitochondria. *Annu. Rev. Microbiol.*, **42**, 465.
69. Steinbüchel, A. and Müller, M. (1986) Anaerobic pyruvate metabolism of *Tritrichomonas foetus* and *Trichomonas vaginalis* hydrogenosomes. *Mol. Biochem. Parasitol.*, **20**, 57.
70. Johnson, P. J. (1993) Metronidazole and drug resistance. *Parasitol. Today*, **9**, 183.
71. Lahti, C. J. and Johnson, P. J. (1991) *Trichomonas vaginalis* hydrogenosomal proteins are synthesized on free polyribosomes and may undergo processing upon maturation. *Mol. Biochem. Parasitol.*, **46**, 307.
72. Kulda, J., Nohynkova, E., and Ludvik, J. (1986) Basic structure and function of the trichomonad cell. *Acta Univ. Carol. Biol.*, **30**, 181.
73. Johnson, P. J., d'Oliveira, C. E., Gorrell, T. E., and Müller, M. (1990) Molecular analysis of the hydrogenosomal ferredoxin of the anaerobic protist Trichomonas vaginalis. *Proc. Natl. Acad. Sci. USA*, **87**, 6097.
74. Lahti, C. J., d'Oliveira, C. E., and Johnson, P. J. (1992) Beta-succinyl coenzyme A syn-

thetase from *Trichomonas vaginalis* is a soluble hydrogenosomal protein with an amino terminal sequence that resembles mitochondrial presequences. *J. Bacteriol.*, **174**, 6822.

75. Lahti, C. J., Bradley, P. J., and Johnson, P. J. (1994) Molecular characterization of the α-subunit of the *Trichomonas vaginalis* hydrogenosomal succinyl CoA synthetase. *Mol. Biochem. Parasitol.*, **66**, 309.

76. Lange, S., Rozario, C., and Müller, M. (1994) Primary structure of the hydrogenosomal adenylate kinase of *Trichomonas vaginalis* and its phylogenetic relationships. *Mol. Biochem. Parasitol.*, **66**, 297.

77. Hrdy, I. and Müller, M. (1995) Primary structure and eubacterial relationships of the pyruvate:ferredoxin oxidoreductase of the amitochondriate eukaryote *Trichomonas vaginalis*. *J. Mol. Evol.* in press.

78. Hrdy, I. and Müller, M. (1995) Primary structure of the hydrogenosomal malic enzyme of *Trichomonas vaginalis* and its relationship to homologous enzymes. *J. Eukaryot. Microbiol.* (in press).

79. Hendrick, J. P., Hodges, P. E., and Rosenberg, L. E. (1989) Survey of amino terminal proteolytic cleavage sites in mitochondrial precursor proteins: leader peptides cleaved by two matrix proteases share a three amino acid motif. *Proc. Natl. Acad. Sci. USA*, **86**, 4056.

80. von Heijne, G. (1986) Mitochondrial targeting sequences may form amphiphilic alpha helices. *EMBO J.*, **5**, 1335.

81. Johnson, P. J., Bradley, P. J., and Lahti, C. J. (1995) Cell biology of trichomonads: protein targeting to the hydrogenosome. In *Molecular approaches to parasitology*, (ed. J. C. Boothroyd and R. Komuniecki) Vol. 12, p. 399. Wiley-Liss, New York.

82. Cavalier-Smith, T. (1987) The simultaneous symbiotic origin of mitochondria, chloroplasts and microbodies. *Ann. NY Acad. Sci.*, **503**, 55.

83. Yarlett, N., Orpin, C. G., Munn, E. A., Yarlett, N. C., and Greenwood, C. A. (1986) Hydrogenosomes in the rumen fungus *Neocallimastix patriciarum*. *Biochem. J.*, **236**, 729.

84. O'Fallon, J. V., Wright, R. W., and Calza, R. E. (1991) Glucose metabolic pathways in the anaerobic rumen fungus *Neocallimastrix frontalis* EB188. *Biochem. J.*, **274**, 595.

85. Marvin-Sikkema, F. D., Lahpor, G. A., Kraak, M. N., Gottschal, J. C., and Prins, R. A. (1992) Characterization of an anaerobic fungus from llama faeces. *J. Gen. Microbiol.*, **138**, 2235.

86. Lloyd, D., Hillman, K., Yarlett, N., and Williams, A. G. (1989) Hydrogen production by rumen holotrich protozoa: Effects of oxygen and implications for metabolic control by in situ conditions. *J. Protozool.*, **36**, 205.

87. Paul, R. G., Williams, A. G., and Bulter, R. D. (1990) Hydrogenosomes in the rumen entodiniomorphid ciliate *Polyplastron multivesiculatum*. *J. Gen. Microbiol.*, **136**, 1981.

88. Yarlett, N., Hann, A. C., Lloyd, D., and Williams, A. (1981) Hydrogenosome in the rumen protozoan *Dasytricha ruminantium* Schuberg. *Biochem. J.*, **200**, 365.

89. van Bruggen, J. J. A., Stumm, C. K., and Vogels, G. D. (1983) Symbiosis of methanogenic bacteria and sapropelic protozoa. *Arch. of Microbiol.*, **136**, 89.

90. Fenchel, T. and Finlay, B. J. (1991) The biology of free-living anaerobic ciliates. *Eur. J. Protistol.*, **26**, 201.

91. Broers, C. A. M., Stumm, C. K., Vogels, G. D., and Brugerolle, G. (1990) *Psalteriomonas lanterna* gen. nov./sp. nov., a free-living amoeboflagellate isolated from freshwater anaerobic sediments. *Eur. J. Protistol.*, **25**, 369.

92. Marvin-Sikkema, F. D., Kraak, M. N., Veenhuis, M., Gottschal, J. C., and Prins, R. A. (1993) The hydrogenosomal enzyme hydrogenase from anaerobic fungus *Neocallimastix* sp. L2 is recognized by antibodies, directed against the C-terminal microbody protein targeting signal SKL. *Eur. J. Cell Biol.*, **61**, 86.

10 | Mechanisms of drug resistance in protozoan parasites

DYANN F. WIRTH and ALAN COWMAN

1. Emergence of drug resistance and its impact

Drug resistance in microorganisms is an increasing problem in world health today. Commonly used antimicrobial agents are no longer effective and the rate at which resistance develops, often to multiple drugs, has increased the urgency for a solution to this problem. Drug-resistant strains have been found among almost all infectious organisms from viruses to multicellular parasitic nematodes. The impact of drug-resistant infectious diseases on world health is just beginning to be assessed. Certain antimicrobial agents are no longer usable as resistance is so widespread. An example of this trend is the antimalarial drug chloroquine; in many malaria-endemic areas, chloroquine resistance is so prevalent that the drug is no longer recommended. Yet, antimalarials are the most commonly purchased drugs in the world. Resistance to other antimalarials is emerging faster than new drugs are being developed. Similar trends are apparent in other infectious diseases and the problem of drug resistance promises to be a major challenge facing the world in the later part of this decade and into the 21st century.

Since chemotherapeutic agents remain the first-line defence for the treatment of parasitic infections, the problem of drug resistance complicates both therapeutic and prophylactic regimens. In the case of diseases with high mortality, such as malaria, the necessity for early identification of resistant infections is particularly acute. Similarly, the identification of alternative drugs is particularly important. One of the imperatives of the field of chemotherapy of protozoan parasites is the need to discover and develop new drugs that will be effective against drug resistant organisms. Malaria is leading the field in the emergence of drug-resistance but other diseases are displaying reduced drug susceptibility, a clear precursor to the emergence of resistant organisms.

A problem which is particularly manifest in protozoan parasites is the number of drugs for which neither the mechanism of action nor the mechanism of resistance is well understood. Many drugs have been discovered empirically and only recently has there been a significant focus on drug targets and mechanisms of resistance.

The goals of this chapter are to delineate what is known about drug targets and resistance mechanisms and to illuminate future directions for research.

2. Role of intrinsic genetic and biological components

One important consideration regarding the emergence and spread of drug resistance in protozoan parasites is the life cycle of the organism and its genetic properties. For example, human malaria does not have an animal reservoir and therefore there is the potential for the entire parasite population to be exposed to drug while in the human host. Furthermore, once a resistant organism emerges it can be readily transmitted to another person. In contrast, zoonotic diseases such as leishmaniasis have a major animal reservoir, and in many areas of the world human infection derives primarily from the animal reservoirs. In this case, the exposure of the parasite population to drug is limited and even in the case where a resistant organism emerges in a treated person, the probability of that organism being transmitted to another human is low. Another factor in the rate of emergence of drug resistant organisms is the number of parasites exposed to drug. In a single person infected with *Plasmodium falciparum* the parasite burden can reach as high as 10^{11} organisms, whereas in infections such as leishmaniasis the parasite burden is typically four to six logs lower. A final consideration is the ploidy of the organism. For haploid organisms such as *Plasmodium* spp., point mutations are immediately phenotypically expressed, at least for traits which involve single genes. In contrast, in diploid organisms such as *Leishmania* spp., point mutations may result in no phenotypic change if the trait is recessive. Thus, many factors including genetic, biological, and ecological components influence the emergence and spread of drug resistance. Analysis of current trends indicates that the prevalence of drug resistance in protozoan parasites parallels the situation in other micro-organisms and is increasing rather than decreasing.

3. Impact of drug resistance on public health and disease

The most striking example of the impact of drug-resistant protozoan parasites on public health and disease is the development of resistance by the human malaria parasite, *P. falciparum*, to most antimalarial agents. The combined use of chloroquine, a safe, cheap, and highly effective antimalarial, and DDT to control the mosquito vector, was the strategy for malaria eradication (1). The emergence of vector resistance to DDT and *P. falciparum* resistance to chloroquine dashed the premature optimism. Chloroquine resistance has spread to most parts of the world where malaria is endemic (2) and this necessitates the use of alternative drugs that are effective against parasites increasingly resistant to chloroquine (3). The development of resistance to chloroquine has not yet resulted in a major increase in deaths from the disease, at least in areas where there are only moderate levels of resistance, but it is having an important impact on the cost of treatment (4). In the case of drugs used in the treatment of other parasitic infections, for example leishmaniasis, the limited number of existing drugs, including pentavalent antimony, are both toxic and require long treatment regimens (5–7). Decreased sensitivity of the infection to these drugs will mean even longer treatment regimens with increased side effects in the patient (8, 9). One concern is that with decreased drug efficacy and increased cost, the number of infections which go untreated or undertreated will increase resulting in more serious disease, increased resistance, and a significant public health impact.

4. Common themes of resistance in protozoan parasites

The ability of protozoan parasites to develop resistance to many of the drugs used to combat them is due to their genetic adaptability that enables the selection of appropriate strategies to circumvent the lethal effects of antiprotozoal compounds. Gene amplification is a strategy that is very common among protozoan parasites. This can involve amplification and overexpression of either a gene encoding the target enzyme of the particular drug or a gene that encodes a protein capable of transporting the drug or in some way affecting its concentration in the cell. A second common mechanism of resistance can occur when antiprotozoal drugs act as competitive inhibitors of enzyme activity. Mutations in the enzyme can decrease its affinity for the drug, providing the parasite with a degree of resistance. The ability of protozoan parasites to adapt to the selective pressures provided by chemotherapeutic attack is a major challenge and a greater understanding of the targets and resistance mechanisms used by protozoan parasites is clearly required.

5. Amplification and drug resistance

The amplification of genes in drug resistance most commonly involves amplification of the target enzyme, but an equally important mechanism is the amplification of genes which encode molecules involved in drug transport. The best characterized examples of gene amplification are in *Leishmania* spp (10). This parasite has the capacity to amplify genes as extrachromosomal circles and linear molecules, thereby increasing the gene copy number and, correspondingly, the expression of the proteins encoded by the amplified DNA segments. In contrast, in the closely related organism, *Trypanosoma brucei*, there are not yet examples of amplification associated with drug resistance. Whether this relates to differences in the ability of these organisms to produce and replicate extrachromosomal elements or to fundamental differences in the biochemical targets remains to be elucidated. Further information on amplicon structure and formation is covered in Chapter 2.

The association of amplified extrachromosomal circles with drug resistance implicated these circles in conferring drug resistance, but definitive proof was provided by transfection analysis (11–17). This type of analysis is particularly important in the case of the large amplicons where several genes may be amplified. In the case of *Leishmania* spp., shuttle plasmid expression vectors were developed which remained as extrachromosomal elements in transfected parasites. These plasmids are present in multiple copies and thus can be used to directly test the functional role of any sequence from the amplified circles. In this way, the functional elements from the amplified DNA can be directly tested by transfection of sensitive parasites with expression vectors containing either single genes or segments of the amplified circle and subsequent testing for the drug resistance phenotype. In a parallel development, using the method of targeted gene disruption, the test gene can be 'knocked-out' in the amplicon and the remainder of the amplicon can be tested for contributory or overlapping roles in conferring resistance (18). Combining these approaches has proven to be extremely powerful and has allowed the functional identification of those genes necessary for conferring drug resistance.

The first identified example of gene amplification in drug resistance was the amplification of dihydrofolate reductase/thymidylate synthase (DHFR/TS) in methotrexate resistant *L. major* (10). Parasites which were selected in a stepwise manner for decreased sensitivity to killing by the antifolate methotrexate displayed an amplification of two regions of DNA as extrachromosomal circles. These were termed the R-circle and the H-circle. Subsequent analysis demonstrated that the R-circle contained the DHFR/TS gene and transfection experiments have demonstrated that the amplification of the DHFR/TS gene can confer methotrexate resistance. Thus, amplification of the R-circle alone is sufficient to confer drug resistance. It is interesting that recent analysis of the H-circle has identified at least two genes involved in conferring drug resistance, one encodes a short-chain dehydrogenase involved in methotrexate resistance and the other, *ltpgpA*, is a member of the *mdr*-related gene family and implicated in oxyanion and arsenite resistance (11, 12, 19, 20).

Selection of parasites with several other drugs has resulted in the amplification of target enzymes. *Leishmania* parasites selected for resistance to tunicamycin, an inhibitor of glycosyltransferase, have been shown to possess an amplified extrachromosomal circle which contains the glycosyltransferase coding region (21, 22). The amplified DNA is derived from a chromosomal location distinct from both the R-region and the H-region. As in the case of the R-region, the amplified circle contains several additional transcribed regions but subsequent transfection analysis has demonstrated that only overexpression of the glycosyltransferase gene is required for tunicamycin resistance (22). Selection of *Leishmania* in increasing concentrations of α-difluoromethylornithine results in increased ornithine decarboxylase activity, a phenotype which is unstable in the absence of drug (23). Two types of amplification elements are found, a linear 140kb element which contains the ODC gene and a circular element of 70kb which lacks the ornothine decarboxylase gene (24). Selection of *Leishmania* with mycophenolic acid has resulted in a cell line with an amplified inosine monophosphate-dehydrogenase. In this case, the amplified DNA is in the form of a linear extrachromosomal element amplified 10–20 times (25, 26).

6. Mutations in target enzymes

Many antiprotozoal drugs act on a specific enzyme in a metabolic pathway and a common mechanism to evade the action of some drugs is mutation of the target enzyme. This usually results in reduced affinity of binding of the antiprotozoal compound without greatly affecting the normal function of the enzyme and the ability of the parasite to survive. While there are several examples of drug resistant organisms in which this mechanism is implicated, only a few have been characterized at the molecular level.

7. Folate antagonists

The antifolate drugs inhibit enzymes of the protozoan folic acid biosynthesis pathway (see Fig. 1). High-level resistance to antifolate drugs is now very common in malaria parasites and it is clear that the major mechanism by which resistance is acquired is by accumulation of mutations in target enzymes (27, 28). The DHFR inhibitors pyrimethamine and proguanil are the most important antifolate drugs used to treat malaria, but resistance to both is now widespread (29, 30). The effect of the DHFR inhibitors on the malaria parasite is to reduce the biosynthesis of folate resulting in decreased synthesis of pyrimidines and blockage of DNA replication (31), as well as decreased conversion of glycine to serine and decreased methionine synthesis (32).

Early studies of pyrimethamine resistance in rodent malaria models showed that the resistance involved a single gene (33–35) and kinetic studies on the DHFR enzyme suggested that the major mechanism of resistance was reduced affinity of binding of the competitive inhibitor to DHFR (36). Cloning and sequence analysis of the *P. chabaudi* DHFR gene in resistant parasites identified an alteration at amino acid 106 (from the wild-type serine to asparagine) suggesting that this was important in the generation of resistance (37).

Kinetic studies of the DHFR enzyme in pyrimethamine-resistant isolates of *P. falciparum* originally obtained from the field showed that the enzyme had a reduced affinity for the drug (38, 39). Isolation of the DHFR gene from *P. falciparum* showed that it was a bifunctional enzyme with thymidylate synthetase (27, 40), as is the case with most protozoa so far examined (41). A genetic cross between pyrimethamine-resistant and pyrimethamine-sensitive parents showed that resistance segregated with the DHFR allele from the resistant parent (28). Sequence analysis of the DHFR gene from the parents and progeny of the cross and from other *P. falciparum* isolates showed that the key amino acid whose mutation results in resistance is located at position 108 (27, 28, 42, 43). This is homologous to the change at amino acid 106 noted above. Additional mutations towards the N-terminus of DHFR confer greatly increased levels of pyrimethamine resistance. Expression of the different *P. falciparum* DHFR alleles in *Escherichia coli* has allowed direct analysis of the effect of the different mutations on the enzyme kinetics and binding of pyrimethamine (44).

It has been shown that field isolates of *P. falciparum* can be resistant to

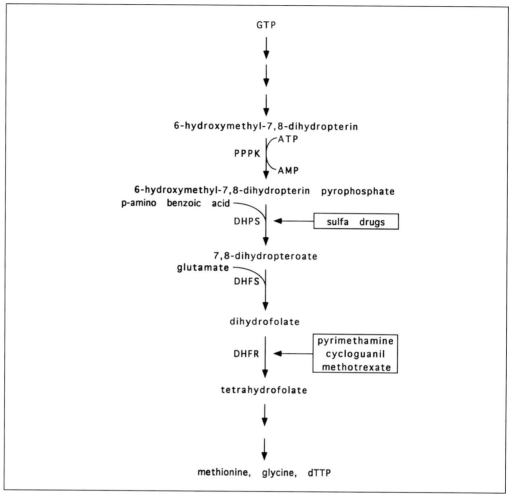

Fig. 1 Scheme of the possible folate biosynthetic pathway in protozoan parasites. The potential sites of action of particular drugs (boxed) are shown. PPPK, 6-hydroxymethyl-7,8-dihydropterin pyrophosphokinase; DHPS, dihydropteroate synthetase; DHFS, dihydrofolate synthetase; DHFR, dihydrofolate reductase. DHFS has not yet been identified in *P. falciparum*.

pyrimethamine and cycloguanil (45, 46). In some cases, the isolates represent a mixture of parasites that are resistant to pyrimethamine or cycloguanil alone, and in other cases the *P. falciparum* lines show a true cross-resistance phenotype and are resistant to both compounds simultaneously (47, 48). Sequence analysis of the DHFR gene in these isolates has shown that amino acid 108 in DHFR is important in resistance to cycloguanil as well as to pyrimethamine. Interestingly, a subsequent mutation at position 164 can confer high-level cross-resistance.

The sulphone and sulphonamide (sulfa) drugs act on the folic acid biosynthetic pathway either by directly inhibiting the enzyme dihydropteroate synthase (DHPS) or by inhibiting DHFS (the next enzyme in the pathway), following conversion to the dihydropteroate analogue by DHPS. Analysis of the mechanism of sulfa drug resistance has shown that the metabolism of sulfadoxine to the dihydropteroate analogue is reduced (49). The gene for DHPS has recently been cloned (50, 51) and shown to encode a bifunctional enzyme with 6-hydroxymethyl-7,8-dihydropterin pyrophosphokinase, the enzyme that mediates the previous step in the biosynthetic pathway. Sequence analysis of the DHPS gene from sulfadoxine-resistant and sulfadoxine-sensitive *P. falciparum* has identified amino acid differences that may be involved in the mechanism of resistance.

8. Drug transport

Transport across membranes is essential for drugs to reach their intracellular targets. Drugs are often recognized and transported in the same way as natural substrates and one common mechanism of resistance in bacterial systems is a defect in this transport. Similarly, examples of reduced transport of certain drugs have been found in parasite systems. *Leishmania mexicana* strains resistant to toxic nucleosides display reduced purine accumulation (52) and trypanosomes resistant to melarsen lack one of the high-affinity adenosine transporters found in wild-type cells, suggesting that resistance to arsenicals is due to loss of uptake (53). As discussed below, there remains significant controversy as to the role of drug transport in chloroquine resistant *Plasmodium falciparum*.

A second major mechanism of drug resistance that involves drug transport is increased drug efflux. The phenomenon of multidrug resistance (MDR) was first described in mammalian neoplastic cells derived either through step-wise selection in various anti-neoplastic drugs *in vitro* or from tumours which became refractory to chemotherapy (54). Neoplastic cells subjected to treatment with a single drug became resistant to multiple, unrelated drugs through a common mechanism of drug efflux mediated by overexpression of a membrane transport protein, the P-glycoprotein. The primary role of this P-glycoprotein mediated mechanism of drug resistance has been demonstrated by transfection of drug sensitive cells with the cDNA of the *mdr 3* (*mdr1a*) gene (55). *mdr* homologues have been found in other organisms, including mouse (56–59), rat (60), hamster (61), yeast (62, 63), and some protozoans (15, 64–67). They all share the same predicted structures, namely an internal duplication with each half having six transmembrane domains and one highly conserved nucleotide-binding motif. It is postulated that the P-glycoproteins originated from the duplication of the bacterial membrane transporters genes (68).

Although it is well established that the P-glycoprotein is responsible for multidrug resistance in mammalian cells, the role of this gene in normal cells has yet to be firmly established. However, recent experiments in transgenic mice have investigated the putative functions of both the *mdr2* and *mdr1/3* genes. Neither gene is essential for mouse development, but defects have been identified in phospholipid transport in *mdr2* knockout mice, while in *mdr3* knockout mice the integrity of the blood–brain barrier has been disrupted (69, 70). Evidence from analysis of drug efflux, and the similarity of the P-glycoprotein structure with bacterial transport proteins, has led to the hypothesis that the P-glycoprotein is directly involved in the transport of drugs from the cell. In support of this hypothesis, recent work has demonstrated that the mouse *mdr* gene can functionally complement *ste6* in yeast for the transport of pheromone (63). Second, reconstitution experiments have shown that the P-glycoprotein has a drug-inducible ATPase activity (71). Recently, it has also been shown to regulate a chloride-selective channel activity which is volume regulated and ATP dependent (72). The chloride ion channel activity can be dissociated from the drug-binding activity of the P-glycoprotein (73).

mdr-like genes have been studied recently in some parasitic protozoa, including *Plasmodium falciparum* (64, 65, 74), *Entamoeba histolytica* (66), *Leishmania tarentolae* (67), and *L. major* (11). In *L. tarentolae* and *L. major*, an *mdr*-like gene (*ltpgpA* and *lmpgpA*, respectively) was found to be amplified on an extra-chromosomal circle (H-circle) following selection with various agents such as methotrexate or arsenite. In *L. major*, transfection experiments showed that the *lmpgpA* is responsible for the resistance to arsenite and some antimonial drugs but not to the hydrophobic drugs like vinblastine or puromycin to which mammalian MDR cells are resistant (11).

A second *mdr*-like gene has been identified in both *Leishmania donovani* and *Leishmania enrietti*. Parasites were selected in increasing concentrations of vinblastine and the resulting resistant cloned parasites displayed an amplification of an extra-chromosomal circle termed the V-circle (13, 15). This amplicon contains a gene with a high degree of homology with the mammalian *mdr 1* and *mdr 3* genes and was

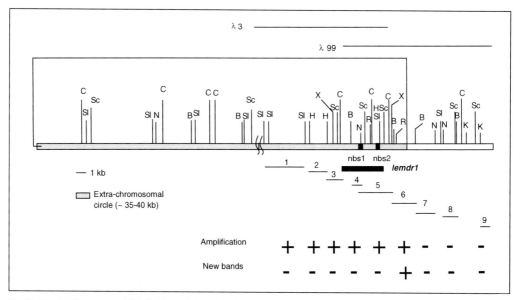

Fig. 2 Restriction map of λ3, λ99, and the V-circle. Regions covered by λ3 and λ99 are represented as bars on top of the diagram. The two nucleotide binding sites of *lemdr1* are represented as black boxes (nbs1 and nbs2). The coding region of *lemdr1* is represented by a dark bar underneath the restriction map. The stippled region represents the region which is amplified as an extra-chromosomal circle. Fragments 1–9 were used as probes for hybridization analysis. + and - signs indicate the presence or absence of amplification or new bands in the resistant cells. Sc, *Sac*I; S1, *Sal*I; B, *Bam*HI; R, *Eco*RI; K, *Kpn*I; N, *Not*I; C, *Cla*I; X, *Xho*I; H, *Hinc*II.

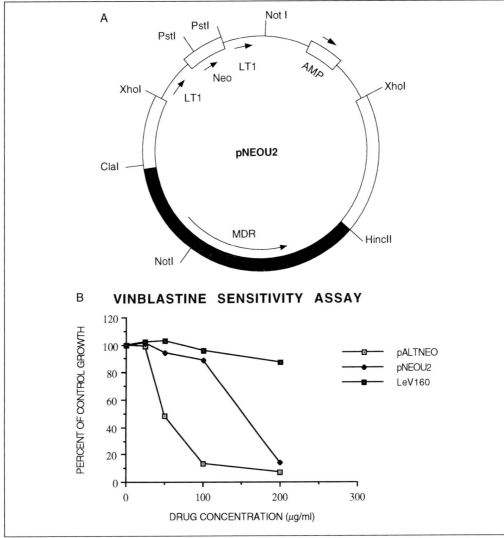

Fig. 3. Transfection of wild-type *L. enriettii* with *lemdr1* confers vinblastine resistance. A: Restriction map of pNEOU2, was prepared by subcloning the 7-kb *Xho*I fragment (see Fig. 2) into pALTNEO. B: Stably transfected cell lines of pALTNEO and pNEOU2 and the positive control LeV160 were assayed for vinblastine sensitivity at 0, 25, 50, 75, 100, and 200 µg ml^{-1}.

termed *ldmdr1* or *lemdr1*. The role of these genes was tested using direct transfection analysis in which *ldmdr1* or *lemdr1* was expressed on an amplified extrachromosomal circle. While expression of the coding region conferred low-level resistance and nonsense mutations introduced into the coding region ablated this resistance, the coding region alone did not confer high-level vinblastine resistance and little or no cross-resistance to puromycin was detected (See Fig. 3). In contrast, when the entire V-circle was first converted to a shuttle vector (using the method of homologous integration) and reintroduced into sensitive parasites, high level vinblastine resistance was observed. These data indicate that additional factors on the V-circle are necessary for high level resistance (18).

An important question remains regarding the role of these genes in the normal situation. Important insights into the function of mammalian *mdr* genes have been provided by gene 'knockout' experiments as described above. In similar experiments in *Leishmania*, the *lemdr1* gene has been disrupted by targeted deletion. Similar to the observations made in mammalian cells, *L. enriettii* that does not express the *lemdr1* gene is extremely sensitive to low levels of drug and other toxins, implying a role for these genes in protecting cells from toxic effects of drugs and other compounds.

9. Quinine-like antimalarials

Peruvian Indians in South America chewed the bark of the Chinchona tree to combat 'bad fevers' which were usually caused by attacks of malaria. The 4-aminoquinolines (chloroquine, amodiaquine, mepacrine and sontaquine) are analogues of quinine and were first synthesized in the 1930s and 40s. More recently developed related compounds include mefloquine and halofantrine. The quinine-like antimalarials have their major effect on the mature stages of the asexual life cycle of the parasite. Studies on the effect of drugs such as chloroquine on the parasite have shown that the first morphological changes are swelling of the food vacuole, vesiculation, and accumulation of undigested haemoglobin (75–78). Similar parasite organellar alterations have also been seen with mefloquine, suggesting that the major action of these drugs is to block the function of the food vacuole (79, 80). There have been a number of hypotheses to explain the action of chloroquine including alteration of vacuolar pH (81) and inhibition of enzymatic functions in the organelle (82). However, the most likely mechanism is the decoupling of ferriprotoporphyrin IX (FP) sequestration into haemozoin resulting in the accumulation of toxic amounts of FP (83–88). A haeme polymerase-like activity has been identified in *P. falciparum* that enhances the polymerization of haeme to haemozoin (89), suggesting a possible target for chloroquine and other antimalarials. The exact mechanism of killing by chloroquine has not yet been definitively determined but current evidence suggests that inhibition of the sequestration of FP is an important part.

Chloroquine is the most important antimalarial used to combat human malaria but unfortunately the malaria parasites, in particular *P. falciparum*, have been able to develop mechanisms to evade the toxic action of this drug. It is clear that chloroquine-resistant *P. falciparum* parasites concentrate less drug than do sensitive parasites (90–93) suggesting that the mode of action of the drug is independent of the mechanism of resistance. Decreased accumulation of chloroquine in resistant parasites could occur by two mechanisms: first, efflux of the drug; and second, alterations that reduce the uptake of the drug in the parasite. It is also possible that both mechanisms may be operating simultaneously.

Some studies have found that uptake of chloroquine into resistant and sensitive *P. falciparum* is the same (93, 94), whereas others found a marked difference in uptake of the drug between resistant and sensitive parasites (95–100). An explanation that may reconcile these conflicting results is that genetic heterogeneity of the parasite lines studied may influence the expression of the chloroquine-resistance phenotype.

Chloroquine concentrates several hundred-fold more in malaria-infected erythrocytes than it does in uninfected erythrocytes (101, 102). It was originally suggested that this was due to a 'receptor' such as FP (90); however, it is more likely to be a result of the weak base properties of the compound that allows its accumulation in acidic vesicles such as the food vacuole (103, 104). The amount of accumulation is therefore predicted to be dependent on the pH gradient across the food vacuole membrane and this has been confirmed by experiments showing that the level of chloroquine uptake is dependent on the pH of the external medium (92, 105, 106). Some studies show that the rate of energy-dependent uptake of chloroquine is reduced in chloroquine-resistant parasites and, although rapid efflux of chloroquine can occur, it is equivalent in both resistant and sensitive parasites (99, 100). Experimental evidence for this model has been obtained by showing that chloroquine-resistant *P. falciparum* are more sensitive to inhibition with the vacuolar ATPase inhibitor bafilomycin A1 (100). Two subunit genes of the H^+-ATPase from *P. falciparum* have been cloned and the proteins characterized but no alterations have been found in either of these subunits that could explain chloroquine resistance (107, 108).

The demonstration that the P-glycoprotein homologue 1 (Pgh1) is primarily localized to the membrane of the food vacuole (109) and that it appears to be involved in regulation of pH (110, 111) suggests it could play a role in accumulation of chloroquine. Expression of this protein in Chinese hamster ovary (CHO) cells has shown that it confers a chloroquine-sensitive phenotype on the cells (111). This occurs by an increase in acidity of the CHO cell lysosome due to the expression of the Pgh1 protein on the lysosome membrane and a resulting increased accumulation of the drug (110).

It has been suggested that chloroquine is transported out of resistant parasites at a rate 40–50 times greater than in chloroquine-sensitive parasites (93). This efflux was shown to be an energy-dependent process and could be inhibited by a number of different drugs such as verapamil and vinblastine. The chloroquine-resistant phenotype could also be reversed by an array of different drugs including verapamil and desipramine (112, 113). It has been demonstrated that desipramine in combination with chloroquine clears a chloroquine-resistant *P. falciparum* infection of owl monkeys (113). Resistance to other antimalarials such as quinine can also be modulated by verapamil and some other compounds (114). The ability of chloroquine-resistant *P. falciparum* to export chloroquine (93) and the demonstration that verapamil can modulate this process has suggested similarities with the MDR phenotype of mammalian tumour cells (see above).

10. Genetic characterization of chloroquine resistance

Widespread use of chloroquine began in the late 1940s and chloroquine resistance was reported almost simultaneously in two areas in 1960—first in South America and, soon after, in Southeast Asia. From these two foci, chloroquine-resistance appears to have spread across the malarious world. Now chloroquine-resistant malaria can be found in all areas where malaria is endemic. The long time required to develop resistance, the geographical spread, and the inability to select chloroquine resistance *in vitro* suggest that chloroquine resistance in *P. falciparum* has a complex phenotype involving a number of genes.

The demonstration that the chloroquine-resistance phenotype has similarities to the MDR phenotype of mammalian tumour cells suggested that P-glycoprotein homologues in *P. falciparum* might be involved. This led to the search for mdr gene homologues in *P. falciparum*; the first isolated was termed the *pfmdr1* gene (65, 64). It encodes Pgh1, which has the typical structure shared by many members of this protein family, with six transmembrane regions and one nucleotide binding site repeated in tandem (109). A second gene (*pfmdr2*) has also been isolated (64) and it encodes a protein with 10 putative transmembrane regions and only one nucleotide-binding fold (115, 116). The pfmdr2 protein has strong homology with the HMT1 protein of *Schizosaccharomyces pombe* (117) which is involved in heavy metal tolerance. This suggests that pfmdr2 may be involved in metal homeostasis in the parasite.

Isolation of two *mdr* gene homologues from *P. falciparum* allowed testing of the hypothesis that they were involved in the chloroquine-resistance phenotype. A genetic cross performed between a chloroquine-resistant and a chloroquine-sensitive strain of *P. falciparum* showed no linkage of the *pfmdr1* gene with the chloroquine-resistance phenotype (74); however, the chloroquine efflux phenotype segregated as a single locus on chromosome 7 (118). The locus was linked to a 400kb region of the chromosome, however; the gene has not yet been isolated.

This is in contrast to analysis of a large number of alleles of the *pfmdr1* gene, which has identified amino acid differences that appear to be linked to chloroquine resistance (119). However, sequence analysis of *pfmdr1* of more recent isolates has shown that the amino acid differences do not occur in the same pattern (120) suggesting either that the isolates used in the original study were more closely related than expected or that additional mechanisms of chloroquine resistance have developed that are independent of the *pfmdr1* gene. There is strong evidence that the amino acid differences are important in the function of the Pgh1 protein, as one common mutant allele, when expressed in CHO cells, does not cause a chloroquine-sensitive phenotype whereas the wild-type gene does (110, 111). Also, the same mutant allele when expressed in *Saccharomyces cerevisiae* does not complement the *Ste 6* mutation whereas the wild type gene does (121). *Ste 6* encodes a protein that has a similar structure to the Pgh1 protein, and it is necessary for the secretion of the a peptide, a yeast mating type factor (122). These results suggest that the amino acid differences found in the Pgh1 protein are functionally important. While it is

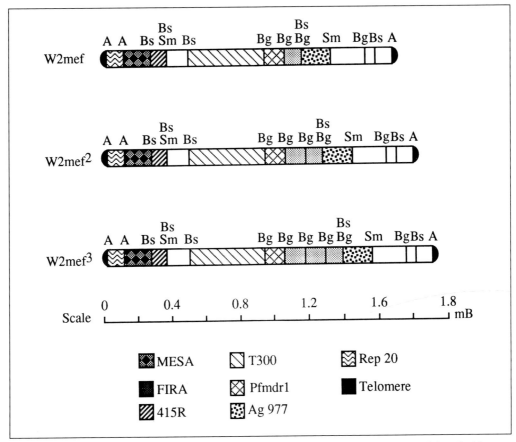

Fig. 4 Mefloquine selection of cloned lines of *P. falciparum* is linked to the amplification of the *pfmdr1* gene. The *pfmdr1* gene is localized on chromosome 5 (65) and a region of 100kb of DNA surrounding the gene is amplified in tandem to up to four copies causing an increase in size (123). W2mef was selected on increasing concentrations of mefloquine to generate the parasite lines W2mef² and W2mef³. Shaded regions indicate restriction fragments that hybridize to the probes in the key. Restriction enzyme sites are: A, *Apa*I; Bg, *Bgl*I; Bs, *Bss*HII; Sm, *Sma*I.

clear that the *pfmdr1* gene is not the only gene involved in chloroquine resistance, mutations and differences in the level of expression appear to be able to affect the level of chloroquine resistance.

Selection of *P. falciparum* mutants *in vitro* that are more resistant to mefloquine has resulted in amplification and overexpression of the *pfmdr1* gene (64, 123, 124; see Fig. 4). This usually results in the amplification of large segments of DNA arrayed in tandem along chromosome 5 (123). The increased mefloquine resistance is also accompanied by an increase in halofantrine and quinine resistance (123, 124). An analysis of field isolates from Thailand has shown a strong linkage between amplification of the *pfmdr1* gene with resistance to mefloquine and cross-resistance to halofantrine (120). These results have suggested that the amplification of the *pfmdr1* gene in field isolates has occurred as a result of quinine drug pressure. Analysis of the breakpoints of a large number of drug-resistant *P. falciparum* isolates has shown that they do not share the same breakpoint, suggesting that the amplification of *pfmdr1* that is found in many isolates has arisen in multiple independent events (125). It is clear from these results that this area of the *P. falciparum* genome is under strong selective pressure, probably as a result of heavy antimalarial drug use in endemic areas. It was also noted in some selection experiments that the increased mefloquine resistance was accompanied by a decrease in chloroquine resistance (120, 123, 124). This is consistent with the earlier observation that selection for increased chloroquine resistance was linked to decreased mefloquine resistance and deamplification and decreased expression of the *pfmdr1* gene (126). These results are consistent with drug sensitivity data that show a linkage between resistance to mefloquine and cross-resistance to quinine in field isolates (127). It is clear that there is a strong correlation of amplification of *pfmdr1* and overexpression of the Pgh1 protein with mefloquine, halofantrine, and, in some cases, quinine resistance (120, 123, 124, 126). This linkage suggests that over-expression of Pgh1 is involved in the mechanism of resistance to these drugs but the molecular basis for this is not yet understood.

11. Drug resistance in *Toxoplasma gondii*

Drug resistance is becoming a problem in *Toxoplasma gondii* with emergence of resistant organisms in clinical settings. Laboratory research has predicted this outcome for some time since several-drug resistant organisms have been selected *in vitro* (128). Although mutagenesis was used to enhance the frequency of drug-resistant organisms, these studies foreshadowed the recent clinical failures of anti-*Toxoplasma* drugs. Of particular interest are parasites resistant to atovaquone, a drug which is effective against both *Toxoplasma gondii* and *Plasmodium falciparum* and is hypothesized to affect the mitochondrial bc1 complex (129). Mutants selected in the laboratory for resistance to atovaquone are cross-resistant to deoquinate. The mechanism of resistance remains to be elucidated but experiments rule out an inhibition of *de novo* pyrimidine synthesis. Recent developments in the transfection of *Toxoplasma gondii* should greatly facilitate the determination of mechanisms of drug resistance (130–132; see Chapter 4).

12. Metronidazole resistance

Nitroheterocyclic compounds are the preferred drugs for the treatment of anaerobic protozoal diseases, in particular human giardiasis, amoebiasis, and trichomoniasis infections. Metronidazole is the most commonly used of this group of compounds and is currently used to treat *Trichomonas vaginalis* (133), *Giardia intestinalis* (134), and *Entamoeba histolytica* (135). Clinical resistance to metronidazole has been reported in *Trichomonas* (136) and *Giardia* (137). Although it is still proving to be a useful drug for the treatment of these infections, the increased incidence of resistance and treatment failures suggests that this efficacy may be compromised in the future (138).

In *Trichomonas*, metronidazole can be reduced to a toxic radical in a specialized organelle termed the hydrogenosome (139; see Chapter 9). This organelle is a carbohydrate metabolizing structure that contains the electron transport components ferredoxin, hydrogenase and pyruvate:ferredoxin oxidoreductase that are required for drug activation (see Fig. 5). In *Giardia*, it has been shown that ferredoxin can reduce metronidazole in a reaction coupled to pyruvate dehydrogenase (140); as *Giardia* does not contain hydrogenosomes, however, the activation is likely to be cytosolic. The major effects of the activation of metronidazole to short-lived radicals are likely to be inactivation of the cell's enzymatic machinery, and interaction with various cellular components such as DNA and membranes which would result in damage and death of the parasite.

The mechanism of resistance by *Trichomonas* and *Giardia* has been studied primarily with parasite lines induced in the laboratory by drug pressure (138, 140–144). In *Trichomonas*, pyruvate:ferredoxin oxidoreductase and hydrogenase activities are reduced, and the morphology of the hydrogenosome altered. Other lines of *T. vaginalis* selected for metronidazole resistance have greatly decreased levels of ferredoxin (145). Metronidazole resistance in *Giardia* has been induced *in vitro* (144, 146) and reduction of the level of pyruvate:ferredoxin oxidoreductase has been confirmed as a mechanism used by the parasite (147, 148). The exact mechanisms used by anaerobic protozoa to evade the action of metronidazole have yet to be elucidated. However, most resistant strains so far examined are reduced in their ability to activate the drug. This generally occurs by a decrease in activity of one of the proteins involved in metronidazole activation, which include pyruvate:ferredoxin oxidoreductase, ferredoxin, and hydrogenase.

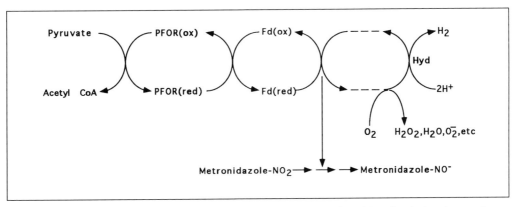

Fig. 5 The proposed mechanism for converting metronidazole to the activated form by electron transport components of *T. vaginalis*. Metronidazole is activated by the transfer of electrons from ferredoxin to the nitro group of the drug. PFOR, pyruvate:ferredoxin oxidoreductase; Fd, ferredoxin; Hyd, hydrogenase.

13. Conclusions and perspectives

One of the imperatives in the field of chemotherapy of protozoan parasites is the need to discover and develop new drugs that will be effective against drug-resistant organisms. Malaria is leading the field in the emergence of drug resistance but other diseases are displaying reduced drug susceptibility, a clear precursor to the appearance of resistant organisms. New approaches must be taken that will prevent or delay this outcome. For other microbial infections, alternating the use of drugs with different mechanisms of action or combining two or more drugs with independent and synergistic effects have been viable strategies, but for most parasitic infections, this approach is limited by the availability of drugs with different targets. Clearly, as more drugs specific to unique parasite pathways are discovered, such a combinatorial and alternating approach becomes a viable option. However, even when drug combinations are used, resistance eventually emerges. Thus, alternative strategies are also necessary. One intriguing possibility is to develop approaches which are targeted at the organism's ability to amplify its DNA. A common theme in many of the resistance mechanisms described in this chapter is the amplification of either target enzyme genes or amplification of the P-glycoprotein-like transporters. While little is known about the mechanisms of gene amplification in these organisms, inhibition of these pathways could prevent resistance to a large number of drugs. In the case of resistance due to P-glycoprotein overexpression, inhibition of drug efflux using channel blockers is in the early stages of clinical development both in cancer chemotherapy and for treatment of drug resistant *P. falciparum*. This approach could be applied to drug efflux-mediated resistance in other organisms. Finally, with detailed information on specific mutations in target enzymes, it may be possible to use rational drug design approaches to select drug candidates that could inhibit resistant enzymes.

References

1. WHO (1955) *WHA8.30 malaria eradication* from the 9th Plenary Meeting. WHO Official Records No. **63**, 31.
2. WHO. (1991) *WHO Weekly Epidemiol. Rec.*, **66**, 157.
3. Wernsdorfer, W. H. (1991) The development and spread of drug-resistant malaria. *Parasitol. Today*, **11**, 297.
4. Evans, D. B. and Jamison, D. T. (1994) Economics and the argument for parasitic disease control. *Science*, **264**, 1866.
5. Baum, K. F, Berens, R. L (1994) Successful treatment of cutaneous leishmaniasis with allopurinol after failure of treatment with ketoconazole. *Clin. Infect. Dis.*, **18**,: 813.
6. Baumann, R. J, Hanson, W. L, McCann, P. P, Sjoersdma, A., and Bitonti, A. J. (1990) Suppression of both antimony-susceptible and antimony-resistant *Leishmania donovani* by a bis(benzyl)polyamine analog. *Antimicrob. Agents Chemother.*, **34,** 722.
7. Croft, S. L, Hogg, J, Gutteridge, W. E, Hudson, A. T, and Randall, A. W. (1992) The activity of hydroxynaphthoquinones against *Leishmania donovani*. *J. Antimicrob. Chemother.*, **30,** 827.

8. Gramiccia, M, Gradoni, L., and Orsini, S. (1992) Decreased sensitivity to meglumine antimoniate (Glucantime) of *Leishmania infantum* isolated from dogs after several courses of drug treatment. *Ann. Trop. Med. Parasitol.*, **86,** 613.
9. Jackson, J. E, Tally, J. D, Ellis, W. Y, Mebrahtu, Y. B, Lawyer, P. G, Were, J. B, Reed, S. G, Panisko, D. M, and Limmer, B. L. (1990) Quantitative *in vitro* drug potency and drug susceptibility evaluation of *Leishmania* spp. from patients unresponsive to pentavalent antimony therapy. *Am. J. Trop. Med. Hyg.*, **43,** 464.
10. Beverley, S.M (1991) Gene amplification in *Leishmania*. *Annu. Rev. Microbiol.*, **45,** 417.
11. Callahan, H. L, Beverley, and S. M. (1991) Heavy metal resistance: a new role for P-glycoproteins in *Leishmania*. *J. Biol. Chem.*, **266,** 8427.
12. Callahan, H. L, Beverley, S. M (1992) A member of the aldoketo reductase family confers methotrexate resistance in *Leishmania*. *J. Biol. Chem.*, **267,** 24165.
13. Chow, L. M, Wong, A. K, Ullman, B, Wirth, D. F. (1993) Cloning and functional analysis of an extrachromosomally amplified multidrug resistance-like gene in *Leishmania enriettii*. *Mol. Biochem. Parasitol.*, **60,** 195.
14. Gamarro, F., Chiquero, M. J, Amador, M. V, Legare, D., Ouellette, M., and Castanys, S. (1994) P-glycoprotein overexpression in methotrexate-resistant *Leishmania tropica*. *Biochem. Pharmacol.*, **47,** 1939.
15. Henderson, D. M, Sifri, C. D., Rodgers, M., Wirth, D. F., Hendrickson, N., Ullman, B. (1992) Multidrug resistance in *Leishmania donovani* is conferred by amplification of a gene homologous to the mammalian *mdr1* gene. *Mol. Cell. Biol.*, **12,** 2855.
16. Lee, M. G and Van der Ploeg, L. H. (1990) Homologous recombination and stable transfection in the parasitic protozoan *Trypanosoma brucei*. *Science*, **250,** 1583.
17. Papadopoulou, B., Roy, G., Dey, S., Rosen, B. P, Ouellette, M. (1994) Contribution of the Leishmania P-glycoprotein-related gene *ltpgpA* to oxyanion resistance. *J. Biol. Chem.*, 269: 11980.
18. Wong, A. K, Chow, L. M, and Wirth, D. F. (1994) A homologous recombination strategy to analyze the vinblastine resistance property of the V-circle in *Leishmania*. *Mol. Biochem. Parasitol.*, **64,** 75.
19. Ouellette, M. and Borst, P. (1991) Drug resistance and P-glycoprotein gene amplification in the protozoan parasite *Leishmania*. *Res. Microbiol.*, **142,** 737.
20. Papadopoulou, B., Roy, G., and Ouellette, M. (1992) A novel antifolate resistance gene on the amplified H circle of Leishmania. *Embo J*, **11,** 3601.
21. Katakura, K., Peng, Y., Pithawalla, R., Detke, S., and Chang, K. P. (1991) Tunicamycin-resistant variants from five species of *Leishmania* contain amplified DNA in extrachromosomal circles of different sizes with a transcriptionally active homologous region. *Mol. Biochem. Parasitol.*, **44,** 233.
22. Liu, X. and Chang, K. P. (1992) The 63-kilobase circular amplicon of tunicamycin-resistant *Leishmania amazonensis* contains a functional N-acetylglucosamine-1-phosphate transferase gene that can be used as a dominant selectable marker in transfection. *Mol. Cell. Biol.*, **12,** 4112.
23. Coons, T., Hanson, S., Bitonti, A. J., McCann, P. P., and Ullman, B. (1990) Alpha-difluoromethylornithine resistance in Leishmania donovani is associated with increased ornithine decarboxylase activity. *Mol. Biochem. Parasitol.*, **39,** 77.
24. Hanson, S., Beverley, S. M., Wagner, W., Ullman, B. and (1992) Unstable amplification of two extrachromosomal elements in alpha-difluoromethylornithine-resistant *Leishmania donovani*. *Mol. Cell. Biol.*, **12,** 5499.
25. Wilson, K., Collart, F. R., Huberman, E., Stringer, J. R., and Ullman, B. (1991) Amplifica-

tion and molecular cloning of the IMP dehydrogenase gene of *Leishmania donovani*. *J. Biol. Chem.*, **266**, 1665.
26. Wilson, K., Beverley, S. M., and Ullman, B. (1992) Stable amplification of a linear extrachromosomal DNA in mycophenolic acid-resistant *Leishmania donovani*. *Mol. Biochem. Parasitol.*, **55**, 197.
27. Cowman, A. F., Morry, M. J., Biggs, B. A., Cross, G. A. M., and Foote, S. J. (1988) Amino acid changes linked to pyrimethamine resistance in the dihdrofolate reductase-thymidylate synthase gene of *Plasmodium falciparum*. *Proc. Natl. Acad. Sci. USA*, **85**, 9109.
28. Peterson, D. S., Walliker, D., and Wellems, T. E. (1988) Evidence that a point mutation in dihydrofolate reductase-thymidylate synthase confers resistance to pyrimethamine in falciparum malaria. *Proc. Natl. Acad. Sci. USA*, **85**, 9114.
29. Watkins, M. W. and Mosobo, M. (1993) Treatment of *Plasmodium falciparum* malaria with pyrimethamine-sulfadoxine: selective pressure for resistance is a function of long elimination half-life. *Trans. Roy. Soc. Trop. Med. Hyg.*, **87**, 75.
30. Roche, J., Benito, A., Ayecaba, S., Amela, C., Molina, R., and Alvar, J. (1993) Resistance of *Plasmodium falciparum* to antimalarial drugs in equatorial Guinea. *Ann. Trop. Med. Parasitol.*, **87**, 443.
31. Schellenberg, K. A. and Coatney, G. R. (1961) The influence of antimalarial drugs on nucleic acid synthesis in *Plasmodium gallinaceum* and *Plasmodium berghei*. *Biochem. Pharmacol.* **6**, 143.
32. Gutteridge, W. E. and Trigg, P. I. (1971) Action of pyrimethamine and related drugs against *Plasmodium knowlesi in vitro*. *Parasitology*, **62**, 431.
33. Bishop, A. (1962) An analysis of the development of resistance to proguanil and pyrimethamine in *Plasmodium gallinaceum*. *Parasitology*, **52**, 495.
34. Walliker, D. and Carter, T. A. S. (1975) Genetic studies on *Plasmodium chabaudi*: recombination between enzyme markers. *Parasitology* **70**, 19.
35. Knowles, G., Sanderson, A., and Walliker, D. (1981) *Plasmodium yoelii*: genetic analysis of crosses between two rodent malaria subspecies. *Exp. Parasitol.*, **52**, 243.
36. Diggens, S. M., Gutteridge, W. E., and Trigg, P. I. (1970) An altered dihydrofolate reductase associated with a pyrimethamine resistant *P. berghei* produced in a single step. *Nature*, **228**, 579.
37. Cowman, A. F. and Lew, A. M. (1989) Antifolate drug selection results in duplication and rearrangement of chromosome 7 in *Plasmodium chabaudi*. *Mol. Cell. Biol.* **9**, 5182.
38. McCutchan, T. F., Welsh, J. A., Dame, J. B., Quakyi, I. A., Graves, P. M., Drake, J. C., and Allegra, C. J. (1984) Mechanism of pyrimethamine resistance in recent isolates of *Plasmodium falciparum*. *Antimicrob. Agents Chemother.*, **26**, 656.
39. Chen, G. X. and Zolg, J. W. (1987) Purification of the bifunctional thymidylate synthase-dihydrofolate reductase complex from the human malaria parasite *Plasmodium falciparum*. *Mol. Pharmacol.* **32**, 723.
40. Bzik, D. J., Li, W.-B., Horii, T., and Inselburg, J. (1987) Molecular cloning and sequence analysis of the *Plasmodium falciparum* dihydrofolate reductase-thymidylate synthase gene. *Proc. Natl. Acad. Sci. USA*, **84**, 8360.
41. Garrett, C. E., Coderre, J. A., Meek, T. D., Garvey, E. P., Claman, D. M., Beverley, S.M., and Santi, D. V. (1984) A bifunctional thymidylate synthetase-dihydrofolate reductase in protozoa. *Mol. Biochem. Parasitol.* **11**, 257.
42. Snewin, V. A., England, S. M., Sims, P. F., and Hyde, J. E. (1989) Characterisation of the dihydrofolate reductase-thymidylate synthetase gene from human malaria parasites highly resistant to pyrimethamine. *Gene*, **76**, 41.

43. Zolg, J. W., Plitt, J. R., Chen, G. X., and Palmer, S. (1989) Point mutations in the dihydrofolate reductase–thymidylate synthase gene as the molecular basis for pyrimethamine resistance in *Plasmodium falciparum*. *Mol. Biochem. Parasitol.* **36,** 253.
44. Sirawaraporn, W., Sirawaraporn, R., Cowman, A. F., Yuthavong, Y., and Santi, D. V. (1990) Heterologous expression of active thymidylate synthase–dihydrofolate reductase from *Plasmodium falciparum*. *Biochemistry,* **29,** 10779.
45. Schapira, A., Bygbjerg, I. C., Jepsen, S., Flachs, H., and Bentzon, M. W. (1986) The susceptibility of *Plasmodium falciparum* to sulfadoxine and pyrimethamine: correlation of *in vivo* and *in vitro* results. *Am. J. Trop. Med. Hyg.* 35, 239.
46. Watkins, W. M., Sixsmith, D. G., and Chulay, J. D. (1984) The activity of proguanil and its metabolites, cycloguanil and *p*-chlorophenylbiguanide, against *Plasmodium falciparum in vitro*. *Am. J. Trop. Med. Hyg.* **78,** 273.
47. Foote, S. J., Galatis, D., and Cowman, A.F. (1990) Amino acids in the dihydrofolate reductase-thymidylate synthase gene of *Plasmodium falciparum* involved in cycloguanil resistance differ from those involved in pyrimethamine resistance. *Proc. Natl. Acad. Sci. USA,* **87,** 3014.
48. Peterson, D. S., Milhous, W. K., and Wellems, T. E. (1990) Molecular basis of differential resistance to cycloguanil and pyrimethamine in *Plasmodium falciparum* malaria. *Proc. Natl. Acad. Sci. USA,* **87,** 3018.
49. Dieckmann, A. and Jung, A. (1986) Mechanisms of sulfadoxine resistance in *Plasmodium falciparum*. *Mol. Biochem., Parasitol.* **19,** 143.
50. Triglia, T. and Cowman, A. F. (1994) Primary structure and expression of the dihydropteroate synthetase gene of *Plasmodium falciparum*. *Proc. Natl. Acad. Sci. USA,* **91,** 7149.
51. Brooks, D., Wang, P., Read, M., Watkin, W., Sims, P., and Hyde, J. (1994) Sequence variation in the hydroxymethyldihydropterin pyrophosphokinase-dihydropteroate synthase gene in lines of the human malaria parasite, *Plasmodium falciparum*, with differing resistance to sulfadoxine. *Eur. J. Biochem.* **224,** 397.
52. Kerby, B. R. and Detke, S. (1993) Reduced purine accumulation is encoded on an amplified DNA in *Leishmania mexicana amazonensis* resistant to toxic nucleosides. *Mol. Biochem. Parasitol,* **60,** 171.
53. Carter, N. S. and Fairlamb, A. H. (1993) Arsenical-resistant trypanosomes lack an unusual adenosine transporter. *Nature,* **361,** 173. (Erratum appears in *Nature,* **361,** 374).
54. Endicott, J. A. and Ling, V. (1989) The biochemistry of P-glycoprotein-mediated multidrug resistance. *Annu. Rev. Biochem.* **58,** 137.
55. Shen, D.-W., Fojo, A., Roninson, I. B., Chin, J. E., Soffir, R., Pastan, I., and Gottesman, M. M. (1986) Multidrug-resistance in DNA-mediated transformants is linked to transfer of the human *MDR1* gene. *Mol. Cell. Biol.* **6,** 4039.
56. Gros, P., Croop, J. M., and Housman, D. (1986) Mammalian multi-drug resistance gene: complete cDNA sequence indicates strong homology to bacterial transport proteins. *Cell,* **47,** 371.
57. Gros, P., Raymond, M., Bell, J., and Housman, D. (1988) Cloning and characterization of a second member of the mouse *mdr* gene family. *Mol. Cell. Biol.* **8,** 2770.
58. Devault, A. and Gros, P. (1990) Two members of the mouse MDR gene family confer multidrug resistance with overlapping but distinct drug specificities. *Mol. Cell Biol.* **10,** 1652.
59. Hsu, S. I. H., Cohen, D., Kirschner, L. S., Lothstein, L., Hartstein, M., and Horwitz, S. B. (1990) Structural analysis of the mouse *mdr1a* (P-glycoprotein) promoter reveals the basis for differential transcript heterogeneity in multidrug resistant J774.2 cells. *Mol. Cell. Biol.* **10,** 3596.

60. Silverman, J. A., Raunio H., Gant, T. W., and Thorgeirsson, S. S. (1991) Cloning and characterisation of a member of the rat multidrug resistance (*mdr*) gene family. *Gene* **106**, 221.
61. De Bruijn, M. L. H., Van der Bliek, A. M., Biedler J. L., and Borst, P. (1986) Differential amplification and disproportionate expression of five genes in three multidrug-resistant Chinese hamster lung cell lines. *Mol. Cell. Biol.* **6**, 4717.
62. McGrath, J. P. and Varshavsky, A. (1989) The yeast *STE6* gene encodes a homologue of the mammalian multidrug resistance P-glycoprotein. *Nature*, **340**, 400.
63. Raymond, M., Gros, P., Whiteway, M., and Thomas, D. Y. (1992) Functional complementation of yeast *ste6* by a mammalian multidrug resistance *mdr* gene. *Science*, **256**, 232.
64. Wilson, C. M., Serrano, A. E., Wasley, A., Bogenschutz, M. P., Shankar, A. H., and Wirth, D. F. (1989) Amplification of a gene related to mammalian mdr genes in drug-resistant falciparum malaria. *Science*, **244**, 1184.
65. Foote, S. J., Thompson, J. K., Cowman, A. F., and Kemp, D. J. (1989) Amplification of the multidrug resistance gene in some chloroquine-resistant isolates of *P. falciparum*. *Cell*, **57**, 921.
66. Samuelson, J., Ayala, P., Orozco, E., and Wirth D. F. (1990) Emetine-resistant mutants of *Entamoeba histolytica* over-express mRNAs for multidrug resistance. *Mol. Biochem. Parasitol.* **38**, 281.
67. Ouellette, M., Fase-Fowler, F., and Borst, P. (1990) The amplified H-circle of methotrexate-resistance *Leishmania tarentolae* contains a novel P-glycoprotein gene. *EMBO J.* **9**, 1027.
68. Chen, C., Chin, J. E., Ueda, K., Clark, D. P., Pastan, I., Gottesman, M. M., and Roninson, I. B. (1986) Internal duplication and homology with bacterial transport proteins in the mdr1 (P-glycoprotein) gene from multidrug-resistant human cells. *Cell*, **47**, 381.
69. Smit, J. J. M., Schinkel, A. H., Oude Elferink, R. P. J., Groen, A. K., Wagenaar, E., van Deemter, L., Mol, C. A. A. M., Ottenhoff, R., van der Lugt, N. M. T., van Roon, M. A., van der Valk, M. A., Offerhaus, G. J. A., Berns, A. J. M., and Borst, P. (1993) Homozygous disruption of the murine *mdr2* P-glycoprotein gene leads to a complete absence of phospholipid from bile and to liver disease. *Cell*, **75**, 451.
70. Schinkel, A. H., Smit, J. J. M., van Tellingen, O., Beijnen, J. H., Wagenaar, E., van Deemter, L., Mol, C. A. A. M., van der Valk, M. A., Robanus-Maandag, E. C., te Riele, H. P. J., Berns, A. J. M., and Borst, P. (1994) Disruption of the mouse *mdr1a* P-glycoprotein gene leads to a deficiency in the blood–brain barrier and to increased sensitivity to drugs. *Cell*, **77**, 491.
71. Ambudkar, S. V., Lelong, I. H., Zhang, J., Cardarelli, C. O., Gottesman, M. M., and Pastan, I. (1992) Partial purification and reconstitution of the human multidrug resistance pump: characterization of the drug-stimulatable ATP hydrolysis. *Proc. Natl. Acad. Sci. USA*, **89**, 8472.
72. Valverde, M. A., Dia, Z. M., Sepulveda, F., Gill, D. R., Hyde, S. C., and Higgins, C. F. (1992) Volume-regulated chloride channels associated with multi-drug resistance P-glycoprotein. *Nature*, **350**, 830.
73. Gill, D. R., Hyde, S. C., Higgins, D. F., Valvede, M. A., Mintenig, G. M., and Sepulveda, F. V. (1992) Separation of drug transport and chloride channel functions of the human multidrug resistance P-glycoprotein. *Cell*, **71**, 23.
74. Wellems, T. E., Panton, L. J., Gluzman, I. Y., do Rosario, R. V., Gwadz, R. W., Walker, J. A., and Krogstad, D. J. (1990) Chloroquine resistance not linked to *mdr*-like genes in a *Plasmodium falciparum* cross. *Nature*, **345**, 253.

75. Macomber, P. B. and Sprinz, H. (1967) Morphological effects of chloroquine on *Plasmodium berghei* in mice. *Nature*, **214**, 937.
76. Warhurst, D. C. and Hockley, D. J. (1967) Mode of action of chloroquine on *Plasmodium berghei* and *P. cynomolgi*. *Nature*, **214**, 935.
77. Langreth, S. G., Nguyen, D. P., and Trager, W. (1978) *Plasmodium falciparum*: merozoite invasion *in vitro* in the presence of chloroquine. *Exp. Parasitol.*, **46**, 235.
78. Sinden, R. E. (1982) Gametocytogenesis of *Plasmodium falciparum in vitro*: ultrastructural observations on the lethal action of chloroquine. *Ann. Trop. Med. Parasitol.* **76**, 15.
79. Peters, W., Portus, J., and Robinson, B. L. (1977) The chemotherapy of rodent malaria, XXVIII. The development of resistance to mefloquine (WR 142,490). *Ann. Trop. Med. Parasitol.*, **71**, 419.
80. Jacobs, G. H., Aikawa, M., Milhous, W. K., and Rabbege, J. R. (1987) An ultrastructural study of the effects of mefloquine on malaria parasites. *Am. J. Trop. Med. Hyg.* **36**, 9.
81. Peters, W. (1970) *Chemotherapy and drug resistance in malaria*. Academic Press; London.
82. Vander, J. D., Hunsaker, L. A., and Campos, N. M. (1986) Characterization of a hemoglobin-degrading, low molecular weight protease from *Plasmodium falciparum*. *Mol. Biochem. Parasitol.*, **18**, 389.
83. Chou, A. C. and Fitch, C. D. (1980) Hemolysis of mouse erythrocytes by ferriprotoporphyrin IX and chloroquine. *J. Clin. Invest.*, **66**, 856.
84. Chou, A. C. and Fitch, C. D. (1981) Mechanism of hemolysis induced by ferriprotoporphyrin IX. *J. Clin. Invest.*, **68**, 672.
85. Chou, A. C., Chevli, R., and Fitch, C. D. (1980) Ferriprotoporphyrin IX fulfills the criteria for identification as the chloroquine receptor of malaria parasites. *Biochemistry*, **19**, 1543.
86. Orjih, A. U., Banyal, H. S., Chevli, R., and Fitch, C. D. (1981) Hemin lyses malaria parasites. *Science*, **214**, 667.
87. Fitch, C. D., Chevli, R., Kanjananggulpan, P., Dutta, P., Chevli, K., and Chou, A. C. (1983) Intracellular ferriprotoporphyrin IX is a lytic agent. *Blood*, **62**, 1165.
88. Fitch, C. D. (1983) Mode of action of antimalarial drugs. *Ciba Found. Symp.*, **94**, 222.
89. Slater, A. F. and Cerami, A. (1992) Inhibition by chloroquine of a novel haem polymerase enzyme activity in malaria trophozoites. *Nature*, **355**, 167.
90. Fitch, C. D. (1970) *P. falciparum* in owl monkeys: drug resistance and chloroquine binding capacity. *Science*, **169**, 289.
91. Verdier, F., Le Bras, J., Clavier, F., Hatin, I., and Blayo, M. (1985) Chloroquine uptake by *Plasmodium falciparum*-infected human erythrocytes during *in vitro* culture and its relationship to chloroquine resistance. *Antimicrob. Agents Chemother.* **27**, 561.
92. Yayon, A., Cabantchik, Z. I., and Ginsburg, H. (1985) Susceptibility of human malaria parasites to chloroquine is pH dependent. *Proc. Natl. Acad. Sci. USA*, **82**, 2784.
93. Krogstad, D. J., Gluzman, I. Y., Kyle, D. E., Oduola, A. M. and Martin, S. K. (1987) Efflux of chloroquine from *Plasmodium falciparum*: mechanism of chloroquine resistance. *Science*, **238**, 1283.
94. Krogstad, D. J., Gluzman, I. Y., Herwaldt, B. L., Schlesinger, P. H. and Wellems, T. E. (1992) Energy dependence of chloroquine accumulation and chloroquine efflux in *Plasmodium falciparum*. *Biochem. Pharmacol.*, **43**, 57.
95. Geary, T. G., Divo, A. D., Jensen, J. B., Zangwill, M., and Ginsburg, H. (1990) Kinetic modelling of the response of *Plasmodium falciparum* to chloroquine and its experimental testing *in vitro*. Implications for mechanism of action and resistance to the drug. *Biochem. Pharmacol.*, **40**, 685.
96. Ginsburg, H. and Stein, W. D. (1991) Kinetic modelling of chloroquine uptake by

malaria-infected erythrocytes. Assessment of the factors that may determine drug resistance. *Biochem. Pharmacol.*, **41**, 1463.
97. Ferrari, V. and Culter, D. J. (1991) Simulation of kinetic data on the influx and efflux of chloroquine by erythrocytes infected with *Plasmodium falciparum*. Evidence for a drug-importer in chloroquine-sensitive strains. *Biochem. Pharmacol.*, **42**, 69.
98. Ginsburg, H. and Krugliak, M. (1992) Quinoline-containing antimalarials-mode of action, drug resistance and its reversal. An update with unresolved puzzles. *Biochem. Pharmacol.*, **43**, 63.
99. Bray, P. G., Howells, R. E. and Ward, S. A. (1992) Vacuolar acidification and chloroquine sensitivity in *Plasmodium falciparum*. *Biochem. Pharmacol.*, **43**, 1219.
100. Bray, P. G., Howells, R. E., Ritchie, G. Y., and Ward, S. A. (1992) Rapid chloroquine efflux phenotype in both chloroquine-sensitive and chloroquine-resistant *Plasmodium falciparum*. A correlation of chloroquine sensitivity with energy-dependent drug accumulation. *Biochem. Pharmacol.*, **44**, 1317.
101. Macomber, P. B., O'Brien, R. L., and Hahn, F. E. (1966) Chloroquine: physiological basis of drug resistance in *Plasmodium berghei*. *Science*, **152**, 1374.
102. Polet, H. and Barr, C. F. (1968) Chloroquine and dihydroquinine. *In vitro* studies on their antimalarial effect on *P. knowlesi*. *J. Pharmacol. Exp. Ther.*, **164**, 380.
103. Homewood, C. A., Warhurst, D. C., Peters, W., and Baggaley, V. C. (1972) Lysosomes, pH and the antimalarial action of chloroquine. *Nature*, **235**, 50.
104. De Duve, C., De Barsy, T., Poole, B., Trouet, A., Tulkens, P., and Van Hoof, F. (1974) Lysosomotropic agents. *Biochem. Pharmacol.*, **23**, 2495.
105. Yayon, A., Cabantchik, Z. I., and Ginsburg, H. (1984) Identification of the acidic compartment of *Plasmodium falciparum*-infected human erythrocytes as the target of the antimalarial drug chloroquine. *EMBO J.*, **3**, 2695.
106. Ginsburg, H. and Geary, T. G. (1987) Current concepts and new ideas on the mechanism of action of quinoline-containing antimalarials. *Biochem. Pharmacol.*, **36**, 1567.
107. Karcz, S. R., Herrmann, V. R. and Cowman, A. F. (1993) Cloning and characterization of a vacuolar ATPase A subunit homologue from *Plasmodium falciparum*. *Mol. Biochem. Parasitol.*, **58**, 333.
108. Karcz, S. R., Herrmann, V. R., Trottein, F., and Cowman, A. F. (1994) Cloning and characterization of the vacuolar ATPase B subunit from *Plasmodium falciparum*. *Mol. Biochem. Parasitol.*, **65**, 123.
109. Cowman, A. F., Karcz, S., Galatis, D., and Culvenor, J. G. (1991) A P-glycoprotein homologue of *Plasmodium falciparum* is localized on the digestive vacuole. *J. Cell Biol.*, **113**, 1033.
110. Van Es, H. H. G., Renkema, H., Aerts, H., and Schurr, E. (1994) Enhanced lysosomal acidification leads to increased chloroquine accumulation in CHO cells expressing the *pfmdr1* gene. *Mol. Biochem. Parasitol.*, **68**, 209.
111. Van Es, H. H. G., Karcz, S., Chu, F., Cowman, A. F., Vidal, S., Gros, P., and Schurr, E. (1994) Expression of the plasmodial *pfmdr1* gene in mammalian cells is associated with increased susceptibility to chloroquine. *Mol. Cell. Biol.*, **14**, 2419.
112. Martin, S. K., Oduola, A. M., and Milhous, W. K. (1987) Reversal of chloroquine resistance in *Plasmodium falciparum* by verapamil. *Science*, **235**, 899.
113. Bitonti, A. J., Sjoerdsma, A., McCann, P. P., Kyle, D. E., Oduola, A. M., Rossan, R. N., Milhous, W. K., and Davidson, D. E. J. (1988) Reversal of chloroquine resistance in malaria parasite *Plasmodium falciparum* by desipramine. *Science*, **242**, 1301.
114. Kyle, D. E., Oduola, A. M., Martin, S. K., and Milhous, W. K. (1990) *Plasmodium falci-*

parum: modulation by calcium antagonists of resistance to chloroquine, desethylchloroquine, quinine, and quinidine *in vitro*. *Trans. R. Soc. Trop. Med. Hyg.*, **84**, 474.

115. Zalis, M. G., Wilson, C. M., Zhang, Y., and Wirth, D. F. (1993) Characterization of the pfmdr2 gene for *Plasmodium falciparum*. *Mol. Biochem. Parasitol.*, **62**, 83.
116. Rubio, J. P. and Cowman, A. F. (1994) *Plasmodium falciparum*: the pfmdr2 protein is not overexpressed in chloroquine-resistant isolates of the malaria parasite. *Exp. Parasitol.*, **79**, 137.
117. Ortiz, D. F., Kreppel, L., Speiser, D. M., Scheel, G., McDonald, G., and Ow, D. W. (1992) Heavy metal tolerance in the fission yeast requires an ATP-binding cassette-type vacuolar membrane transporter. *EMBO J.*, **11**, 3491.
118. Wellems, T. E., Walker, J. A., and Panton, L. J. (1991) Genetic mapping of the chloroquine-resistance locus on *Plasmodium falciparum* chromosome 7. *Proc. Natl. Acad. Sci. USA*, **88**, 3382.
119. Foote, S. J., Kyle, D. E., Martin, R. K., Oduola, A. M., Forsyth, K., Kemp, D. J., and Cowman, A. F. (1990) Several alleles of the multidrug-resistance gene are closely linked to chloroquine resistance in *Plasmodium falciparum*. *Nature*, **345**, 255.
120. Wilson, C. M., Volkman, S. K., Thaithong, S., Martin, R. K., Kyle, D. E., Milhous, W. K., and Wirth, D. F. (1993) Amplification of *pfmdr1* associated with mefloquine and halofantrine resistance in *Plasmodium falciparum* from Thailand. *Mol. Biochem. Parasitol.*, **57**, 151.
121. Volkman, S. K., Cowman, A. F., and Wirth, D. F. (1995) Functional complementation of the *ste6* gene of *Saccharomyces cerevisiae* with the *pfmdr1* gene of *Plasmodium falciparum*. *Proc. Natl. Acad. Sci. USA* (in press).
122. Berkower, C. and Michaelis, S. (1991) Mutational analysis of the yeast a-factor transporter STE6, a member of the ATP binding cassette (ABC) protein superfamily. *EMBO J.*, **10**, 3777.
123. Cowman, A. F., Galatis, D. and Thompson, J. K. (1994) Selection for mefloquine resistance in *Plasmodium falciparum* is linked to amplification of the *pfmdr1* gene and cross-resistance to halofantrine and quinine. *Proc. Natl. Acad. Sci. USA*, **91**, 1143.
124. Peel, S. A., Bright, P., Yount, B., Handy, J., and Baric, R. S. (1994) A strong association between mefloquine and halofantrine resistance and amplification overexpression and mutation in the P-glycoprotein gene homolog (*pfmdr*) of *Plasmodium falciparum in vitro*. *Am. J. Trop. Med. Hyg.*, **51**, 648.
125 Triglia, T., Foote, S. J., Kemp, D. J., and Cowman, A. F. (1991) Amplification of the multi-drug resistance gene pfmdr1 in *Plasmodium falciparum* has arisen as multiple independent events. *Mol. Cell. Biol.*, **11**, 5244.
126. Barnes, D. A., Foote, S.J., Galatis, D., Kemp, D.J. and Cowman, A.F. (1992) Selection for high-level chloroquine resistance results in deamplification of the *pfmdr1* gene and increased sensitivity to mefloquine in *Plasmodium falciparum*. *EMBO J.*, **11**, 3067.
127. Brasseur, P., Kouamouo, J., Moyou-Somo, R., and Druilhe, P. (1992) Multi-drug resistant falciparum malaria in Cameroon in 1987–1988. II. Mefloquine resistance confirmed *in vivo* and *in vitro* and its correlation with quinine resistance. *Am. J. Trop. Med. Hyg.*, **46**, 8.
128. Pfefferkorn, E. R., Borotz, S. E., Nothnagel, R. F. (1992): *Toxoplasma gondii*: characterization of a mutant resistant to sulfonamides. *Exp. Parasitol.*, **74**, 261.
129. Pfefferkorn, E. R., Borotz, S. E., Nothnagel, R. F. (1993): Mutants of *Toxoplasma gondii* resistant to atovaquone (566C80) or decoquinate. *J. Parasitol.*, **79**, 559.
130. Donald, R. G. and Roos, D. S. (1993) Stable molecular transformation of *Toxoplasma*

gondii: a selectable dihydrofolate reductase-thymidylate synthase marker based on drug-resistance mutations in malaria. *Proc. Natl. Acad. Sci. USA*, **90**, 11703.

131. Donald, R. G. and Roos, D. S. (1994) Homologous recombination and gene replacement at the dihydrofolate reductase-thymidylate synthase locus in *Toxoplasma gondii*. *Mol. Biochem. Parasitol.*, **63**, 243.
132. Kim, K., Soldati, D., and Boothroyd, J. C. (1993) Gene replacement in *Toxoplasma gondii* with chloramphenicol acetyltransferase as selectable marker. *Science*, **262**, 911.
133. Lossick, J. G. and Kent, H. L. (1991) Trichomoniasis: trends in diagnosis and management. *Am. J. Obstet. Gynecol.*, **165**, 1217.
134. Boreham, P. F. L. (1991) Giardiasis and its control. *Pharm. J.*, **234**, 271.
135. Knight, R. (1980) The chemotherapy of amebiasis. *J. Antimicrob. Chemother.*, **6**, 577.
136. Robinson, S. C. (1962) Trichomonal vaginitis resistant to metronidazole. *Can. Med. Assoc. J.*, **86**, 665.
137. Boreham, P. F. L., Phillips, R. E., and Shepherd, R. W. (1984) The sensitivity of *Giardia intestinalis* to drugs *in vitro*. *J. Antimicrob. Chemother.*, **14**, 449.
138. Upcroft, J. A. and Upcroft, P. (1993) Drug resistance and *Giardia*. *Parasitol. Today*, **9**, 187.
139. Muller, M. (1980) The hydrogenosome. *Symp. Soc. Gen. Microbiol.*, **30**, 127.
140. Townson, S. M., Upcroft, J. A., and Upcroft, P. (1994) Reduction of metronidazole in *Giardia*. In *Giardia - from molecules to disease and beyond* (ed. R. C. A. Thompson, J. A. Reynoldson, and A. J. Lymbery), p. 376. CAB International, Wallingford, Oxon.
141. Yarlett, N., Yarlett, N. C., and Lloyd, D. (1986) Ferredoxin dependent reduction of nitroimidazole derivatives in drug-resistant and susceptible strains of *Trichomonas vaginalis*. *Biochem. Pharmacol.*, **35**, 1703.
142. Cerkasovova, A., Cerkasov, J., and Kulda, J. (1988) Resistance of trichomonads to metronidazole. *Acta Univ. Carol. BiologicaI*, **30**, 485.
143. Kabickova, H., Kulda, J., Cerkasovova, A., and Peckova, H. (1988) Metronidazole-resistant *Tritrichomonas foetus*: activities of hydrogenosomal enzymes in course of development of anaerobic resistance. *Acta Univ. Carol. Biologica*, **30**, 513.
144. Townson, S. M., Laqua, H., Upcroft, P., Boreham, P. F. L. and Upcroft, J. A. (1992) Induction of metronidazole and furazolidone resistance in *Giardia*. *Trans. Roy. Soc. Trop. Med. Hyg.*, **86**, 521.
145. Johnson, P. J. (1993) Metronidazole and drug resistance. *Parasitol. Today*, **9**, 183.
146. Boreham, P. F. L., Phillips, R. E., and Shepherd, R. W. (1988) Altered uptake of metronidazole in vitro by stocks of *Giardia intestinalis* with different drug sensitivities. *Trans. R. Soc. Trop. Med. Hyg.*, **82**, 104.
147. Smith, N. C., Bryant, C., and Boreham, P. F. L. (1988) Possible role for pyruvate:ferredoxin oxidoreductase and thiol-dependent peroxidase and reductase activities in resistance to nitroheterocyclic drugs in *Giardia intestinalis*. *Int. J. Parasitol.* **18**, 991.
148. Ellis, J. E., Wingfield, J. M., Cole, D., Boreham, P. F. L., and Lloyd, D. (1993) Oxygen activities of metronidazole-resistant and -sensitive stocks of *Giardia intestinalis*. *Int. J. Parasitol.*, **23**, 35.

11 | Glycosyl-phosphatidylinositols and the surface architecture of parasitic protozoa

MALCOLM J. McCONVILLE

1. Introduction

A functionally diverse family of eukaryotic cell surface proteins are anchored to the plasma membrane via glycosyl-phosphatidylinositol (GPI) glycolipids (for recent reviews see Refs 1 and 2). These glycolipids constitute an alternative anchoring mechanism to the transmembrane polypeptide domain of type I membrane proteins. GPI-anchored proteins make up a variable proportion of the protein repertoire on the cell surface of complex metazoan organisms, but are usually less abundant than proteins with a transmembrane polypeptide domain. In contrast, GPI anchors are found on the most abundant surface proteins of many parasitic protozoa, including the African trypanosomes, *Trypanosoma cruzi*, *Leishmania* spp., *Plasmodium* spp. and *Toxoplasma gondii* (2) and are thus essential in maintaining the surface architecture of these organisms.

There is increasing evidence that GPI anchors offer particular advantages to the parasitic protozoa and may enhance or be essential for the function of the major surface proteins. Many protozoan parasites also synthesize a diverse range of GPI glycolipid structures which are not linked to protein. These glycolipids may be extremely abundant (especially in the trypanosomatids) and represent a novel class of virulence factors. The parasitic protozoa have also proved to be useful model systems for elucidating the steps involved in GPI biosynthesis. These studies have highlighted the possibility of designing drugs to inhibit these pathways which might be useful in the treatment of a variety of parasitic diseases.

2. Structure of protozoan GPI protein anchors

The GPI anchors of a variety of parasite proteins have been partially or completely characterized (Fig. 1). They contain a glycan core, which is conserved throughout

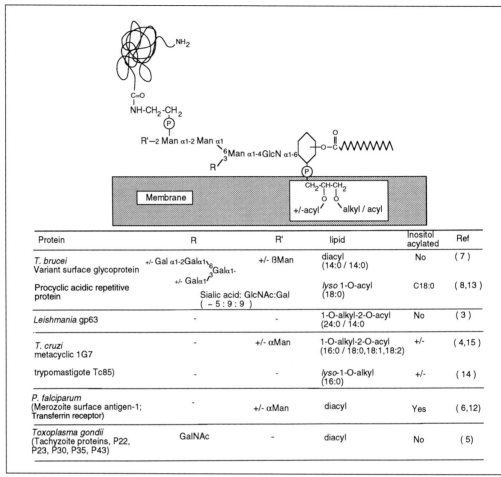

Fig. 1 Structures of the protozoan GPI protein anchors. Species-specific differences occur in the nature of the glycan side-chains (R and R') which substitute the conserved glycan core, in the lipid composition of the PI moiety and in the presence of an additional acyl group on the inositol ring. These structures were deduced from analysis of the proteins as well as putative GPI precursors.

eukaryotic evolution, that links the carboxyl terminus of the protein to an inositolphospholipid. The glycan backbone of the parasite anchors may be unmodified (3) or substituted with monosaccharides (αMan or GalNAc) (4–6) and/or extensive oligosaccharide side-chains (7, 8). The most elaborate side-chains occur on the GPI anchors of the variable surface glycoprotein (VSG) and the procyclic acidic repetitive protein (PARP), which are the major surface proteins of African trypanosome bloodstream and insect stages, respectively. While the VSG anchor contains side-chains of αGal residues (7; Fig. 1), the side-chains of the PARP anchor have been partially characterized as branched chains of poly-N-acetyllactosamine which may contain terminal sialic acid residues (8). The sialic acid residues on the PARP anchor are probably added on the cell surface by a parasite *trans*-sialidase that uses sialylated host glycoconjugates as the sugar donor (9). The presence of distinct GPI anchors on the proteins of the procyclic and bloodstream trypanosome forms may reflect differences in the expression of particular transferases (such as the *trans-sialidase* which is not expressed in the bloodstream form) as well as the nature of the proteins themselves. Interestingly, these side-chains are unique to the African trypanosomes and may represent specialized adaptations that are required for parasite survival (see below). Finally, none of the parasite anchors are substituted with extra ethanolamine-phosphate residues, which commonly occurs on the anchors of higher eukaryotes (2) and on the free GPIs of these organisms (10, 11).

The lipid moieties of the parasite GPI anchors are either diacyl-, *lyso* acyl-glycerol (5–7, 13) or alkylacyl-, *lyso* alkyl-glycerol (4, 14, 15). The fatty acid and alkyl chain compositions of these anchors are variable and frequently unlike those in the general phosphatidylinositol (PI) pool. For example, while the general PI pool has a high content of unsaturated fatty acids, the fatty acids (and alkyl chains) of the GPI anchors are predominantly saturated. To date, none of the parasite anchors have been found to contain ceramide, although ceramide occurs in the free GPIs of some of these parasites (see below) and in the anchors of other unicellular eukaryotes, such as yeast and slime moulds (2). Acylation of the inositol ring with palmitate or other acyl groups occurs frequently in mammalian GPI anchors (2) but may be less common in the parasites (Fig. 1). This modification renders the protein anchors resistant to PI-specific phospholipase C cleavage and is developmentally regulated in the African trypanosomes (13).

3. Function of GPI protein anchors in the protozoa

GPI-anchored proteins tend to dominate the cell surface of unicellular eukaryotes (both free-living and parasitic) and are also highly enriched in the apical membrane of many mammalian epithelial cells (2, 16). This distribution suggests that GPI anchors may be used preferentially on cell surfaces that are exposed to harsh extracellular environments which may reflect the particular properties a GPI anchor confers on the protein, or the major functions of the proteins that occur on the surfaces of these cells, or both. In this respect, it may be important that GPI-anchored proteins are unable to interact directly with the cytoskeleton or other cytoplasmic components. This may be advantageous for cells which are exposed to a harsh external envionment (such as the parasitic protozoa) where the frequent use of this type of anchor may buffer the cytoplasm from chemical or physical insult to surface proteins. On the other hand, this property means that GPI-anchored proteins cannot be directly involved in transmembrane signaling. While this may preclude the frequent use of GPI anchors in metazoan organisms, where many surface proteins are involved in transmembrane signalling, it may not be so critical in the protozoa, where the major surface proteins are involved in cellular nutrition (hydrolases), surface protection (coat proteins) or cell adhesion, all functions that primarily involve the ectoplasmic domain of the protein.

GPI anchors may offer other advantages over a transmembrane polypeptide domain that account for their prevalent use in the protozoa and less common, but none the less significant, use in metazoan organisms. While some of these functions may be common to all eukaryotes, there is evidence that some protozoan species have evolved specialized functions for their GPI anchors or, alternatively, have expanded upon the GPI biosynthetic pathway to generate novel classes of free GPIs as major surface components (see below). Some of the functions for GPI anchors which are thought to be important in the parasitic protozoa include a role in subcellular trafficking, control of the half-life of proteins, the selective release of surface proteins, the formation of tightly packed surface coats, and the possible modulation of host signal-transduction pathways.

3.1 Subcellular trafficking

There is increasing evidence that GPI-anchored proteins can associate with specific membrane microdomains in mammalian cells, and that this property may affect both the intracellular trafficking of these proteins and their distribution on the cell surface (reviewed in Ref. 16). These microdomains were identified on the basis of their (surprising) insolubility in cold, non-ionic detergents, such as Triton X-100, and are enriched in GPI-anchored proteins, sphingolipids (glycosphingolipids, sphingomyelin), cholesterol, and a number of non-GPI-anchored integral membrane proteins (17). GPI-anchored proteins appear to associate with these domains once they enter the Golgi, possibly because their glycerolipid moieties contain saturated aliphatic chains, which prefer the relatively non-fluid environment of the sphingolipid-enriched domains (17, 18). It has been proposed that this association may provide a mechanism for the apical trafficking of GPI-anchored proteins in epithelial cells (16), for their association with specialized plasma membrane domains such as caveolae and other non-clathrin coated pits (19), and transmembrane signaling in lymphocytes after antibody-mediated clustering of GPI-anchored proteins (20).

It is likely that GPI-anchored proteins in lower eukaryotes also associate with specific membrane microdomains although the extent to which this occurs may vary markedly in different organisms. The association of GPI-anchored proteins with sphingolipid enriched domains in yeast is strongly suggested by the finding that inhibition of ceramide synthesis specifically reduces the rate of transport of GPI-anchored proteins from the endoplasmic reticulum to the Golgi, without affecting transport of soluble or transmembrane proteins (21). Thus, as in mammalian cells, a GPI anchor may act as a sorting signal and affect the rate of transport through the secretory pathway. There is less evidence for the association of GPI-anchored proteins with similar membrane microdomains in the protozoa, as abundant GPI-anchored proteins, such as trypanosome VSG and leishmanial gp63, are extracted in high yield with cold Triton X-100. This lack of detergent insolubility is not surprising given that sphingolipids are much less abundant in these parasites. However, it is possible that these proteins may still associate with particular microdomains, but are more soluble in Triton X-100 because of the distinct lipid composition of these domains. Indirect evidence for this type of association comes from recent studies on the transferrin receptor of *T. brucei* bloodstream forms. This receptor is a dimer or oligomeric complex of two different proteins: a 50–60kDa polypeptide with a GPI-anchor; and a related 40kDa polypeptide with an unmodified carboxyl terminus (22, 23). This complex is only found in the flagellar pocket of the bloodstream-form trypanosomes, and, unlike VSG, requires zwitterionic detergents for complete solubilization (22, 23). Interestingly, when both proteins are transfected into the insect (procyclic) stage which normally lacks this receptor, a transferrin-binding complex is formed which is no longer restricted to the flagellar pocket and is soluble in cold Triton X-100 (23). These data suggest that bloodstream form trypanosomes have a mechanism for selectively retaining some GPI-anchored proteins within the flagellar pocket, which may involve both oligomerization and association with membrane microdomains. The former mechanism may be analogous to the sequestration of GPI-anchored proteins in caveolae in mammalian cells after their oligomerization with cross-linking antibody (24). The localization of the transferrin receptor to the flagellar pocket may reduce the chance of this invariant protein being targeted by the immune system and thus be critical for trypanosome survival in the bloodstream.

A number of adhesion and receptor proteins in both mammalian (LFA-3, FcγPIII, N-CAMs) and protozoal (leishmanial gp63) cells are expressed in either a GPI-anchored form or an alternate transmembrane or soluble form as a result of either differential mRNA splicing of a single gene transcript or expression of different genes (2). This in turn may modulate the function and possibly the cellular location of that protein. For example, in *L. mexicana* and *L. chagasi*, the major surface protease, gp63, is encoded by three distinct classes of genes (25, 26). Two of these gene classes encode a protein with a GPI attachment signal, while the third class of genes encodes a soluble protein with a C-terminal extension. The expression of these genes is developmentally regulated which affects both the subcellular location and presumably the function of this protease. In the promastigote stage, the GPI-anchored forms of gp63 predominate and most of the protease is found on the cell surface (25, 26). However, in the intracellular amastigote stage, the hydrophilic form (which is constitutively expressed in both developmental stages) predominates and the protein is targeted to the extended lysosomes, called megasomes (27). This provides an example of a GPI anchor acting as a signal to sort proteins to the cell surface, and as a way of 'fine tuning' protein function.

3.2 Protein turnover

A variety of studies have shown that the half-life of GPI-anchored proteins on the surface of mammalian cells and some lower eukaryotes, such as slime moulds, is relatively long, possibly due to their slow rate of endocytosis via clathrin-coated pits (28–30). In contrast, the rate of endocytosis of GPI-anchored proteins in a number of kinetoplastid parasites can be extremely rapid (31–33). High rates of endocytosis are typically observed after opsonization of these parasites with antibody or Fc fragments. In the case of bloodstream form trypanosomes, it was found that a monolayer of surface antibody directed against the VSG coat could be internalized within 20 minutes (32). Endocytosis of the antibody–VSG complexes occurs within the flagellar pocket, the sole site of endocytosis in these parasites. It is still not clear whether this rate of endocytosis occurs constitutively or is induced by the bound protein, and whether all VSG molecules are internalized. Interestingly, most of the endocytosed VSG is apparently recycled back to the cell surface after removal of opsonic antibody (31, 32). Thus, despite a high rate of endocytosis, the overall half-lives of the GPI-anchored proteins in trypanosomes may be comparable with those in mammalian cells. It is likely that this recycling is important for internalization of nutrient receptors and for the removal of surface-bound immune complexes. However, as yet, nothing is known about the signal(s) that is required for recycling in these parasites and whether it is specific for GPI-anchored proteins.

3.3 Lipase-mediated release

The presence of a GPI anchor could allow the selective release of some surface proteins via the action of an endogenous phospholipase. While this is an attractive proposal, clear evidence for this type of release has been demonstrated in only a few cases, both in higher eukaryotes (2) and in the protozoa. PI-specific phospholipase C-mediated release of surface proteins has been reported in *T. cruzi* (34, 35) and in *Plasmodium* spp. (36). Cleavage of the 76kDa serine protease of *Plasmodium* spp. results in the activation of this enzyme which is required for merozoite invasion of erythrocytes (36). Activation of proteases following lipase-mediated release also occurs with some mammalian dipeptidases (37) and may provide a novel mechanism for regulating the activities of specific enzymes. In contrast, there is no evidence for lipase-mediated release in *Leishmania* or in the African trypanosomes, despite the presence of a well characterized GPI-phospholipase C in the latter (38, 39). The *T. brucei* lipase occurs on the cytoplasmic face of intracellular vesicles (39), suggesting that it is not involved in the shedding and turnover of VSG molecules. Although the role of this enzyme remains obscure, it is possible that it may be involved in the catabolism of excess GPI precursors.

Release of GPI-anchored molecules may also occur via non-enzymatic processes. For example, the *Leishmania* lipophosphoglycan spontaneously partitions out of the plasma membrane as monomers or small micelles by virtue of containing a *lyso* alkyl-PI lipid moiety (40), and rapid shedding of membrane vesicles containing GPI-anchored proteins has been reported in *T. cruzi* (41).

3.4 Maintenance of surface coats

The presence of a GPI anchor may allow some protozoal proteins to be expressed on the cell surface at such high densities that they can form protective coats. Examples of such coat proteins include the VSG of African trypanosomes and the circumsporozoite protein of *Plasmodium* sporozoites (7, 42). These proteins form a densely packed monolayer over the cell surface (about 10^7 copies per cell) and are thought to protect the parasite from both non-specific and specific components of the host immune system. The GPI anchor would facilitate the tight packing of these proteins with minimum perturbation to the plasma membrane. It may also allow proteins to be packed more closely around, and even within, the lipid-filled spaces of large integral membrane proteins (such as the glucose transporter) which would otherwise produce 'holes' in the protein coat.

In some cases, the GPI anchor may itself contribute to the barrier properties of the protein coat. Three dimensional modelling studies of the αGal-substituted VSG anchor suggest that the glycan moiety has approximately the same cross-sectional area as the N-terminal domain of this protein (43) and may thus form a glycocalyx between the lipid bilayer and the protein. The αGal side-chains of this anchor are added while VSG dimers are being packaged into coat arrays in the Golgi for transport to the cell surface (44) and it is thought that the galactosyltransferases could act as spatial probes, filling the space between the protein and the membrane. The major protein, PARP, on the trypanosome insect stage also forms a protective coat. Although the PARP coat is less dense than the VSG coat, the side-chains on the PARP anchor are much larger and may constitute the main protective glycocalyx over the plasma membrane in the absence of other surface glycolipids (8).

3.5 Modulation of host signal transduction pathways

There is increasing evidence that protozoal GPI glycolipids can modulate signal transduction pathways in a variety of host cells, including macrophages, adipocytes, and lymphocytes. GPI glycolipids, derived from *P. falciparum* proteins or organic solvent extracts of whole cells, stimulate macrophages to produce tumour necrosis factor (TNF) and interleukin-1 (IL-1), and appear to be insulin-mimetic in adipocytes (45). They can also cause some of the symptoms associated with malarial infections when injected into mice, such as transient pyrexia, hypoglycemia, and lethal cachexia (45). The plasmodial GPIs have a simple glycan backbone and are only unusual in having a myristate (C14:0) chain on the inositol headgroup of dipalmitoyl-PI (Gerold, P., Schmidt, A., and Schwarz, R., unpublished). Both the lipid and glycan moieties are required for TNF stimulation (45, 46). The GPIs from African trypanosomes cause similar effects, although there are quantitative differences in the degree to which TNF and IL-1 production is stimulated in macrophages (47). Recent studies suggest that these parasite GPIs are directly activating host non-receptor tyrosine kinase(s) and protein kinase C (46, 47). It is possible that subtle differences in the structure of the GPIs from different species may modulate the activities of specific kinases in different ways. For example, in contrast to the plasmodial GPIs, the structurally distinct protein-free GPIs of *Leishmania* strongly inhibit macrophage protein kinase C (48). These parasites may be exploiting a GPI-based signal transduction pathway in mammalian cells, to either disable the microbicidal responses of the host cell (in the case of *Leishmania*) or modify the host immune response (in the case of *Plasmodium* and *T. brucei*). Antibodies against these GPI toxins can neutralize some of these effects and may have therapeutic potential (46, 49).

4 Protein-free GPI glycolipids in the trypanosomatids

The trypanosomatid parasites differ from most other eukaryotes in having free GPIs as their major class of glycolipid (2). While some of these glycolipids have the same backbone structure as the protein anchors and can act as anchor precursors (as in *T. brucei*), they frequently diverge from the protein anchors beyond the core sequence (Manα1–4GlcNα1–6*myo*inositol-1-PO$_4$-lipid) and are extremely abundant metabolic end-products. These novel glycoinositolphospholipids (GIPLs) were first identified in *T. cruzi* and have now been identified in a number of non-kinetoplastid parasites such as *Trichomonas* (50) and *Entamoeba histolytica* (51). However, the most extensively characterized, in terms of their structure and function, are the GIPLs and related lipophosphoglycans (LPGs) from leishmanial parasites (for reviews see Refs 2 and 52).

4.1 Structure of the GIPLs and LPG in *Leishmania* parasites

LPG forms a dense glycocalyx (about 5 million molecules per cell) over the entire surface of the leishmanial promastigote (insect) stage, but is expressed at very low levels or is absent from the intracellular amastigote (mammalian) stage (30, 53). The polymorphic phosphoglycan chains of LPG contain a conserved backbone of phosphorylated disaccharide repeat units and are anchored to the membrane by a novel GPI glycolipid (Fig. 2). Inter-specific polymorphism occurs in terminal cap structures and in the nature of the highly variable side-chains that branch off the disaccharide repeat units (40, 54–56) (Fig. 2). These side-chains may be ligands for host receptors and may also profoundly affect the extended helical configuration of the phosphoglycan backbone depending on their site of attachment (56, 57). The structure of LPG can also be developmentally regulated during the parasite life cycle (58–60). In particular, the differentiation of promastigotes from an avirulent 'procyclic' form to a virulent 'metacyclic' form in the insect midgut is associated in several species with a pronounced increase in the average chain length of the LPG (58, 60) and, in *L. major*, with the capping of the βGal-terminating side chains of LPG with βAra residues (58).

In contrast to the LPGs, the GIPLs are present at very high levels (about 50 million copies per cell) in all developmental stages of the parasite (53, 61). Three distinct series of GIPL structures have been characterized which are expressed in a species- and stage-specific manner (10, 53, 61, 62). These glycolipids may be structurally related to the GPI protein anchors, to the LPG anchors, or contain motifs in common with both anchors (Fig. 3). They may also contain additional substituents, such as ethanolamine-phosphate and β-galactose residues, which are not found in the anchors and generally have an alkylacyl-PI lipid moiety that is distinct from the lipid moieties of the protein or LPG anchors.

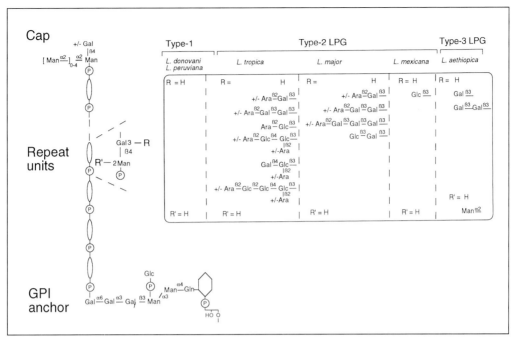

Fig. 2 Polymorphism in leishmanial LPG. The LPGs from different species have a common backbone structure, which includes a conserved GPI anchor and a phosphoglycan moiety made up of repeating phosphorylated disaccharide units. These repeat units can be substituted with species-specific side-chains which are attached to either the galactose (R) or the mannose (R') residues. The LPGs have been classified as being type-1 (unsubstituted), type-2 (side-chains are linked to galactose), or type-3 (side-chains are linked to mannose) (56).

4.2 Function of *Leishmania* LPG and GIPLs

There is compelling evidence that LPG is required for promastigote survival in both the sandfly vector and the mammalian host. LPG-deficient strains are unable to survive in either host, although virulence may be re-established if they revert back to an LPG-expressing phenotype or if exogenous LPG is incorporated into their surface membrane (52, 63). A variety of studies have shown that these polymorphic molecules are able to fulfill a number of distinct functions in both the insect and mammalian host.

4.2.1 LPG function in the sandfly vector

The glycocalyx formed by LPG appears to protect the plasma membrane and cell surface proteins from hydrolases in the sandfly digestive tract as well as from components of the mammalian immune system that are taken up with the bloodmeal (64). LPG may also be involved in mediating the attachment of rapidly dividing procyclic promastigotes to epithelial cells along the sandfly midgut (65, 66). This

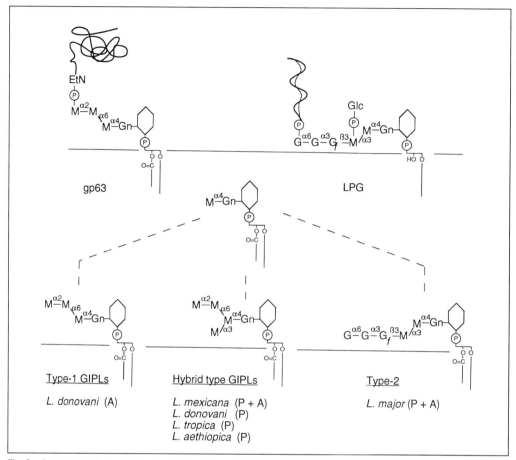

Fig. 3 Structures of the leishmanial GIPLs. Three distinct GIPL series have been characterized which are structurally related to the GPI protein anchors (as exemplified by gp63) or the LPG anchor. Only the most polar GIPL in each of the series is shown. M, Man; G, Gal_p; G_f, Gal_f; Gn, GlcN; P, phosphate; EtN, ethanolamine. The distributions in the promastigote (P) or amastigote (A) stages are indicated.

attachment is essential for the establishment of sandfly infection, as unattached promastigotes are lost when the remains of the bloodmeal are excreted. Interestingly, different domains of LPG appear to be involved in this attachment process depending on the parasite–vector combination. For example, attachment of *L. major* procyclic forms to the midgut of its natural sandfly vector, *Phlebotomus papatasi*, is mediated by a receptor that recognizes the clustered galactose-terminating side chains of procyclic LPG (66; Fig. 4). Most of these side-chains become capped with arabinose residues during metacyclogenesis, which may allow the selective release and anterior migration of infective metacyclic promastigotes for subsequent transmission at the next bloodmeal (58). These β-Gal side-chains appear to be essential for infection of *P. papatasi*, as other species of *Leishmania* that lack these side-chains

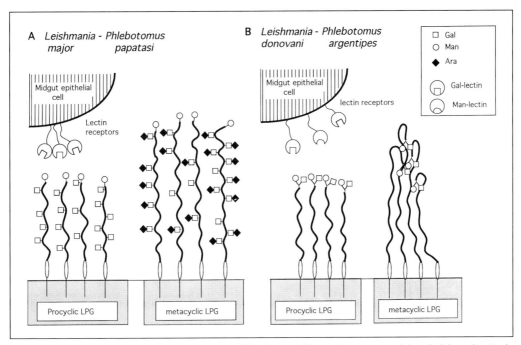

Fig. 4 Model showing how the developmental modification of LPG structure may modulate *Leishmania* attachment to midgut cells of the sandfly vector. A: *L. major* procyclic forms express an LPG with clustered β-Gal sidechains that are recognized by a lectin receptor(s) on the midgut of *P. papatasi* (58, 66). As the parasites differentiate to the metacyclic stage, new LPG molecules are expressed on the cell surface which contain mainly βAra-terminating side-chains, allowing midgut detachment. B: In contrast, the binding of *L. donovani* procyclic forms to the midgut cells of *P. argentipes* is mediated by a receptor that recognizes the LPG cap structures (60). These cap structures are expressed on the longer LPG chains of metacyclics but are not accessible to the lectin receptors (possibly due to a conformation change in the LPG (60)), thereby allowing parasite detachment.

on their LPGs are unable to establish an infection in this sandfly (65). In contrast, *L. donovani* promastigotes bind to the midgut of *P. argentipes* sandflies via one or more sugar-binding receptors which recognize the terminal cap structures of LPG (60; (Fig. 4). Similar cap structures are present on all the leishmanial LPGs, consistent with the finding that this sandfly is a permissive vector for a number of *Leishmania* species (65). Detachment of *L. donovani* promastigotes from the midgut may be mediated by a developmentally regulated change in LPG conformation which appears to make the cap structures cryptic to protein probes and receptors (60). The LPG chains of these metacyclic promastigotes are longer and their ultrastructural organization on the surface of metacyclic promastigotes suggests that they may undergo some form of self-association, which in turn may lead to the loss of accessibility of the terminal cap structures (Fig. 4). These studies suggest that much of the species- and stage-specific polymorphism in LPG structure is due to the presence of different sugar-binding receptors in these sandfly vectors and the role of LPG in modulating promastigote attachment and detachment to the sandfly midgut.

4.2.2 LPG function in the mammalian host

LPG is thought to protect promastigotes from complement-mediated lysis and to be required for promastigote attachment to mammalian macrophages as well as their survival within the phagolysosome. Resistance to complement-mediated lysis is acquired during the differentiation of rapidly dividing procyclic promastigotes to the non-dividing metacyclic promastigotes and is associated with an approximate doubling in the average length of the LPG chains and an increase in the thickness of the surface glycocalyx. This coat may form a more effective barrier, preventing insertion of the complement membrane attack complex into the parasite membrane and cell lysis (67). Opsonization of the surface LPG coat with complement components such as C3 or C3b appears to be important in mediating the attachment and uptake of promastigotes by macrophages via the complement receptors CR1 and CR3 (2, 52). This route of entry is preferentially used by metacyclic promastigotes and may avoid activation of the macrophage oxidative burst. It is possible that other serum proteins may also bind the LPG coat and either activate complement or act as ligands for macrophage receptors (68). Some macrophage receptors appear to be able to bind directly to LPG epitopes (69), although the nature of the receptor and the physiological significance of this interaction in macrophage invasion is unclear. Once the promastigotes have been taken into the phagolysosomes, LPG may protect the plasma membrane from lysosomal hydrolases and oxygen metabolites generated by the macrophage (70). LPG also appears to inhibit specific signal transduction pathways in the macrophage which are dependent on protein kinase C (48). These pathways are involved in the activation of the oxidative burst, macrophage chemotaxis, lymphokine production, and apoptosis, all of which are impaired to some extent by intracellular *Leishmania* parasites or purified LPG (48, 71). While there is strong experimental evidence for these functions both *in vitro* and *in vivo*, it is not clear how the LPG, which is located on either the macrophage surface or on the lumenal side of the phagolysosome membrane, interacts with the cytoplasmically disposed protein kinase C.

4.2.3 Function of the GIPLs

The GIPLs are abundant surface components on all developmental stages of *Leishmania*, although they may only be accessible to macromolecular probes in the amastigote stage which has little or no LPG and few detectable surface proteins (2). The GIPLs may thus be the major cell surface components of amastigotes and potentially cover the entire plasma membrane with a dense hydrogen-bonded network which could protect the amastigote membrane from hydrolases in the macrophage phagolysosome (53, 61). Amastigotes use a variety of receptors to gain entry into macrophages, some of which could bind directly to the surface GIPLs or to serum proteins that recognize the GIPLs (68). Although the GIPLs have been shown to inhibit protein kinase C *in vitro*, they are not shed from the parasite membrane as rapidly as LPG (72) and their ability to interact with host kinase *in vivo* has not been established. Finally, the GIPLs are also abundant in the membrane of the

4.3 Distribution of free GPIs in other protozoan parasites

GIPLs are the major cellular glycoconjugates of several other kinetoplastid parasites. Interestingly, the structures of the GIPLs from genera which are closely related to *Leishmania*, such as *Endotrypanum* (11), *Leptomonas* (73), and *Crithidia* (P. Schneider, K. Milne, A. Treumann, N. Zitmann, and M. A. J. Ferguson, unpublished data), share a high degree of homology with the *Leishmania* LPGs and type-2 GIPLs. These GIPLs contain glycan structures—the core sequence Manα1–3Manα1–4GlcN—that are not found in the GIPLs of more distantly related genera such as *T. cruzi* (74; Fig. 5), suggesting that they may have appeared after these groups diverged from

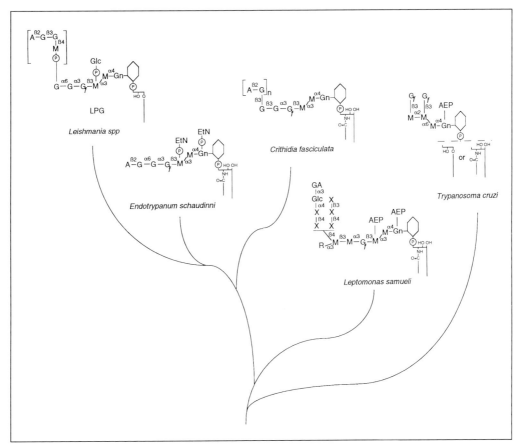

Fig. 5 Structures of the protein-free GIPLs from different genera of the Trypanosomatidae. The relationship between GIPL structure and evolutionary relatedness, as determined by nuclear ribosomal RNA analysis (98), is indicated. *T. cruzi* (14, 69), *L. samueli* (68), *E. schaudinni* (11), *C. fasciculata* (P. Schneider, K. Milne, A. Treumann, N. Zitmann, and M. Ferguson, unpublished). A, Ara; G, Gal; G$_f$, Gal$_f$; GA, GlcA; Gn, GlcN; M, Man; R, Rha; X, Xyl; AEP, aminoethylphosphonate.

each other. Some of these GIPLs may be extended with polysaccharide chains which contain completely novel sequences (as in *Leptomonas samueli*) or related sequences to those found in the leishmanial LPGs (as in *Crithidia fasciculata*) and may also be substituted with the phosphorylated residues aminoethylphosphonate or ethanolamine phosphate (Fig. 5). It is notable that most of these GIPLs contain an inositol-phosphoceramide lipid moiety instead of the alkylacyl-PI lipids that occur on the protein anchors of the same species. Ceramide lipids are introduced into the GPI protein anchors of yeast by a novel form of lipid remodelling that involves the exchange of glycerolipid for ceramide (2), and it is possible that a similar process may occur in the parasites (75).

5. Metabolism and biosynthesis of protozoan GPIs

5.1 Protein anchor biosynthesis

The GPI anchor is pre-assembled before its attachment to the carboxyl terminus of the protein, although both the glycan and lipid moieties may be modified after protein attachment. The biosynthesis of the GPI anchor precursor was first elucidated in African trypanosomes, which synthesize more than ten times the number of GPI precursors required for protein anchoring (76, 77, reviewed in Refs 1, 2, and 78). The main steps in this pathway are summarized in Fig. 6. Recent studies suggest that all the intermediates in this pathway are assembled on the cytoplasmic leaflet of the endoplasmic reticulum and that, following the addition of the ethanolamine phosphate group, they are translocated to the lumenal compartment (79, 80). While the main features of this pathway are broadly similar in all eukaryotes, there are frequently species- and even stage-specific differences in the nature of the initial PI precursor and the way in which the final lipid composition of the GPI anchor is achieved. In the bloodstream trypanosomes, a series of fatty acid remodelling steps converts the disteroyl-PI lipid of polar GPI intermediates to the unique dimyristoyl-PI (7, 81; Fig. 6). Myristate exchange may continue after the anchor has been attached to VSG, possibly after the internalization of surface VSG (82), and appears to be functionally important because first, the same remodelling steps occur in all African typanosomes; second, these parasites have an absolute requirement for this fatty acid and preferentially direct it into GPI anchors if supply is reduced (83); and third, analogues of myristate which are incorporated into GPI anchors are toxic and cause disruption of the cellular membranes (84). In the procyclic stage of *T. brucei*, only the first of these fatty acid remodelling steps occurs, leading to the formation of a mature precursor with a *lyso* 1-*O*-steroyl-PI lipid (85). While fatty acid remodelling of GPI precursors may occur in both *Toxoplasma gondii* and *Plasmodium falciparum*, the distinctive alkyl chain composition of the *Leishmania* GPI precursors is probably achieved by selection of a minor subpopulation of PI molecular species by the enzymes that catalyse the formation of GlcN-PI (72).

Species- and stage-specific variability in the GPI biosynthetic pathway also occurs in the extent and timing of inositol acylation. In mammalian cells and in

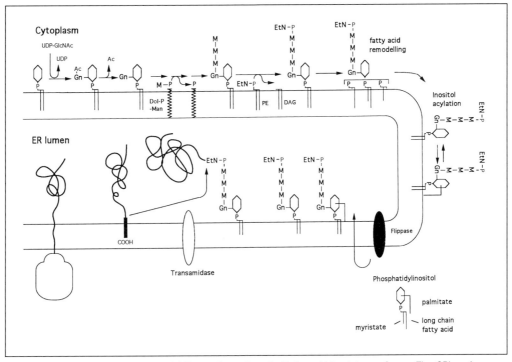

Fig. 6 Model for the biosynthesis of GPI protein anchors in *T. brucei* bloodstream forms. The GPI anchors are preformed on the cytoplasmic face of the endoplasmic reticulum by the sequential addition of monosaccharides and an ethanolamine-phosphate residue to PI (78–80, 99). The lipid moities of these glycolipids are remodelled by fatty acid (C14:0) exchange, while the inositol residue may be reversibly acylated with palmitate (81, 86). Both the acylated and non-acylated GPI species may be flipped to the lumenal face of the endoplasmic reticulum but only the non-acylated form is attached to the newly formed C-terminus of the protein generated by a putative transamidase. M, Man; Gn, GlcN; EtN, ethanolamine phosphate; P, phosphate.

some parasites (*Plasmodium* and procyclic trypanosomes), most of the early intermediates are inositol acylated (1, 2, 6). In contrast, only the mature GPI precursor (called glycolipid A or P2) is inositol acylated in bloodstream trypanosomes, to form glycolipid C (or P3) (76). Glycolipids A and C are both relatively abundant in bloodstream trypanosomes, but only glycolipid A is attached to protein *in vivo* (78). Glycolipids A and C are in dynamic equilibrium with each other, through the actions of an acyltransferase and a deacylase (86) and it is thought that glycolipid C could act as a reserve for the protein anchor precursor. It is notable that the enzyme involved in inositol acylation in trypanosomes is completely inhibited by phenylmethylsulphonyl fluoride (PMSF), whereas the equivalent enzymes in mammalian cells are not, suggesting that it may be possible to selectively inhibit the parasite GPI biosynthetic pathway (86). However, such inhibitors will have to be extremely effective, as neither PMSF or another inhibitor of GPI biosynthesis, mannosamine (which decreases GPI biosynthesis by 90 and 80%, respectively) deplete the precursor pool of bloodstream trypanosomes sufficiently to disrupt the integrity of the VSG coat (76, 77).

The timing of addition of glycan side-chains may also vary in different species. The extensive side chains of trypanosome VSG and PARP are added in the endoplasmic reticulum and the Golgi (44, 85), while the monosaccharide side-chains (Man and GalNAc) of the *Plasmodium* and *Toxoplasma* anchors are apparently added before the precursors are attached to protein (5, 6).

Only one of the enzymes from this pathway, the de-*N*-acetylase from *T. brucei*, has been partially purified (87). However, several genes which are required for GPI biosynthesis have now been cloned from mammalian cells by complementation of mutant cell lines that are defective in GPI biosynthesis (reviewed in Ref. 88). Several different complementation classes of thymoma mutants have been characterized that are defective in the synthesis of GlcNAc-PI, the addition of the third mannose residue, and the addition of the ethanolamine phosphate. cDNAs which overcome the defect in some of these mutants have now been cloned and sequenced. Although the sequence of these genes suggests that they might have the right orientation to be glycosyltransferases, they have not yet been shown to have enzymatic activity. It is possible that some of these gene products may be structural elements in an enzymatic complex as it has been shown that at least three gene products (not including those involved in the synthesis of PI and UDP-GlcNAc) are required for the biosynthesis of GlcNAc-PI (88).

Attachment to protein is directed by a C-terminal signal and involves a coupled reaction in which the signal sequence is proteolytically removed and then replaced with a GPI glycolipid. This reaction occurs after the nascent protein has been translocated into the lumen of the endoplasmic reticulum and is catalysed by an as yet uncharacterized membrane-bound transamidase. The GPI signal sequence generally involves a highly degenerate hydrophobic domain at the C terminus (about 12–20 amino acids long), a short hydrophilic spacer sequence (10–12 amino acids long), and a short cleavage or attachment site (89). In the two mammalian proteins, decay accelerating factor and alkaline phosphatase, the amino acid residues on either side of the cleavage site tend to have small side-chains, such as Ser, Gly or Ala. There is some evidence that these residues, plus the next residue on the carboxyl side of the cleavage site are accomodated within the catalytic site of the transamidase and that the overall size of these three amino acids determines whether they will be substrates for this enzyme (90). Interestingly, the cleavage/attachment sites of trypanosome VSG and the *Plasmodium* circumsporozoite protein are poorly recognized by the putative transamidase in mammalian cells (90). The parasite proteins have slightly bulkier amino acids around their cleavage site (for example, Asp-Ser-Ser in VSG), suggesting that protozoan transamidases have a larger binding site than the mammalian enzyme (90).

5.2 Biosynthesis of protein-free GPIs in *Leishmania*

The biosynthesis of the LPG anchor and the related type-2 GIPLs has been studied in *L. major*, in which this pathway is up regulated (72), and in LPG-deficient strains of *L. donovani* (91; Fig. 7). This biosynthetic pathway contains the same initial steps

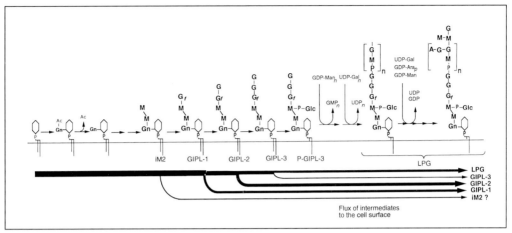

Fig. 7 Model for the biosynthesis of the leishmanial LPG and relationship with the type-2 GIPLs of *L. major*. The LPG anchor is built up by the sequential addition of monosaccharides to PI. Several of these precursor species (GIPLs 1–3) are synthesized in enormous excess over the requirement for LPG synthesis in *L. major* and are eventually transported to the cell surface where they have a long life-time (72). A small proportion of these precursors are extended with disaccharide repeat units (92, 93) which may be further substituted with glycan branches after chain elongation (94).

as the protein anchor pathway and involves the sequential addition of monosaccharides to PI (72). There is evidence that the glycosyl-transferases along this pathway select for molecular species with long (for example, C24:0) alkyl chains (10, 62, 72), which may ensure that the mature LPG molecules are stably attached to the plasma membrane. Two GIPL species have been identified (GIPL-3 and P-GIPL-3) which, based on their structure and kinetics of labeling, are likely to be the immediate precursors to LPG (72; Fig. 7). The phosphoglycan backbone is built up on these GIPLs through the repeated action of a mannosyltransferase and a galactosyl transferase that sequentially add Man-1-PO$_4$ followed by Gal from their respective nucleotide-sugar donors, GDP-Man and UDP-Gal (92, 93; Fig. 7). The distinctive βGal side-chains of *L. major* LPG are preferentially added to unsubstituted LPG chains in a microsomal system, suggesting that they are normally added late in the synthesis of the phosphoglycan moiety (94). These side-chains are capped with arabinopyanose residues in metacyclic promastigotes by a transferase that uses a novel GDP-βArap nucleotide sugar (95). This developmental modification, and possibly the activation of the arabinotransferase, is stimulated by the depletion of haemoglobin from the medium (96) which may be an important signal for parasite development in the sandfly digestive tract. By analogy with the protein anchor biosynthetic pathway, it is likely that the initial steps in type-2 GIPL and LPG anchor biosynthesis occur on the cytoplasmic side of the endoplasmic reticulum. However, there is indirect evidence that the more polar intermediates in this pathway can flip across to the lumenal compartment (80). Subsequent steps in LPG biosynthesis, including the formation of the phosphoglycan chains, probably occur in the Golgi.

None of the glycosyltransferases involved in LPG and GIPL biosynthesis have been purified, although one of the genes has recently been isolated and cloned by genetic complementation of an LPG-deficient *L. donovani* strain (97). The complementing gene (named *LPG1*) codes for a putative type-II protein which has the same overall structure, as, but no homology with, other glycosyltransferases. This mutant is thought to have a defect in LPG core synthesis at the point of addition of galactofuranose (91) and it is possible that the product of *LPG1* codes for a novel galactofuranosyl-transferase or an enzyme involved in nucleotide-sugar synthesis. Given that a number of different LPG-deficient mutants are available, it is likely that this approach will be useful in identifying a range of genes involved in the biosynthesis and regulation of LPG.

6. Conclusion

GPI anchors appear to play a crucial role in maintaining the surface architecture of many parasitic protozoa by acting as membrane anchors for some of the predominant surface proteins. The frequent use of GPI anchors in these organisms suggests that they offer advantages over the use of transmembrane polypeptide anchors. It is likely that they reduce the turnover rate of these proteins and possibly insulate the cytoplasm from the external environment. They are also required for the formation of protein coats, of which they may be an integral part, for membrane targeting and for the release, with activation of activity in some cases of specific proteins. Free GPI glycolipids that are not linked to protein are important components in the surface membrane of the trypanosomatids and other parasites. These glycolipids are vital constituents in the protective surface glycocalyx and may be involved in mediating host–parasite interactions. Studies on the biosynthesis of GPI glycolipids in these organisms have already identified the presence of novel enzymes as well as differences in the catalytic and regulatory sites of GPI biosynthetic enzymes common to both parasites and mammalian cells. It is anticipated that new approaches in the treatment of these parasitic diseases will be developed through an understanding of the function and biosynthesis of these molecules.

References

1. Englund, P. T. (1993) The structure and biosynthesis of glycosyl phosphatidylinositol protein anchors. *Annu. Rev. Biochem.*, **62,** 121.
2. McConville, M. J. and Ferguson, M. A. J. (1993) The structure, biosynthesis and function of glycosylated phosphatidylinositols in the parasitic protozoa and higher eukaryotes. *Biochem. J.*, **294,** 305.
3. Schneider, P., Ferguson M. A. J., McConville, M. J., Mehlert, A., Homans, S.W., and Bordier, C. (1990) Structure of the glycosyl-phosphatidylinositol membrane anchor of the *Leishmania major* surface protease. *J. Biol. Chem.*, **265,** 16955.
4. Güther, M. L. S., Cardoso de Almeida, M. L., Yoshida, N., and Ferguson, M. A. J. (1992) Structural studies on the glycosylphosphatidylinositol membrane anchor of *Trypanosoma cruzi* 1G7-antigen. *J. Biol. Chem.*, **267,** 6820.
5. Tomavo, S., Dubremetz, J-.F., and Schwarz, R. T. (1992) A family of glycolipids from

Toxoplasma gondii. Identification of candidate glycolipid precursor(s) for *Toxoplasma gondii* glycosylphosphatidylinositol membrane anchors. *J. Biol. Chem.*, **267**, 11721.

6. Gerold, P., Dieckmann-Schuppert, A., and Schwarz, R. T. (1994) Glycosyl phosphatidylinositols synthesized by asexual erythrocytic stages of the malarial parasite, *Plasmodium falciparum*. Candidates for plasmodial glycosylphosphatidylinositol membrane anchor precursors and pathogenicity factors. *J. Biol. Chem.*, **269**, 2597.

7. Ferguson, M. A. J., Homans, S. W., Dwek, R. A. and Rademacher, T. W. (1988) Glycosylphosphatidylinositol moiety that anchors *Trypanosoma brucei* variant surface glycoprotein to the membrane. *Science*, **239**, 753.

8. Ferguson, M. A. J., Murray, P., Rutherford, H., and McConville, M. J. (1993) A simple purification of the procyclic acidic repetitive protein and demonstration of a sialylated glycosylphosphatidylinositol membrane anchor. *Biochem. J.*, **290**, 51.

9. Pontes de Carvalho, L. C., Tomlinson, S., Vandekerckhove, F., Bienen, E. J., Clarkson, A. B., Jiang, M.-S., Hart, G. W., and Nussenzweig, V. (1993) Characterization of a novel *trans*-sialidase of *Trypanosoma brucei* procyclic trypomastigotes and identification of procyclin as the major sialic acid acceptor. *J. Exp. Med.*, **177**, 465.

10. McConville, M. J., Collidge, T., and Schneider, P. (1993) The glycoinositolphospholipids of *Leishmania mexicana* promastigotes. Evidence for the presence of three distinct pathways of glycolipid biosynthesis. *J. Biol. Chem.*, **268**, 15595.

11. Wait, R., Jones, C., Previato, J. O., and Menonca-Previato, L. (1994) Characterization of phospholipid oligosaccharides from parasitic protozoa by fast-atom bombardment and collisional activation mass-spectrometry. *Braz. J. Med. Biol. Res.* **27**, 203.

12. Haldar, K., Ferguson, M. A. J., and Cross, G. A. M. (1985) Acylation of a *Plasmodium falciparum* merozoite surface antigen via *sn*-1,2-diacyl glycerol. *J. Biol. Chem.*, **260**, 4969.

13. Field, M. C., Menon, A. K., and Cross, G. A. M. (1991) A glycosylphosphatidylinositol protein anchor from procyclic stage *Trypanosoma brucei*: lipid structure and biosynthesis. *EMBO J*,. **10**, 2731.

14. Couto, A. S., de Lederkremer, R. M., Colli, W., and Alves, M. J. M. (1993) The glycosylphosphatidylinositol anchor of the trypomastigote-specific Tc-85 glycoprotein from *Trypanosoma cruzi*. Metabolic-labeling and structural studies. *Eur. J. Biochem.*, **217**, 597.

15. Heise, N., Cardoso de Almeida, M. L., and Ferguson, M. A. J. (1995) Characterization of the lipid moiety of the glycosylphosphatidylinositol anchor of *Trypanosoma cruzi* IG7-antigen. *Mol. Biochem. Parasitol.*, **70**, 71.

16. Brown, D. A. (1992) Interactions between GPI-anchored proteins and membrane lipids. *Trends Cell Biol.*, **2**, 338.

17. Brown, D. A. and Rose, J. K.(1992) Sorting of GPI-anchored proteins to glycolipid-enriched membrane subdomains during transport to the apical cell surface. *Cell*, **68**, 533.

18. Schroeder, R., London, E., and Brown, D. (1994) Interactions between saturated acyl chains confer detergent resistance on lipids and glycosylphosphatidylinositol (GPI)-anchored proteins: GPI anchored proteins in liposomes and cells show similar behaviour. *Proc. Natl. Acad. Sci. USA*, **91**, 12130.

19. Anderson, R. G. W. (1993) Caveolae: where incoming and outgoing messengers meet. *Proc. Natl. Acad. Sci. USA*, **90**, 10909.

20. Robinson, P. J. (1991) Phosphatidylinositol membrane anchors and T-cell activation. *Immunol. Today*, **12**, 35.

21. Horvath, A., Sütterlin, C., Manning-Krieg, U., Movva, N. R, and Riezman, H. (1994) Ceramide synthesis enhances transport of GPI-anchored proteins to the Golgi apparatus in yeast. *EMBO J.*, **13**, 3687.

22. Steverding, D., Stierhof, Y.-D., Chaudhri, M., Ligtenburg, M., Schell, D., Beck-Sickinger, A. G., and Overath, P. (1994) ESAG 6 and 7 products of *Trypanosoma brucei* form a transferrin binding protein complex. *Eur. J. Cell Biol.*, **64**, 78.
23. Ligtenberg, M. J. L., Bitter, W., Kieft, R., Steverding, D., Janssen, H., Calafat, J., and Borst, P. (1994) Reconstruction of a surface transferrin binding complex in the insect form *Trypanosoma brucei*. *EMBO J.*, **13**, 2565.
24. Mayor, S., Rothberg, K. G., and Maxfield, F. R. (1994) Sequestration of GPI-anchored proteins in caveolae triggered by cross-linking. *Science*, **264**, 1948.
25. Ramamoothy, R., Donelson, J. E., Paetz, K. E., Maybodi, M., Roberts, S. C.and Wilson, M.E. Three distinct RNAs for the surface protease gp63 are differentially expressed during development of *Leishmania chagasi* promastigotes to an infectious form. *J. Biol. Chem.*, **267**, 1888.
26. Medina-Acosta, E., Karess, R. E., and Russell, D. G (1993) Structurally distinct genes for the surface protease of *Leishmania mexicana* are developmentally regulated. *Mol. Biochem. Parasitol.*, **57**, 31.
27. Bahr, V., Stierhof, Y.-D., Ilg, T., Demar, M., Quniten, M., and Overath, P. (1993) Expression of lipophosphoglycan, high molecular weight phosphoglycan and glycoprotein 63 in promastigotes and amastigotes of *Leishmania mexicana*. *Mol. Biochem. Parasitol.*, **58**, 107.
28. Lemansky, P., Fatemi, S. H., Gorican, B., Meyale, S., Rossero, R., and Tartakoff, A. M. (1990) Dynamics and longevity of the glycolipid-anchored membrane protein, Thy-1. *J. Cell Biol.*, **110**, 1525.
29. Keller, G.-A., Siegel, M. W. and Caras, I. W. (1992) Endocytosis of glycophospholipid-anchored and transmembrane forms of CD4 by different endocytic pathways. *EMBO J.*, **11**, 865.
30. Barth, A., Müller-Taubenberger, A., Taranto, P., and Gerisch, G. (1994) Replacement of a phospholipid-anchor in the contact site A glycoprotein of *D. discoideum* by a transmembrane region does not impede cell adhesion but reduces residence time on the cell surface. *J. Cell Biol.*, **124**, 205.
31. Seyfang, A., Mecke, D.. and Duszenko, M. (1990) Degradation, recycling, and shedding of *Trypanosoma brucei* variant surface glycoprotein. *J. Protozool.*, **37**, 546.
32. Webster, P. Russo, D. C. W., and Black, S. J. (1990) The interaction of *Trypanosoma brucei* with antibodies to variant surface glycoproteins. *J. Cell Sci.*, **96**, 249.
33. Teixera, A. R. L. and Santana, J. M. (1989) *Trypanosoma cruzi*: endocytosis and degradation of specific antibodies by parasite forms. *Am. J. Trop. Med. Hyg.*, **40**, 165.
34. Andrews, N. W., Robbins, E. S., Ley, V., Hong, K. S., and Nussenzweig, V. (1988) Developmentally regulated phospholipase C-mediated release of the major surface glycoprotein of amastigotes of *Trypanosoma cruzi*. *J. Exp. Med.*, **167**, 300.
35. Hall, B. F., Webster, P., Ma, A. K., Joiner, K. A. and Andrews, N. W. (1992) Desialylation of lysosomal membrane glycoproteins by *Trypanosoma cruzi*: a role for the surface neuraminidase in facilitating parasite entry into the host cell cytoplasm. *J. Exp. Med.*, **176**, 313.
36. Braun-Breton, C., Rosenberry, T. L., and Da Silva, L. H. P. (1988) Induction of the proteolytic activity of a membrane protein in *Plasmodium falciparum* by phosphatidylinositol-specific phospholipase C. *Nature*, **332**, 457.
37. Brewis, I. A., Turner, A. J., and Hooper, N. M. (1994) Activation of a glycosyl-phosphatidylinositol-anchored membrane dipeptidase upon release from pig membranes by phospholipase C. *Biochem. J.*, **303**, 633.
38. Fox, J. A., Duszenko, M., Ferguson, M. A. J., Low, M. G., and Cross, G. A. M. (1986)

Purification and characterization of a novel glycan-phosphatidylinositol-specific phospholipase C from *Trypanosoma brucei*. *J. Biol. Chem.*, **261**, 15767.

39. Bülow, R., Griffiths, G., Webster, P., Stierhof, Y-D., Opperdoes, F. R., and Overath, P. (1989) Intracellular localization of the glycosyl-phosphatidylinositol-specific phospholipase C of *Trypanosoma brucei*. *J. Cell Sci.*, **93**, 233.

40. Ilg, T., Etges, R., Overath, P., McConville, M. J., Thomas-Oates, J. E., Homans, S. W. and Ferguson, M. A. J. (1992) Structure of *Leishmania mexicana* lipophosphoglycan. *J. Biol. Chem.*, **267**, 6834.

41. Goncalves, M. F., Umezawa, E. S., Katzin, A. M., de Souza, W., Alves, M. J. M., Zingales, B., and Colli, W. (1991) *Trypanosoma cruzi*: shedding of surface antigens as membrane vesicles. *Exp. Parasitol.*, **72**, 43.

42. Ozaki, L. S., Svec, P., Nussenzweig, V., and Godson, G. N. (1983) Structure of the *Plasmodium knowlesi* gene coding for the circumsporozoite protein. *Cell*, **34**, 815.

43. Homans, S. W., Edge, C. J., Ferguson, M. A. J., Dwek, R. A., and Rademacher, T. W. (1989) Solution structure of the glycosylphosphatidylinositol membrane anchor glycan of *Trypanosoma brucei* variant surface glycoprotein. *Biochemistry*, **28**, 2881.

44. Bangs, J. D., Doering, T. L., Englund, P. T. and Hart, G. W. (1988) Biosynthesis of a variant surface glycoprotein of *Trypanosoma brucei*. Processing of the glycolipid membrane anchor and N-linked oligosaccharides. *J. Biol. Chem.*, **263**, 17697.

45. Schofield, L. and Hackett, F. (1993) Signal transduction in host cells by a glycosyl phosphatidylinositol toxin of malaria parasites. *J. Exp. Med.*, **177**, 145.

46. Schofield, L., Vivas, L., Hackett, F., Gerold, P., Schwarz, R. T., Tachado, S. (1994) Neutralizing monoclonal antibodies to glycosylphosphatidylinositol, the dominant TNF-α-inducing toxin of *Plasmodium falciparum*: prospects for the immunotherapy of severe malaria. *Ann. Trop. Med. Parasitol.*, **87**, 617.

47. Tachado, S. D. and Schofield, L. (1994) Glycosylphosphatidyinositol toxin of *Trypanosoma brucei* regulates IL-1α and TNF-α expression in macrophages by protein tyrosine kinase mediated signal transduction. *Biochem. Biophys. Res. Commun.*, **205**, 984.

48. Descoteaux, A. and Turco, S. J. (1993) The lipophosphoglycan of Leishmania and macrophage protein kinase C. *Parasitol. Today*, **9**, 468.

49. Bate, C. A. W. and Kwiatkowski, D. (1994) A monoclonal antibody that recognizes phosphatidylinositol inhibits induction of tumor necrosis factor alpha by different strains of *Plasmodium falciparum*. *Infect. Immun.*, **62**, 5261.

50. Singh, B. N. (1994) The existence of lipophosphoglycan-like molecules in Trichomonads. *Parasitol.Today*, **10**, 152.

51. Bhattacharya, A., Prasad, R., and Sacks, D. L. (1992) Identification and partial characterization of a lipophosphoglycan from a pathogenic strain of *Entamoeba histolytica*. *Mol. Biochem. Parasitol.*, **56**, 161.

52. Turco, S. J. and Descoteaux, A. (1992) The lipophosphoglycan of *Leishmania* parasites. *Annu. Rev. Microbiol.*, **46**, 65.

53. McConville, M. J. and Blackwell, J. M. (1991) Developmental changes in the glycosylated phosphatidylinositols of *Leishmania donovani*: characterization of the promastigote and amastigote glycolipids. *J. Biol. Chem.*, **266**, 15170.

54. McConville, M. J., Thomas-Oates, J. E., Ferguson, M. A. J. and Homans, S. W. (1990) Structure of the lipophosphoglycan from *Leishmania major*. *J. Biol Chem.*, **265**, 19611.

55. Thomas, J. R., McConville, M. J., Thomas-Oates, J., Homans, S. W., Ferguson, M. A. J., Greis, K., and Turco, S. J. (1991) Refined structure of the lipophosphoglycan of *Leishmania donovani*. *J. Biol. Chem.*, **267**, 6829.

56. McConville, M. J., Schnur, L. F., Jaffe, C., and Schneider, P. (1995) Structure of *Leishmania* lipophosphoglycan inter and intra-specifc polymorphism in Old World species. *Biochem. J.*, **310**, 807.
57. Homans, S. W., Melhert, A., and Turco, S. J. (1992) Solution structure of the lipophosphoglycan of *Leishmania donovani*. *Biochemistry*, **31**, 654.
58. McConville, M. J., Turco, S. J., Ferguson, M. A. J., and Sacks, D. L. (1992) Developmental modification of the lipophosphoglycan during the differentiation of *Leishmania major* promastigotes to an infectious stage. *EMBO J.*, **11**, 3593.
59. Moody, S., Handman, E., McConville, M. J., and Bacic, A. (1993) The structure of *Leishmania major* amastigote lipophosphoglycan *J. Biol. Chem.*, **268**, 18457.
60. Sacks, D. L., Pimenta, P. F. P., McConville, M. J., Schneider, P., and Turco, S. J. (1994) Stage-specific binding of *Leishmania donovani* to the sand fly vector midgut is regulated by a conformation change in the abundant surface lipophosphoglycan. *J. Exp. Med.*, **181**, 685.
61. Winter, G., Fuchs, M., McConville, M. J., Stierhof, Y.-D., and Overath, P. (1994) Surface antigens of *Leishmania mexicana* amastigotes: characterization of glycoinositol phospholipids and a macrophage-derived glycosphingolipid. *J. Cell Sci.*, **107**, 2471.
62. McConville, M. J., Homans, S. W., Thomas-Oates, J. E., Dell, A., and Bacic, A. (1990) Structures of the glycoinositolphospholipids from *Leishmania major*. A family of galactofuranose-containing glycolipids. *J. Biol. Chem.*, **265**, 7385.
63. Schlein, Y., Schur, L. F., and Jacobson, R. L. (1990) Released glycoconjugates of indigenous *Leishmania major* enhances survival of a foreign *L. major* in *Phlebotomus papatasi*. *Trans. Roy. Soc. Trop. Med. Hyg.*, **84**, 353.
64. Karp, C. L., Turco, S. J., and Sacks, D. L. (1991) Lipophosphoglycan masks recognition of the *Leishmania donovani* promastigote surface by human Kala-azar serum. *J. Immunol.*, **147**, 680.
65. Pimenta, P. F. P., Saraiva, E. M. B., Rowton, E., Modi, G. B., Garraway, L. A., Beverley, S. M., Turco, S. J., and Sacks, D. L. (1994) Evidence that the vectorial competence of phlebotomine sand flies for different species of *Leishmania* is controlled by structural polymorphisms in the surface lipophosphoglycan. *Proc. Natl. Acad. Sci. USA*, **91**, 9155.
66. Pimenta, P. F. P., Turco, S. J., McConville, M. J., Lawyer, P. G., Perkins, P. V., and Sacks, D. L. (1992) Stage-specific adhesion of *Leishmania* promastigotes to the sandfly midgut. *Science*, **256**, 1812.
67. Puentes, S. M., da Silva, R. P., Sacks, D. L., Hammer, C. H., and Joiner, K. A. (1990) Serum resistance of metacyclic stage *Leishmania major* promastigotes is due to release of C5b-9. *J. Immunol.*, **145**, 4311.
68. Green, P. J., Feizi, T., Stoll, M. S., Thiel, S., Prescott, A., and McConville, M. J. (1994) Recognition of the major cell surface glycoconjugates of *Leishmania* parasites by the human serum mannan-binding protein. *Mol. Biochem. Parasitol.*, **66**, 319.
69. Kelleher, M., Bacic, A., and Handman, E. (1992) Identification of a macrophage-binding determinant on lipophosphoglycan from *Leishmania major*. *Proc. Natl. Acad. Sci. USA*, **89**, 6.
70. Chan, J., Fujiwara, T., Brennan, P., McNeil, M., Turco, S. J., Sibelle, J.-C., Snapper, M., Aisen, P., and Bloom, B. R. (1989) Microbial glycolipids: possible virulence factors that scavenge oxygen radicals. *Proc. Natl. Acad. Sci. USA*, **86**, 2453.
71. Moore, K. J. and Matlashewski, G. (1994) Intracellular infection by *Leishmania donovani* inhibits macrophage apoptosis. *J. Immunol.*, **152**, 2930.
72. Proudfoot, L., Schneider, P., Ferguson, M. A. J., and McConville, M. A. J. (1995) Bio-

synthesis of the glycolipid anchor of lipophosphoglycan and the structurally related glycoinositolphospholipids from *Leishmania major*. *Biochem. J.*, **308**, 45.
73. Previato, J. O., Mendonca-Previato, L., Jones, C., Wait, R., and Fournet, B. (1992) Structural characterization of a novel class of glycophosphosphingolipids from the protozoan *Leptomonas samueli*. *J. Biol. Chem.*, **267**, 24279.
74. de Lederkremer, R. M., Lima, C., Ramirez, M. I., Ferguson, M. A. J., Homans, S. W., and Thomas-Oates, J.E. (1991) Complete structure of the glycan of lipophosphoglycan from *Trypanosoma cruzi* epimastigotes. *J. Biol. Chem.*, **265**, 19611.
75. de Lederkremer, R. M., Lima, C., Ramirez, M. I., Goncalvez, M. F., and Colli, W. (1993) Hexadecylpalmitoylglycerol or ceramide is linked to similar glycophosphoinositol anchor-like structures in *Trypanosoma cruzi*. *Eur. J. Biochem.*, **218**, 929.
76. Ralton, J. E., Milne, K. G., Güther, M. L. S., Field, R. A., and Ferguson, M. A. J. (1993) The mechanism of inhibition of glycosyl-phosphatidylinositol anchor biosynthesis in *Trypanosoma brucei* by mannosamine. *J. Biol. Chem.*, **268**, 24183.
77. Masterson, W. J. and Ferguson, M. A. J. (1991) Phenylmethanesuphonyl fluoride inhibits GPI anchor biosynthesis in the African trypanosome. *EMBO J.*, **10**, 2041.
78. Menon, A. K. (1991) Biosynthesis of glycosyl-phosphatidylinositol. *Cell Biol. Int. Rep.*, **15**, 1007.
79. Vidugiriene, J. and Menon, A. K. (1994) The GPI anchor of cell-surface protein is synthesized on the cytoplasmic face of the endoplasmic reticulum. *J. Cell Biol.*, **127**, 333.
80. Mensa-Wilmot, K., LeBowitz, J. H., Chang, K.-P., Al-Qahtani, A., McGwire, B. S., Tucker, S., and Morris, J. C. (1994) A glycosylphosphatidylinositol (GPI)-negative phenotype produced in *Leishmania major* by GPT phospholipase C from *Trypanosoma brucei*: Topology of two GPI pathways. *J. Cell Biol.*, **124**, 935.
81. Masterson, W. J., Raper, J., Doering, T. L., Hart, G. W., and Englund, P. T. (1990) Fatty acid remodelling: a novel reaction sequence in the biosynthesis of trypanosome glycosyl phosphatidylinositol membrane anchors. *Cell*, **62**, 73.
82. Buxbaum, L. U., Raper, J., Opperdoes, F. R., and Englund, P. T. (1994) Myristate exchange. A second glycosyl phosphatidylinositol myristoylation reaction in African trypanosomes. *J. Biol. Chem.*, **269**, 30212.
83. Doering, T. L., Pessin, M. S., Hoff, E. F., Hart, G. W., Raben, D. M., and Englund, P. T. (1993) Trypanosome metabolism of myristate, the fatty acid required for the variant surface glycoprotein membrane anchor. *J. Biol. Chem.*, **268**, 9215.
84. Doering, T. L., Raper, J., Buxbaum, L. U., Adams, S. P., Gordon, J. I., Hart, G. W., and Englund, P. T. (1991) An analog of myristic acid with selective toxicity for African Trypanosomes. *Science*, **252**, 1851.
85. Field, M. C., Menon, A. K., and Cross, G. A. M. (1992) Developmental variation of glycosylphosphatidylinositol membrane anchors in *Trypanosoma brucei*. *In vitro* biosynthesis of intermediates in the construction of the GPI anchor of the major procyclic surface glycoprotein. *J. Biol. Chem.*, **267**, 5324.
86. Güther, M. L. S., Masterson, W. J., and Ferguson, M. A. J. (1994) The effects of phenylmethylsulphonyl fluoride on inositol acylation and fatty acid remodelling in African trypanosomes. *J. Biol. Chem.*, **269**, 18694.
87. Milne, K. G., Field, R. A., Masterson, W. J., Cottaz, S., Brimacombe, J. S., and Ferguson, M. A. J. (1994) Partial purification and characterization of the N-acetylglucosaminyl-phosphatidylinositol de-N-acetylase of glycosylphosphatidylinositol anchor biosynthesis in African trypanosomes. *J. Biol. Chem.*, **269**, 16403.
88. Kinoshita, T. and Takeda, J. (1994) GPI-anchor synthesis. *Parasitol. Today*, **10**, 139.

89. Caras, I. W. (1991) Probing the signal for glycosylphosphatidylinositol anchor attachment using decay accelerating factor as a model system. *Cell Biol. Int. Rep.*, **15,** 815.
90. Moran, P. and Caras, I. W. (1994) Requirements for glycosylphosphatidylinsotiol attachment are similar but not identical in mammalian cells and parasitic protozoa. *J. Cell Biol.*, **125,** 333.
91. Huang, C. and Turco, S. J. (1993) Defective galactofuranose addition in lipophosphoglycan biosynthesis in a mutant of *Leishmania donovani*. *J. Biol. Chem.*, **268,** 24060.
92. Carver, M. A. and Turco, S. J. (1991) Cell-free biosynthesis of lipophosphoglycan from *Leishmania donovani*. Characterization of microsomal galactosytransferase and mannosyltransferase activities. *J. Biol. Chem.*, **266,** 10974.
93. Carver, M. A. and Turco, S. J. (1992) Biosynthesis of lipophosphoglycan from *Leishmania donovani*: characterization of mannosylphosphate transfer *in vitro*. *Arch. Biochem. Biophys.*, **295,** 309.
94. Ng, K., Handman, E., and Bacic, A. (1994) Biosynthesis of lipophosphoglycan from *Leishmania major*: characterization of (ß1–3)-galactosyltransferase. *Glycobiol.*, **4,** 845.
95. Schneider, P., McConville, M. J., and Ferguson, M. A. J. (1994) Characterization of GDP-α-D-arabinopyranose, the precursor of D-Ara$_p$ in *Leishmania major* lipophosphoglycan. *J. Biol. Chem.*, **269,** 18332.
96. Schlein, Y. and Jacobson, R. L. (1994) Haemoglobin inhibits the development of infective promastigotes and chitanase secretion in *Leishmania major* cultures. *Parasitology*, **109,** 23.
97. Ryan, K. A., Garraway, L. A., Descoteaux, A., Turco, S. J., and Beverley, S. M. (1993) Isolation of virulence genes directing surface glycosyl-phosphatidylinositol synthesis by functional complementation of *Leishmania*. *Proc. Natl. Acad. Sci. USA*, **90,** 8609.
98. Fernandes, A. P., Nelson, K., and Beverley, S. M. (1993) Evolution of nuclear ribosomal RNAs in kinetoplastid protozoa: Perspectives on the age and origins of parasitism. *Proc. Natl. Acad.Sci. USA*, **90,** 11608.
99. Menon, A. K., Eppinger, M., Mayor, S., and Schwarz, R. T. (1993) Biosynthesis of glycosylphosphatidylinositol anchors in *Trypanosoma brucei*: the terminal phosphoethanolamine group is derived from phosphatidylethanolamine. *EMBO J.*, **12,** 1907.

Index

Parasites are referenced according to genus except in the cases of *Trypanosoma brucei* and *Trypanosoma cruzi*.

6kb element, *see also* mitochondria 43–5, 48
35kb element
   expression and function 47
   inheritance 48
   relationship to plastid genomes 45–7
   structure 45–7
   Toxoplasma 57

actin gene 57
adenine arabinoside 62, 67
adenosine transporter 187
adenylate cyclase 90, 103
adenylate kinase 171
African sleeping sickness 3
aldolase 15, 17, 100–1, 166
α-amanitin 91
aminoglycoside phosphotransferase gene, *see* neo
aminoquinolines 190; *see also* chloroquine, mefloquine
animal reservoirs, and drug resistance 182
antifolates 185–7
antigenic variation, *Trypanosoma brucei* 101–7
   Plasmodium 41
antimalarials, *see* pyrimethamine, chloroquine
antimonials 188
anti-parasitic drugs, *see* drug resistance and individual compounds
arabinose, in LPG structure 212, 214–5, 221
arsenicals 184, 187–8
Ascaris, *trans*-splicing 118, 120, 122
atovaquone 194
ATPase 92–3, 191

BAC libraries 62–4
base composition

   Plasmodium 37, 43, 45
   Toxoplasma 57
   trypanosomatids 8
bent DNA 77
ble gene 22, 66–7
Bodo caudatus, RNA editing 153
branch sites, *trans*-splicing 115–18, 124
Caenorhabditis elegans
   genome map 63
   polycistronic transcription 126
cachexia, and malaria 211
calmodulin–ubiquitin associated (CUB) genes 23
cap 4 structure 121–2, 125
cat gene/reporter activity 20, 58, 66–7, 162
ceramide 207–8, 218
Chagas' disease 3
chimeric RNAs, in RNA editing 144–8
chloramphenicol acetyl transferase *see* *cat* gene
chloroquine resistance
   and drug transport 187, 190–1
   and P glycoproteins 191, 192–4
   cost 183
   genetics 192
   and mefloquine resistance 193–4
   spread 182, 192
chloroquine, target of action 190–91
chromatin structure 40, 97, 103
chromosomes
   artificial 25
   in mitosis 6, 36
   organization 11–14, 37–43
chromosome polymorphisms 10–11, 13, 37–42, 193–4
circumsporozoite protein, *Plasmodium* 210, 220
cis-splicing 116, 121, 123, 128, 136–8
cleavage–ligation mechanism, RNA editing 144–5
clonal population structure 10–11, 60
codon usage 8, 57
complement opsonization 216
cosmid vectors 25, 63
CRE1 16, 91

Crithidia
   GIPLs 217–18
   glycosomes 160, 165–6
   kDNA 76–85, 141
   phylogeny 153
   transposon-like elements 16
cycloguanil 186
cytoadherence 41
cytochrome *b* 43–5, 57, 151–3
cytochrome *c* 43, 101
cytochrome oxidase, *Plasmodium* 43, 45
   Toxoplasma 57
   Trypanosoma brucei 138, 143, 151–3
CZAR 16

databases 62–4
DDT 182
debranching enzyme 116
developmental regulation
   glycosomal 160–1
   LPG structure 212–15, 220–1
   mitochondrial 45, 152, 160
   mRNA abundance 45, 92
   RNA editing 139, 152
   VSG expression 92–3, 95, 97, 104–5
DFMO, *see* difluoromethylornithine
DHFR/TS, *see* dihydrofolate reductase/thymidylate synthetase
difluoromethylornithine 184
dihydrofolate reductase/thymidylate synthetase 12, 17, 57
   in drug resistance 184–6
dihydrofolate synthetase 186
dihydropteroate synthase 186–7
DNA polymerase 106
DNA replication, kDNA 78–86
drug activation 171, 195
drug design, potential targets 168, 196
drug resistance
   and drug transport 187–91
   and gene amplification 183–4, 188, 193–4, 196
   genetic analysis 62, 186, 192–4
   and mutation 185–7

drug resistance (*cont.*)
  and public health 182–3
  relationship to parasite genetics and life cycle 182
  spread 181–2, 192
  see also individual diseases and compounds
drug transport, and drug resistance 187–91

electron transport 75, 170
endocytosis, of GPI-anchored proteins 209–10
Endotrypanum, GIPLS 217
enhancer sequences 19, 58, 94
Entamoeba histolytica 1
  P-glycoprotein 188
  GIPLs 212
episomes, *see* extrachromosomal DNAs
ESAGs 90–1, 99
ethanolamine, in GIPLs and LPG structure 212, 214, 217, 218
ether-lipid biosynthesis 167
euglenoids, *trans*-splicing 128
evolution
  hydrogenosomes 174
  parasites 1–2, 153
  RNA editing 139, 152–3
  trans-splicing 128–9
  VSG gene 106–7
expressed sequence tags (ESTs) 62–4
expression site–associated genes, *see* ESAGs
extrachromosomal DNAs 13, 43–8
  and drug resistance 183–4, 188–9
  see also shuttle vectors

fatty acids
  in GPI anchors 207, 211, 218
  β-oxidation 161, 167–8
ferredoxin 170, 171, 195
ferriprotoporphyrin IX 190, 191
flagellar pocket 209
fluorescence-activated cell sorting (FACs) 8
5-fluorouridine 67
folate biosynthesis 186
food vacuole 190–1

G418 (neomycin) 22, 66
galactose, LPG structure 211–13

β-galactosidase 20, 66–7
gene amplification 16
  and drug resistance 183–4, 188, 193–4, 196
gene knockout
  evidence for diploidy 7
  P-glycoprotein 187, 190
gene markers 6
genetic complementation 25, 222
genetic exchange 8–10, 42–3, 55–6, 186, 192–4
genome mapping 11, 61–3
genome organization 6, 14, 16, 17, 37, 42, 57–9
genome size and complexity 6–8, 37, 43, 57
Giardia, drug resistance 1, 194–5
GIPLs 212–14, 216–18, 220–2
glucosephosphate isomerase 165, 166
β-D-glucosyl-hydroxymethyluracil 8, 97
β-glucuronidase (GUS) 20
glyceraldehyde phosphate dehydrogenase 15, 17, 165
glycocalyx 211–12, 216, 222
glycoinositolphospholipids, *see* GIPLs
glycolysis 159–60
glycosomes
  biochemical pathways 89, 159, 160–1
  biogenesis 161, 167–8
  developmental regulation 160–1
  function 160–1
  morphology 160
  relationship to peroxisomes 159, 161–3, 166–8
  targeting signals 161–7, 175
glycosyl-phosphatidylinositol, *see* GPI
glycosyltransferases 220–1
Golgi 208, 211, 220
gp63 genes 12, 17, 208–9
gp85 genes 17
GPI anchors 205–11, 217–22
  biosynthesis 219–20
  and host signal transduction 211
  inositol acylation 218–19
  signal sequence 220
  and subcellular trafficking 208–9
gRNAs 134
  expression 152
  genes 141–2
  structure 143–4

  see also RNA editing, and minicircles
gRNA–mRNA chimeras 144–8
gRNA–protein interactions 148–50
7-methyl guanosine cap 121
guide RNAs, *see* gRNAs
G-U base pairing, in RNA editing 143
GUS (β-glucuronidase) 20

haem polymerase 190
haemozoin 190
halofantrine, 190, 193–4
H-circle 184
Herpetomonas, glycosomes 160
hexose transporter gene 16,17
histones 40
HIV 2
housekeeping genes 12–13, 37, 39, 57
hsp70 genes 12,15
hsp83 genes 100
hydrogenase 170, 174, 195
hydrogenosome
  biogenesis 171–4
  comparison to mitochondria 172–4
  and drug activation 171, 195
  evolutionary origin 174
  function 170–1
  in fungi and ciliates 174
  leader sequences 171–3
  metabolic pathways 170
  morphology 169
  peroxisome-like 174
  protein import *in vitro* 172
hydroxymethyl dihydropterin pyrophosphokinase 186–7
5-hydroxymethyl uracil 8
hyg (hygromycin B phosphotransferase) gene 22, 162
hypoxanthine guanine phosphoribosyl transferase 165
hypoxanthine xanthine guanosine phosphoribosyl transferase gene 57

inducible gene expression 25–6
INGI 16
inosine monophosphate dehydrogenase 184
intergenic regions, trypanosomatids 122
interleukin-1 (IL1) 211

introns
  Toxoplasma 57
  SL RNA 115–16, 119–20
isoenzyme variation 7, 10–11, 59–60

KAHRP 39, 41
karyotype (molecular)
  Plasmodium 37
  Leishmania spp 12–14
  Trypanosoma brucei 14
  Trypanosoma cruzi 12, 13
  Toxoplasma 57
kDNA 6
  components and gene organization 75–6, 140–2
  condensation 77
  network structure 76–8
  replication 78–86
  see also RNA editing
kinetoplast DNA, *see* kDNA
knobs 41

lacZ gene 66–7
leader sequences
  hydrogenosomal 171–3, 175
Leishmania
  drug resistance 183–4, 187, 189
  gene amplification 183–4
  genome size and complexity 7
  glycosomes 160–1, 165–6
  life cycle 212
  minicircles 141–2
  mitochondrial RNPs 149
  P-glycoprotein genes 188–90
  species 9
leishmaniasis 3, 182
Leptomonas
  GIPLs 217, 218
  glycosomes 160
  SL RNA 119–20
life cycle
  and drug resistance 182
  Leishmania 212
  Plasmodium 35–7, 42
  Toxoplasma 55–6, 59–60
  Trypanosoma brucei 7–8, 89, 100–1, 209
  see also developmental regulation
lipophosphoglycan, *see* LPG
LPG mutants, complementation 222
LPG structure 212–16, 220–2
luciferase 20, 162, 163–5
lysosomes 209

macrophage, oxidative burst 216
malaria
  morbidity and mortality 3, 35, 41, 211
  vaccines 3
  see also Plasmodium
malate dehydrogenase 160, 166
malic enzyme, hydrogenosomal 171
mannose, LPG structure 213
maxicircles 75–6
  genes 140–1
  replication 85
  see also kDNA
mdr, *see* multidrug resistance and P-glycoproteins
mefloquine 190, 193–4
megasomes 209
meiosis 9, 42–3
melarsen 187
metacyclogenesis
  Leishmania 214
  Trypanosoma brucei 9, 89
methotrexate 184
metronidazole 171, 194–5
microbodies, *see* glycosomes, peroxisomes, and hydrogenosomes
microsporidia 1
mini-chromosomes, *T. brucei* 14, 15
mini-exon RNA, *see* spliced leader (SL) RNA
minicircles 75–6
  function 141
  replication 78–85
  size and heterogeneity 141–2
  topology 76, 78
  valence 77–8, 83
  see also kDNA, gRNAs
mitochondria 1, 194
  developmental regulation 45, 152, 160
  versus hydrogenosomes 172–4
  mitochondrial gene expression 44–5;
  see also RNA editing
mitochondrial genome 43–4, 48, 57, 75–6, 140–2
mitochondrial RNPs 148–50
MSA-1 42
multidrug resistance 187; *see also* P-glycoprotein
mycophenolic acid 184
myristate 211, 218

NADH dehydrogenase 152
neo gene 22

nucleolus 103

organelles, *see* 35kb element, hydrogenosomes, glycosomes, mitochondria
origin of DNA replication 24, 85
ornithine decarboxylase 184

PAGs 91–2
peroxisomes 159, 161–3, 166–8, 174
PfEMP1 41
PFGE 6, 11–14, 57, 63
P-glycoprotein
  and drug resistance 184, 188–90, 191, 192–4, 196
  and food vacuole 191
  function 187–90
  gene amplification 188, 193–4
  structure 187, 192
phagolysosome 216
Phlebotomine spp., *see* sandfly
phleomycin 22, 66–7
phosphoenolpyruvate carboxykinase 160, 162, 165
phosphofructokinase gene 12
phosphoglucose isomerase gene 15, 17
phosphoglycerate kinase 92, 93, 161, 165
phospholipase C 15, 207, 210
phylogeny 1–2, 153
Phytomonas, glycosomes 160
plasmids, *see* extrachromosomal DNAs and shuttle vectors
Plasmodium
  antigenic variation 41
  chromosome structure and polymorphism 37–43, 193–4
  cytoadherence 41
  drug resistance 64, 183, 185–7, 190–4
  35Kb element 45–8
  extrachromosomal DNAs 43–8
  food vacuole 190–1
  genome 37, 42
  GPI anchors 206, 220
  histones 40
  in vitro culture 35
  life cycle 35–7, 42
  mitochondria 43–5, 48
  nucleosomes 39–40
  P-glycoprotein genes 188, 192
  ploidy 37
  recombination 42–3
  repetitive sequences 37–9
  species 35

Plasmodium (cont.)
  transfection 48
  see also malaria
plastid genome, *see* 35kb element
ploidy 6–8, 14, 37, 182
point mutations 105–6, 182, 185–7
polyadenylation 18, 20–1, 92, 99, 101, 123
polycistronic transcription, nuclear 17–18, 90–2, 122
polycistronic transcription, mitochondrial 45, 151
polypyrimidine tract 18, 123–7
population genetics 10–11, 59–60
position effects 97, 103
positional cloning 63
post-translational control 101
pre-mRNA splicing pathways 116
procyclic acidic repetitive protein (PARP), *see* procyclin
procyclin 89, 206, 211
  genes 15, 17, 91, 92–101
  mRNA 100–1
procyclin-associated genes, *see* PAGs
promoter
  α-amanitin sensitivity 91
  chromosomal context 97
  pol I 20, 93
  pol II 17, 92–4
  pol III 94
  procyclin 94–7
  regulation 95
  ribosomal 94–5
  SL RNA genes 122
  and transfection 20
  VSG 90–1, 94–7
protein kinase C 211, 216
pulsed field gel electrophoresis, *see* PFGE
purine salvage 161
puromycin 22
pyrimethamine 66–7, 185–6
pyrimidine biosynthesis 45, 161, 194
pyruvate dehydrogenase 195
pyruvate:ferredoxin oxidoreductase 170, 171, 195
pyruvate kinase gene 12, 15

quinine 190, 194

RAPD technique 60
R-circle 184

recombination 7, 19, 41–3, 90, 103–4
repetitive sequences 37–9, 58
reporter genes 19–20, 58, 66–7, 162–5
restriction fragment length polymorphisms, *see* RFLPs
RFLPs 7, 60–1
ribosomal protein S12 153
ribosomal RNA genes
  35 kb element 46–7
  fragmented 44
  mitochondrial 44, 140–1
  nuclear 12,16,17, 57–8
ribozymes 136
RIME 16, 91
RNA editing
  developmental regulation 139, 152
  discovery of 138
  evolution 139, 152–3
  forms of 134–5
  function 150–2
  G-U base pairing 143
  in vitro system 138
  proposed mechanisms 144–8
  RNPs 148–50
  see also gRNAs
RNA ligase 145, 148, 149
RNA polymerase
  α amanitin-sensitivity 91
  organellar 47
RNA polymerase I 17, 20, 93
RNA polymerase II 17, 92–4
RNA polymerase III 17
RNA processing, *see trans*-splicing, *cis*-splicing, RNA editing, polyadenylation
RNAse P 136–7
RNPs
  in RNA editing 138, 148–50
  in *trans*-splicing 116–18, 121, 126

SAG1 gene 58
sandfly 9, 213–15
selectable markers 22, 62, 66–7, 162
sequence-tagged sites (STSs) 63–4
Shine Dalgarno sequence 1
shuttle vectors 25, 184, 189–90
sialic acid 206
sinefungin 62, 121
SLACS 16
Sm-binding site 117
small nuclear ribonucleoprotein particles, *see* snRNPs
small nuclear RNAs, *see* U-snRNAs

snRNPs 116–18, 121, 126
spherical body 46
sphingolipids 208
splice-acceptor sites 18, 98, 115, 119, 123–4
spliced leader (SL) RNA 18, 115, 119, 120–2
  genes 12, 15
  promoter 94
  RNP 121
spliced leader-associated RNA 118
spliceosome 116
splicing, *see cis*-splicing and *trans*-splicing
stage-regulation, *see* developmental regulation
subtelomeric regions 38–9, 42–3
succinyl-CoA synthetase 170, 171
sulfa drugs 186–7

tandem gene arrays 6, 14
targeting signals, glycosomal 161–7
  hydrogenosomal 172
telomeres 14, 24
  DNA modifications in 97
  Plasmodium falciparum 40–2
  repetitive elements 90
  VSG expression sites 90, 105
terminal uridylyl transferase 143, 147–8, 149
tetracycline repressor 25
6-thioguanosine 67
topoisomerase II 78, 83–4, 85
Toxoplasma
  and AIDS 57, 60
  developmental regulation 59
  drug resistance 194
  gene expression 58
  genetics 55, 59–61
  genome size and ploidy 55–6
  GPI anchors 206, 220
  identification 59–60
  libraries 62–4
  life cycle 55–6, 59–60
  mitochondria 57, 194
  mutagenesis 61, 67
  population structure 60
  pyrimidine metabolism 61, 194
  stage-specific antigens 59
toxoplasmosis 3, 56–7
transcription initiation 20, 92–3, 96–7
  elongation 92, 95, 98
  termination 21, 98

transesterification mechanism, RNA editing 145
transfection
  Toxoplasma 66–7
  trypanosomatids 18–25, 90, 100, 161–5, 184, 189
  Plasmodium 48
transferrin receptor 90, 209
translation initiation 58
translational control 100–2
transposon-like elements 16, 91
trans-sialidase 206–7
trans-splicing
  evolution 128–9
  kinetics 121, 125
  nematodes 115, 118, 120, 122
  and polyadenylation 126
  protozoa 18, 20–1, 92, 98, 115–133
  RNPs 116–18, 121, 126
  stage-regulation 99–100
trichomonads, organelles 168; *see also* hydrogenosomes
Trichomonas 1, 168–9
  drug resistance 194–5
  GIPLs 212
  hydrogenosomes 168–75
trichomoniasis 3, 169, 171
trimethylguanosine cap 117, 121
triosephosphate isomerase gene 15
trisomy 9
Tritrichomonas, hydrogenosomes 169
tRNA processing 137–8

tRNAs, 35kb element 46–7
tRNAs, mitochondrial, 140
Trypanoplasma
  RNA editing 153
  glycosomes 160, 165
Trypanosoma brucei
  cell cycle 78, 89
  developmental regulation 88–107, 209
  genome size and complexity 7
  glycosomes 160–8
  GPI anchors 205–6, 209–10, 218–19
  life cycle 7–8, 89, 100–1, 209
  ploidy 6–8, 14
  population structure 10–11
  RNA editing 134, 138, 140–2, 148–50, 152
  sub-species 88
Trypanosoma cruzi
  genome size and complexity 7
  GIPLs 217
  glycosomes 160–1
  GPI anchors 206, 217
  life cycle 8
  ploidy 6–8
trypanosomatid, phylogeny 153
trypanosomiasis 3
trypanothione reductase 166
tsetse fly (*Glossinia*) 9, 88–9, 97
tubulin genes 12, 15–17, 57, 123–5
tumour necrosis factor (TNF) 211
tunicamycin, resistance 184
TUTase, *see* terminal uridylyl transferase

U-snRNAs 116–18
ubiquitin gene 12,13,17
uracil phosphoribosyl transferase gene 57
UTR, in gene regulation 58, 99, 100–1

Var genes, *see* PfEMP1
variant surface glycoprotein, see VSG
vinblastine 188–90
V-circle 188–90
virulence factor 205
VSG 89, 101–2
  antigenic variation 89, 105
  GPI anchor 206, 208–11, 220
VSG genes 6, 14–15, 17, 104–5
  evolution 106–7
  expression sites 90–1, 92, 93–8, 103
  metacyclic expression sites 91–2, 96–7
  point mutations 105–6
  programming 105
  promoter 90–1, 95
  telomeric 8, 105
  transcription, inhibition by protein factors 98
  see also antigenic variation

YAC libraries 62, 63